W9-CGQ-817

ENVIRONMENT
94/95

Thirteenth Edition

A Library of Information from the Public Press

Editor

John L. Allen
University of Connecticut

John L. Allen is professor of geography at the University of Connecticut. He received his bachelor's degree in 1963 and his M.A. in 1964 from the University of Wyoming, and in 1969 he received his Ph.D. from Clark University. His special area of interest is the impact of contemporary human societies on environmental systems.

Cover illustration by Mike Eagle

The Dushkin Publishing Group, Inc.
Sluice Dock, Guilford, Connecticut 06437

The Annual Editions Series

Annual Editions is a series of over 60 volumes designed to provide the reader with convenient, low-cost access to a wide range of current, carefully selected articles from some of the most important magazines, newspapers, and journals published today. Annual Editions are updated on an annual basis through a continuous monitoring of over 300 periodical sources. All Annual Editions have a number of features designed to make them particularly useful, including topic guides, annotated tables of contents, unit overviews, and indexes. For the teacher using Annual Editions in the classroom, an Instructor's Resource Guide with test questions is available for each volume.

VOLUMES AVAILABLE

Africa
Aging
American Foreign Policy
American Government
American History, Pre-Civil War
American History, Post-Civil War
Anthropology
Biology
Business Ethics
Canadian Politics
Child Growth and Development
China
Comparative Politics
Computers in Education
Computers in Business
Computers in Society
Criminal Justice
Drugs, Society, and Behavior
Dying, Death, and Bereavement
Early Childhood Education
Economics
Educating Exceptional Children
Education
Educational Psychology
Environment
Geography
Global Issues
Health
Human Development
Human Resources
Human Sexuality
India and South Asia
International Business
Japan and the Pacific Rim

Latin America
Life Management
Macroeconomics
Management
Marketing
Marriage and Family
Mass Media
Microeconomics
Middle East and the Islamic World
Money and Banking
Multicultural Education
Nutrition
Personal Growth and Behavior
Physical Anthropology
Psychology
Public Administration
Race and Ethnic Relations
Russia, Eurasia, and Central/Eastern
  Europe
Social Problems
Sociology
State and Local Government
Third World
Urban Society
Violence and Terrorism
Western Civilization,
  Pre-Reformation
Western Civilization,
  Post-Reformation
Western Europe
World History, Pre-Modern
World History, Modern
World Politics

Library of Congress Cataloging in Publication Data
Main entry under title: Annual Editions: Environment. 1994/95.
        1. Environment—Periodicals. 2. Ecology—Periodicals.
I. Allen, John L., *comp.* II. Title: Environment.
ISBN 1–56134–274–2   301.31′05   79–644216

Thirteenth Edition

Manufactured in the United States of America

To the Reader

In publishing ANNUAL EDITIONS we recognize the enormous role played by the magazines, newspapers, and journals of the *public press* in providing current, first-rate educational information in a broad spectrum of interest areas. Within the articles, the best scientists, practitioners, researchers, and commentators draw issues into new perspective as accepted theories and viewpoints are called into account by new events, recent discoveries change old facts, and fresh debate breaks out over important controversies.

Many of the articles resulting from this enormous editorial effort are appropriate for students, researchers, and professionals seeking accurate, current material to help bridge the gap between principles and theories and the real world. These articles, however, become more useful for study when those of lasting value are carefully *collected, organized, indexed,* and *reproduced* in a *low-cost format*, which provides easy and permanent access when the material is needed. That is the role played by *Annual Editions*. Under the direction of each volume's *Editor*, who is an expert in the subject area, and with the guidance of an *Advisory Board*, we seek each year to provide in each *ANNUAL EDITION* a current, well-balanced, carefully selected collection of the best of the public press for your study and enjoyment. We think you'll find this volume useful, and we hope you'll take a moment to let us know what you think.

During the last two decades, and particularly during the late 1980s and early 1990s, the environmental predicament foreseen by scientists has begun to emerge in a number of guises such as population/food imbalances, problems of energy scarcity, acid rain, toxic and hazardous wastes, water shortages, massive soil erosion, global atmospheric pollution, forest dieback and tropical deforestation, and the highest rates of plant and animal extinction the world has known. The last half of the 1980s and the opening years of the 1990s have been characterized by drought and famine in Africa, a major environmental chemical accident in Bhopal, India, the burning and cutting of thousands of square miles of tropical rain forest, a near-meltdown of a nuclear power generator in Chernobyl in Russia, abnormally high temperatures and drought in the United States, several serious oil spills, including the infamous wreck of the Exxon *Valdez,* an energy-related military conflict in the oil-rich Persian Gulf that produced significant environmental disruptions, and unprecedented conflicts between advocates of the economic use of resources and the supporters of environmental protection. Moreover, the last few years have brought scientific validation of the concern that the life-protecting ozone layer is being destroyed, and that the long-term global climate changes (the human-enhanced greenhouse effect) that scientists have warned about may have already begun. These and other problems surfaced in spite of the increased environmental awareness and legislation that characterized the decade of the 1970s. They have resulted, in part, from the misguided environmental counterrevolution that characterized the last dozen years and favored the short-term, expedient approach to problem-solving over longer-term economic and ecological good sense. The drive to produce enough food to support a growing population, for example, has resulted in the use of increasingly fragile and marginal resources in Africa, which has produced the desert expansion that brings famine to that troubled continent. Similar social and economic problems have contributed to massive deforestation in Latin America and Southeast Asia. The economic problems caused by resource scarcity have caused the relaxation of environmental quality standards that have become viewed as too costly. The decrease in standards has been particularly apparent in Third World countries striving to become economically developed, and has contributed to accidents such as that at Bhopal. But even in the more highly developed nations, prolonged recession has created an economic climate favoring the slackening of environmental ideals. For the sake of jobs in the timber industry, for example, some of the last few areas of old-growth forests in the United States are threatened. In addition, concerns over energy availability have created the need for military action to save the developed nations' access to cheap oil and have prompted increasing reliance on technological quick fixes—a Faustian bargain that creates conditions under which a terrifying Chernobyl accident can occur. There are signs, however, that a new environmental consciousness is awakening. The dissolution of the Iron Curtain and the environmental horror stories that have emerged from Eastern Europe and the former Soviet Union have given new incentives to international cooperation. Several major publications have claimed the 1990s to be "The Decade of the Environment," and there is growing public clamor that something must be done about environmental quality before it is too late.

The articles contained in *Annual Editions: Environment 94/95* have been selected for the light they shed on these and other problems and issues. The selection process was aimed at including material that will be readily assimilated by the general reader. Additionally, every effort has been made to choose articles that do not engage in futile vilification of the species *Homo sapiens sapiens* as a fouler of its own nest. Accordingly, the selections in this book have been chosen more for their intellectual content than for their emotional tone.

Readers can have input into the next edition by completing and returning the article rating form at the back of the book.

John L. Allen
Editor

Contents

Unit 1

The Global Environment: An Emerging World View

Four selections provide information on the current state of Earth and the changes we will face.

Unit 2

The World's Population: People and Hunger

Four selections examine the problems the world will have in feeding its ever-increasing population.

The concepts in bold italics are developed in the article. For further expansion please refer to the Topic Guide, the Index, and the Glossary.

Unit 3

Energy: Present and Future Problems

Six articles consider the problems of meeting present and future energy needs. Alternative energy sources are also examined.

The concepts in bold italics are developed in the article. For further expansion please refer to the Topic Guide, the Index, and the Glossary.

Unit 4

Pollution: The Hazards of Growth

Seven selections weigh the environmental impacts of the disposal and control of pollution, unwanted radioactive waste, pesticides, urban landfills, and acid rain.

The concepts in bold italics are developed in the article. For further expansion please refer to the Topic Guide, the Index, and the Glossary.

Unit 5

Resources: Land, Water, and Air

Seven selections discuss the environmental problems
affecting our land, water, and air resources.

The concepts in bold italics are developed in the article. For further expansion please refer to the Topic Guide, the Index, and the Glossary.

Unit
6

Biosphere: Endangered Species

Seven articles examine the problems in the world's biosphere. Not only are plants and animals endangered, but so are many human groups who are disastrously affected by deforestation and primitive agricultural policies.

The concepts in bold italics are developed in the article. For further expansion please refer to the Topic Guide, the Index, and the Glossary.

Topic Guide

This topic guide suggests how the selections in this book relate to topics of traditional concern to students and professionals involved with environmental studies. It is useful for locating articles that relate to each other for reading and research. The guide is arranged alphabetically according to topic. Articles may, of course, treat topics that do not appear in the topic guide. In turn, entries in the topic guide do not necessarily constitute a comprehensive listing of all the contents of each selection.

TOPIC AREA	TREATED IN:	TOPIC AREA	TREATED IN:
Acid Rain	21. Common Threads	Environmental Changes	4. Environment of Tomorrow
Agriculture	16. Place for Pesticides?	Environmental Degradation	1. World Transformed 8. Landscape of Hunger 19. Ravaged Republics
Air Pollution	18. Where the Air Was Clear		
Alternative Energy	11. Here Comes the Sun		
Biodiversity	29. Origin and Function of Biodiversity 30. Rain Forest Entrepreneurs 34. New Species Fever	Environmental Impact	28. Desertification and Climate Change
		Environmental Issues	24. Desktop Farms, Backyard Farms, or No Farms?
Biofuels	12. Energy Crops for Biofuels	Environmental Law	25. 20 Years of the Clean Water Act
Biosphere	29. Origin and Function of Biodiversity	Environmental Planning	18. Where the Air Was Clear
Biotechnology	12. Energy Crops for Biofuels	Environmental Policy	1. World Transformed 31. Deforestation and Public Policy
Carrying Capacity	6. New Strategy for Feeding a Crowded Planet	Environmental Problems	7. Population: The Critical Decade 21. Common Threads 22. 25th Environmental Quality Index 32. Out of the Woods
Clean Water Act	25. 20 Years of the Clean Water Act		
Climate Change	28. Desertification and Climate Change 31. Deforestation and Public Policy		
Deforestation	30. Rain Forest Entrepreneurs	Environmental Protection	2. Mirage of Sustainable Development 3. GATT: Menace or Ally?
Desertification	28. Desertification and Climate Change	Environmental Quality	22. 25th Environmental Quality Index
Developing Countries	32. Out of the Woods	Environmental Regulations	3. GATT: Menace or Ally?
Domestic Animals	35. Barnyard Biodiversity	Environmentalist Groups	33. Killed by Kindness
Ecology	16. Place for Pesticides?		
Economic Development	9. Great Energy Harvest 26. Redeeming the Everglades 32. Out of the Woods	Extinction	35. Barnyard Biodiversity
		Family Planning	7. Population: The Critical Decade
Economic Growth	2. Mirage of Sustainable Development	Floodplain Management	23. Beyond the Ark: A New Approach to U.S. Floodplain Management
Economic Policy	1. World Transformed	Food Production	8. Landscape of Hunger 24. Desktop Farms, Backyard Farms, or No Farms?
Economic Problems	32. Out of the Woods		
Ecosystem	29. Origin and Function of Biodiversity		
Energy	9. Great Energy Harvest	Forests	32. Out of the Woods
Energy Conservation	10. What Would It Take to Revitalize Nuclear Power in the United States? 27. Global Warming on Trial	Fossil Fuels	11. Here Comes the Sun
		Genetic Diversity	35. Barnyard Biodiversity
Energy Consumption	14. All the Coal in China	Genetic Heritage	35. Barnyard Biodiversity
Energy Crops	12. Energy Crops for Biofuels	Genetic Variety	29. Origin and Function of Biodiversity
Energy Policy	13. Tilting Toward Windmills	Geothermal Energy	11. Here Comes the Sun
Energy Production	14. All the Coal in China	Global Energy	9. Great Energy Harvest
Energy Technology	11. Here Comes the Sun	Global Environment	5. How Many Is Too Many?
Environmental Activists	3. GATT: Menace or Ally? 26. Redeeming the Everglades		
Environmental Assessment	22. 25th Environmental Quality Index		

TOPIC AREA	TREATED IN:	TOPIC AREA	TREATED IN:
Global Warming	10. What Would It Take to Revitalize Nuclear Power in the United States? 14. All the Coal in China 21. Common Threads 27. Global Warming on Trial 30. Rain Forest Entrepreneurs	**Public Policy**	5. How Many Is Too Many? 10. What Would It Take to Revitalize Nuclear Power in the United States? 19. Ravaged Republics 26. Redeeming the Everglades
Greenhouse Effect	10. What Would It Take to Revitalize Nuclear Power in the United States? 27. Global Warming on Trial	**Public Works Project**	26. Redeeming the Everglades
		Recycling	17. Stewing the Town Dump in Its Own Juice
Habitat Destruction	33. Killed by Kindness	**Renewable Energy**	9. Great Energy Harvest 11. Here Comes the Sun 13. Tilting Toward Windmills
International Cooperation	2. Mirage of Sustainable Development 4. Environment of Tomorrow	**Renewable Resources**	27. Global Warming on Trial
Landfills	17. Stewing the Town Dump in Its Own Juice		
Natural Hazards	23. Beyond the Ark: A New Approach to U.S. Floodplain Management	**Resource Base**	24. Desktop Farms, Backyard Farms, or No Farms?
Nuclear Disaster	20. Chernobyl's Lengthening Shadow	**Solar Power**	11. Here Comes the Sun
Nuclear Power	10. What Would It Take to Revitalize Nuclear Power in the United States? 15. Facing Up to Nuclear Waste	**Solid Waste**	17. Stewing the Town Dump in Its Own Juice
Nuclear Waste	15. Facing Up to Nuclear Waste	**Species**	29. Origin and Function of Biodiversity 34. New Species Fever
Overpopulation	6. New Strategy for Feeding a Crowded Planet	**Sustainable Agriculture**	16. Place for Pesticides?
Ozone Depletion	21. Common Threads	**Sustainable Development**	2. Mirage of Sustainable Development 8. Landscape of Hunger
Pesticide	16. Place for Pesticides?		
Poaching	33. Killed by Kindness	**Sustainable Future**	4. Environment of Tomorrow
Political Action	5. How Many Is Too Many? 7. Population: The Critical Decade	**Sustainable Systems**	1. World Transformed
Pollution	19. Ravaged Republics	**Tropical Forests**	31. Deforestation and Public Policy
Population Growth	5. How Many Is Too Many? 6. New Strategy for Feeding a Crowded Planet 7. Population: The Critical Decade 8. Landscape of Hunger 27. Global Warming on Trial	**Water Pollution**	17. Stewing the Town Dump in Its Own Juice 25. 20 Years of the Clean Water Act
		Wetlands	26. Redeeming the Everglades
Preservation	34. New Species Fever	**Wildlife**	33. Killed by Kindness
Public Health	19. Ravaged Republics	**Wind Power**	11. Here Comes the Sun 13. Tilting Toward Windmills
Public Opinion	10. What Would It Take to Revitalize Nuclear Power in the United States? 13. Tilting Toward Windmills	**World Hunger**	8. Landscape of Hunger
		World Trade	3. GATT: Menace or Ally?

The Global Environment: An Emerging World View

The celebration of Earth Day 1990, the twentieth anniversary of the original Earth Day, came at a time when public apprehension over the environmental future of the planet reached levels unprecedented even during the activist days of the late 1960s and early 1970s. No longer were those concerned about the environment viewed as "eco-freaks" and "tree-huggers" as many serious scientists joined the rising clamor for environmental protection, as did the more traditional environmentally conscious public interest groups. There are a number of reasons for this increased environmental awareness. Some of these reasons arise from environmental events such as drought, heat wave, fire, and famine. But more arise simply from the increase in information and ideas about the global nature of environmental processes. For example, the raising of the Iron Curtain that has separated East and West since the end of World War II and the fragmentation of the former Soviet Union brought visions of the end of the cold war, a reawakening of the democratic spirit in Eastern Europe, and a hope for a more integrated global economy promising both peace and prosperity for the world's peoples. And that same raising of the barrier to the flow of people, goods, and services, has also allowed information and ideas to pass more freely between East and West. Much of what has been learned through this increased information flow, particularly by Western observers, has been of an environmentally ravaged Eastern Europe and Russia—a chilling forecast of what other industrialized nations will become in the near future unless strict international environmental measures are put in place. As distressing as the pictures and descriptions of forest destruction in eastern Germany and the Czech and Slovak Republics or the devastation of the Aral Sea have been, they have had a positive value. For perhaps the first time ever, countries are beginning to recognize that environmental problems have no boundaries and that international cooperation is the only way to solve them.

The subtitle of this first unit, "An Emerging World View," is an optimistic assessment of the future—a future in which less money is spent on defense and more on environmental protection and cleanup. The authors of the Worldwatch Institute's *State of the World* (a publication that has assumed a near-official status as the annual assessment of the global environment) have recently described a New World Order in which political influence will be based more upon leadership in environmental and economic issues than upon military might. Perhaps it is far too early to make such optimistic predictions, to claim that the decade of the 1990s will, indeed, be "The Decade of the Environment," or to conclude that the world's nations—developed and underdeveloped—will begin to recognize that Earth's environment is a single unit. Nevertheless, there is growing international realization that we are all, as environmental activists have said for decades, inhabitants of "Spaceship Earth" and that, as such, we will survive or succumb together.

The articles selected for this unit have been chosen to illustrate this increasingly global perspective on environmental problems and the degree to which environmental problems and their solutions are linked to political, economic, and social problems and solutions. In the lead piece of the section, Lester Brown, director of the Worldwatch Institute, discusses the developing global perspective and what steps are necessary to solve pressing environmental problems. In "The World Transformed," Brown points out that while the standards of living are falling in many developing nations as a result of environmental degradation, the very processes that are leading to economic and environmental dissolution are becoming more clearly understood. In particular, the concept of sustainable use of land and resources is becoming much more than a gleam in environmentalists' eyes—it is becoming official government policy in some regions, with the corresponding creation of international agencies designed to administer environmental regulations that cut across national boundaries. Not all writers on environmental issues agree with Brown's optimistic assessment of where the world is going, in terms of developing the principles of sustained use, or in terms of the manner in which sustained use should be administered. In "The Mirage of Sustainable Development," Thomas DiLorenzo, an economist, argues that sustainable development and use of resources will never work if guided by the principles that seem to prevail among most proponents of the concept. Brown and other advocates of sustained use have long operated upon the assumption that centralized international bureaucracies and agencies are necessary to carry out the coordinated environmental action upon which the notion of sustainable systems supposedly depends. DiLorenzo, on the other hand, suggests that such an approach (what he calls environmental socialism) would bankrupt the global economy and would, therefore, actually be harmful to the environmental systems themselves as potentially destructive environmental actions would be forced by economic necessity. He suggests that the concept of private property, the maintenance of free

markets, and the enforcement of strong liability laws—all anathema to most proponents of sustained use—are essential for a sustained environmental system and a growing economy. The third article in the unit deals with such international legislation. In "The GATT: Menace or Ally?" Hilary French of the Worldwatch Institute describes the conflict between the proponents of the General Agreement on Tariffs and Trade and those who see international trading agreements as a "GATTzilla" monster capable of wreaking untold damage upon environmental systems. As a point of departure, French uses a recent GATT-driven ruling that countries have a right to export tuna regardless of how the fish were caught. Since at least some tuna fishing poses considerable hazards to dolphin populations, such a ruling in favor of free and unrestricted trade poses the potential for considerable damage to dolphin populations in areas where the fishing nations may be too poor or too unconcerned to engage in the kind of dolphin-friendly fishing activities necessary to preserve this marine mammal. As necessary as they might be, international agreements that blur economic boundaries might also be environmentally disruptive.

The next article in the unit ties together many of the threads in the previous articles, taking an historical approach and, at the same time, focusing on the future. In "The Environment of Tomorrow," Martin Holdgate, director general of the World Conservation Union, notes that people have always adapted to environmental change and that this adaptation has often been imperceptible

except in retrospect. Such unconscious adaptation is no longer possible, given the rapidity of human-induced environmental change. If our species is to have a sustainable future on a diverse planet with many environmental inequalities, Holdgate asserts, it can only be through a process of international cooperation that transcends anything we see today.

Looking Ahead: Challenge Questions

What is the relationship between environmental degradation and economic systems? How has the growing awareness of environmental systems improved the chance for increased levels of environmental protection?

What is the conflict between the two major viewpoints on the value of international cooperation and international bureaucracies in the development and implementation of policies directed toward sustainable use of environments?

What conflicts exist between the proponents of international trade agreements and environmentalist groups? Can apparently worthwhile international endeavors, such as free trade agreements, actually have negative consequences for environmental quality?

How can the process of international cooperation increase significantly the chances for the development of sustainable human society? Are there differences between the goals of developed and developing countries that make the achievement of international cooperation on environmental issues difficult, if not impossible?

The World Transformed

Envisioning an Environmentally Safe Planet

Lester R. Brown

Living standards are falling in many countries due to environmental degradation. Fortunately, the changes needed to halt the decline are becoming clear, and some areas are reporting remarkable successes.

Lester R. Brown is president of the Worldwatch Institute, 1776 Massachusetts Avenue N.W., Washington, D.C. 20036. This article is adapted from State of the World 1993 *by Lester R. Brown et al. (W. W. Norton, 1993), which is available from the Futurist Bookstore for $10.95 ($9.95 for Society members), cat. no. B-1656.*

Many people have long understood, at least intuitively, that continuing environmental degradation would eventually exact a heavy economic toll. Unfortunately, no global economic models incorporate the depletion and destruction of the earth's natural support systems. Now, however, we can begin to piece together information from several recent independent studies to get a sense of the worldwide economic effects of environmental degradation. Among the most revealing are studies on the effects of air pollution and acid rain on forests in Europe, of land degradation on livestock and crop production in the world's dryland regions, of global warming on the U.S. economy, and of pollution on health in Russia.

These reports and other data show that the fivefold growth in the world economy since 1950 and the increase in population from 2.6 billion to 5.5 billion have begun to outstrip the carrying capacity of biological support systems and the capacity of natural systems to absorb waste without being damaged. In country after country, demand for crops and for the products of grasslands, forests, and fisheries are exceeding the sustainable yield of these systems. Once this happens, the resource itself begins to shrink as natural capital is consumed. Overstocking grasslands, overcutting forests, overplowing, and overfishing are now commonplace. Every country is practicing the environmental equivalent of deficit financing in one form or another.

Perhaps the most visible environmental deficit is deforestation, the result of tree cutting and forest clearing that exceeds natural regrowth and tree planting. Each year this imbalance now costs the world some 17 million hectares of tropical forests alone. Over a decade, the destruction of tropical forests clears an area the size of Malaysia, the Philippines, Ghana, the Congo, Ecuador, El Salvador, and Nicaragua. Once tropical forests are burned off or clear-cut, the land rapidly loses its fertility, since most of the nutrients in these ecosystems are stored in the vegetation. Although these soils can be farmed for three to five years before fertility drops and can be grazed for five to 10 years before becoming wasteland, they typically will not sustain productivity over the long term.

Clearing tropical forests is, in effect, the conversion of a highly productive ecosystem into wasteland in exchange for a short-term economic gain. As timber resources are depleted in the Third World, transforming countries that traditionally exported forest products into importers, logging companies are turning to remote temperate-zone forests. Canada, for example, is now losing 200,000 hectares a year as cutting exceeds regeneration by a wide margin. Similarly, as Japanese and Korean logging firms move into Siberia, the forests there are also beginning to shrink.

It is not only the axe and chainsaw that threaten forests, but also emissions from power plant smokestacks and automobile exhaust pipes. In Europe, air pollution and acid rain are damaging and destroying the region's traditionally well-managed forests. Scientists at the International Institute for Applied Systems Analysis (IIASA) in Austria have estimated the effect on forest productivity of sulfur-dioxide emissions from fossil-fuel-burning power plants, factories, and automobiles. They concluded that 75% of Europe's forests are now experiencing damaging levels of sulfur deposition. IIASA estimated the losses associated with the deterioration of Europe's forests at $30.4 billion each year, roughly equal to the annual output of the German steel industry. These losses took account not only from the losses of wood but of the costs of increased flooding, lost soil, and the silting of rivers.

Degraded Lands, Depleted Seas

Land degradation is also taking a heavy economic toll, particularly in the drylands of the earth. Using data for 1990, a United Nations assessment of the earth's dryland regions estimated that the degradation of irrigated cropland, rain-fed cropland, and rangeland now costs the world more than $42 billion a year in lost crop and livestock output, a sum that approximates the value of the U.S. grain harvest.

In Africa, where land degradation is most visible, the annual loss of rangeland productivity is estimated at $7 billion, more than the GNP of Ethiopia and Uganda combined. And lost productivity on Africa's rain-fed cropland, largely from soil erosion, totals $1.9 billion, roughly the same as Tanzania's GNP.

Excessive demand directly threatens the productivity of oceanic fisheries as well. The U.N. Food and Agriculture Organization, which monitors oceanic fisheries, indicates that four out of 17 of the world's fishing zones are now overfished. It also reports that most traditional marine fish stocks have reached full exploitation. For example, Atlantic stocks of the heavily fished bluefin tuna have been cut by a staggering 94%. It will take years for such species to recover, even if fishing were to stop altogether.

Dwindling fish stocks are affecting many national economies. In Canada, for example—where the fishing industry traditionally landed roughly 1.5 million tons of fish a year, worth $3.1 billion—depletion of the cod and haddock fisheries off the coast of Nova Scotia has led to shrinking catches and heavy layoffs in the fishing and fish processing industries. In July 1992, in an unprecedented step, Canada banned all cod fishing off the coast of Newfoundland and Labrador for two years in a bid to save the fishery. To cushion the massive layoffs in the industry, the mainstay of Newfoundland's economy,

Photovoltaic operations, such as this one in Kerman, California, may prove to be a cheaper and much safer alternative to using fossil fuels. This experimental 10-acre spread provides electricity for almost 200 homes.

PACIFIC GAS & ELECTRIC

Ottawa authorized a $400-million aid package for unemployment compensation and retraining.

The rising atmospheric concentration of greenhouse gases is potentially the most economically disruptive and costly change that has been set in motion by our modern industrial society. William Cline, an economist with the Washington-based Institute for International Economics, has looked at the long-term economic effects of global warming. As part of this study, he analyzed the effect of a doubling of greenhouse gases on the U.S. economy, which could come as early as 2025. He estimates that heat stress and drought would cost U.S. farmers $18 billion in output, that the increased electricity for air conditioning would require an additional $11 billion, and that dealing with sea-level rise would cost about $7 billion per year. In total, Cline estimates the cost at nearly $60 billion, roughly 1% of the 1990 U.S. GNP.

Not all countries would be affected equally. Some island countries, such as the Republic of the Maldives in the Indian Ocean, would become uninhabitable. Low-lying deltas, such as Egypt and Bangladesh, would be inundated, displacing millions of people. In the end, rising seas in a warming world would be not only economically costly, but politically disruptive as well.

Pollution of the environment exacts a price from every society. Contamination of air, water, and soil by toxic chemicals and radioactivity, along with increased ultraviolet radiation, is damaging human health, running up health-care costs. One study for the United States estimates that air pollution may cost the nation as much as $40 billion annually in health care and lost productivity.

New data from Russia show all too well the devastating effect of pollution by chemical and organic toxins on human health. At an October 1992 news conference, Vladimir Pokrovsky, head of the Russian Academy of Medical Sciences, shocked the world with his frankness: "We have already doomed ourselves for the next 25 years." He added: "The new generation is entering adult life unhealthy. The Soviet economy was developed at the expense of the population's health." Data released by the Academy show 11% of Russian infants suffering from birth defects. With half the drinking water and a tenth of the food supply contaminated, 55% of school-age children suffer health problems. The Academy reported that the increase in illness and early death among those aged 25–40 was particularly distressing. The bottom line is that Russian life expectancy is now falling.

The environmental deficits and debts that the world has incurred in recent decades are enormous, often dwarfing the economic debts of nations. Perhaps more important is the often overlooked difference between economic deficits and environmental ones. Economic debts are something we owe each other. For every borrower there is a lender; resources simply change hands. But environmental debts, especially those that lead to irreversible damage or losses of natural capital, can often be repaid only in the deprivation and ill health of future generations.

Envisioning a New World

That the existing economic system is slowly beginning to self-destruct as it undermines its environmental support systems is evident. The challenge is to design and build an economic system that is environmentally sustainable. Can we envision what this would look like? Yes. And can we devise a strategy for getting from here to there in the time that is available? Again, the answer is yes.

The basic components of an effort to build an environmentally sustainable global economy are rather straightforward. They include reestablishing climate stability, protecting the stratospheric ozone layer, restoring the earth's tree cover, stabilizing soils, safeguarding the earth's remaining biological diversity, and restoring the traditional balance between births and deaths. Endowed with a certain permanence, this new society will be far more satisfying than the ephemeral, throwaway society we now live in.

In this new world, energy would be used far more efficiently. Traditional incandescent light bulbs, for example, would be replaced by the latest compact fluorescents, which provide the same light but use only a quarter as much electricity. In sector after sector—transportation, manufacturing, housing, agriculture—energy would be used with an efficiency dwarfing that of today.

Once we recognize that fossil fuels are no longer environmentally feasible and that nuclear power is not economically or environmentally viable, our only option is solar energy. This energy source takes many forms: hydropower, wind power, solar thermal power plants, firewood, photovoltaic cells, agricultural

Solar energy can be used in a number of ways. In Memmingen, Germany, this former gas station uses electricity from solar cells to feed current into an electrically powered car.

California-based "wind farms" like this produced 3 billion kilowatt-hours of electricity in 1992, providing power to almost 400,000 homes.

wastes, rooftop solar water heaters, alcohol fuels from sugar cane, and many more. All are likely to play a role. Hydropower already supplies one-fifth of the world's electricity. The world's wind power potential may be far greater.

The costs of photovoltaic cells are coming down fast, and these are likely to play an important role eventually. But for large-scale production during this decade, solar thermal power plants of the sort built in Southern California will probably have the edge. The latest of these converts a phenomenal 22% of sunlight into electricity and produces it for 8¢ per kilowatt hour—well below the 12¢ for electricity from new nuclear plants in the state, but still above the 6¢ for that from coal-fired plants.

As costs continue to drop, the resulting cheap solar electricity permits the production of hydrogen fuel from the simple electrolysis of water. Hydrogen provides a means of both storing solar energy and transporting it efficiently over long distances, either by pipeline or by tanker, much the same way that natural gas is transported.

Fortunately, all the world's major population concentrations are located near areas rich in sunlight. For the densely populated northeastern United States and industrial Midwest, it is the sun-drenched South-

west. Natural-gas pipelines that already link Texas and Oklahoma gas fields to these regions might one day carry hydrogen fuel. For the European Community, the source will be southern Spain (which has a solar regime similar to Southern California's) and the coast of North Africa. The Commonwealth of Independent States can turn to the sun-rich Asian republics; India, to the Thar Desert in the northwest; China, and possibly Japan, to central and northwest China, a largely desert region.

Balancing Births and Deaths

Besides energy efficiency, building a new world means reestablishing the balance between births and deaths that prevailed throughout most of our existence as a species. At least one government, working with the World Conservation Union, has recently undertaken such an exercise. In a 1992 study, the government of Pakistan concluded that the country might be able to support an increase in population from 122 million at present to the 200 million projected by 2013, if resources are managed carefully, but that there is no possibility of sustaining the 400 million projected by 2040 if current population growth trends continue. Their assessment of potential population support capacities by agroecological zones suggests that food shortages will come first in the more ecologically fragile low-rainfall regions, and then spread, achieving massive dimensions and enveloping the entire country.

Raising public understanding of the

Mazda's concept car runs on hydrogen-based fuels, which may soon become a viable alternative to gasoline. Cheap electricity produced from solar energy could make large-scale production possible by using simple water electrolysis.

CARL PURCELL / AID

Family-planning workers such as this one in Bombay, India, can help keep the number of births in check and reduce strains on resources.

population threat, filling the gap in the availability of family-planning services, and creating the social conditions conducive to lower fertility are keys to an environmentally sustainable future.

Glimpses of what our new world would look like can be seen here and there: Japan, for example, set the international standard for energy efficiency and, in the process, strengthened its competitive position in world markets. California, with its rapidly growing solar and wind industries, provides a window on future energy sources. Its solar thermal electric generating capacity of 350 megawatts and its wind generating capacity of 1,600 megawatts are enough to satisfy the residential needs of nearly 2 million people, easily enough for San Francisco, San Diego, and Sacramento combined.

On population, Thailand stands out: It has halved its population growth rate in less than 15 years. In the effort to stabilize soils, the United States is leading the way; its innovative Conservation Reserve Program adopted in 1985 reduced soil erosion by one-third by 1990 and may cut it by another third by 1995. In the effort to reduce garbage, Germany is a model with its comprehensive program designed to force companies to assume responsibility for disposing of the packaging used with their products. With beverage containers, Denmark has gone even further—banning those that are not refillable.

In the end, we need not merely a vision, but a shared vision, one that guides and unites us in our day-to-day decision making. Such a vision, a common blueprint, can infuse society with a sense of purpose as we try to build a new world, one much more attractive than today's. This sense of common purpose and excitement is essential if we are to create an environmentally sustainable economic system.

Facing Up to Change

The deterioration in living conditions for much of humanity during the 1980s and early 1990s will continue if economies are not restructured. The fall in incomes in 49 countries between 1980 and 1990 was not an accident or a statistical quirk. It was too large, involved too many countries and too many people, and occurred over too long a time. The great majority of these countries are poor ones where livelihoods are directly dependent on the productivity of croplands, grasslands, and forests. It is in these largely agrarian economies that the link between deteriorating natural systems and living conditions is most direct, and the effects most visible.

Even a casual survey of the planet's physical condition shows

GERMAN INFORMATION CENTER

German workers show off the "Blue Angel" seal, already used on more than 3,000 products to signify that they are made from recycled plastics. German companies are now held responsible for the disposal of their products' packaging materials.

that the costs of burning fossil fuels are rising on many fronts. At some point, the economic costs of deteriorating forests, dying lakes, damaged crops, respiratory illnesses, increasing temperatures, rising sea level, and other destructive effects of fossil-fuel use become unacceptably high. Basic economics argues for a switch to solar energy. Rather than wondering if we can afford to respond to these threats, policy makers should consider the costs of not responding.

The question is whether we will initiate the changes in time and manage the process, or whether the forces of deterioration and decline will prevail, acquiring a momentum of their own.

It is not clear what a deterioration/decline scenario would look like since we have little experience in the modern era with which to judge it. Will our social institutions be overwhelmed by deterioration and decline? Would this scenario look like Lebanon over the past decades? Mozambique, Peru, Somalia, or Sudan in recent years? Yugoslavia in 1992? Will it be a world where dying ecosystems and rising sea levels generate massive flows of environmental refugees, movements that generate social conflicts and overwhelm national borders? Precisely because this is an out-of-control scenario, it is difficult to visualize the form it will take.

Getting off the deterioration/decline path requires an enormous effort, one akin to mobilizing for war. Turning things around begins with us as individuals. We each can do many things. We can recycle, use energy and water more efficiently, and

limit our families to two children. These individual actions, though necessary, are not sufficient. They will not bring about the basic structural changes in the economy needed to make it sustainable. For this, only citizen action to press governments to adopt policies that will transform the economy will suffice.

The overriding need is for a new view of the world, one that reflects environmental realities and that redefines security by recognizing that the overwhelming threat to our future is not military aggression, but the environmental degradation of the planet. Among the principal policy instruments that can convert an economic system that is slowly self-destructing into one that is environmentally sustainable are regulations and tax policy. Until now, governments have relied heavily on regulation, but the record of the last two decades shows that this is not a winning strategy.

Regulations clearly do have a role to play, however. Environmentally damaging chemicals, such as CFCs, can be banned. Regulations are needed on the handling of toxic waste and radioactive materials—things too dangerous to leave to the marketplace. Energy efficiency standards for automobiles and household appliances cut carbon emissions, air pollution, and acid rain.

To transform the economy quickly, however, by far the most effective instrument is tax policy, specifically the partial replacement of income taxes with environmental taxes. An advantage of using tax policy is that it permits the market to work unimpeded, preserving its inherent effi-

ciency. Today, governments tax income because it is an easy way to collect revenue, even though it serves no particular social purpose. Replacing a portion of income taxes with environmental taxes would help to transform the economy quickly. This shift would encourage work and savings and would discourage environmentally destructive activities. In short, it would foster productive activities and discourage destructive ones, guiding both corporate investments and consumer expenditures.

Among the activities that would be taxed are the burning of fossil fuels, the production of hazardous chemical waste, the generation of nuclear waste, pesticide use, and the use of virgin raw materials. The adoption of a carbon tax, which would discourage the burning of fossil fuels, is now being actively considered in both the European Community and Japan. Even a modest carbon tax would quickly tip the scales away from investment in fossil-fuel production and toward investment in energy efficiency and renewable energy sources. Within a year of adoption of an international carbon tax, literally scores of solar thermal plants could be under construction.

We know what we have to do. And we know how to do it. If we fail to convert our self-destructing economy into one that is environmentally sustainable, future generations will be overwhelmed by environmental degradation and social disintegration. Simply stated, if our generation does not turn things around, our children may not have the option of doing so.

THE MIRAGE OF SUSTAINABLE DEVELOPMENT

HOW DO WE ACHIEVE BOTH ECONOMIC GROWTH AND ENVIRONMENTAL PROTECTION? AN ECONOMIST ARGUES THAT PRIVATE-PROPERTY RIGHTS WILL WORK BETTER THAN INTERNATIONAL BUREAUCRACIES.

THOMAS J. DILORENZO

Thomas J. DiLorenzo is professor of economics in the Sellinger School of Business and Management, Loyola College, Baltimore, Maryland 21210. He has written and lectured extensively on public finance and public-policy-oriented issues and is an editorial referee for 15 academic journals.

A longer version of this article was published by The Center for the Study of American Business (CSAB), a nonpartisan research organization at Washington University in St. Louis. Copies of "The Mirage of Sustainable Development," Contemporary Issues Series 56 (January 1993), are available from CSAB, Washington University, Campus Box 1208, One Brookings Drive, St. Louis, Missouri 63130-4899, telephone 314/935-5630.

There is no precise definition of sustainable development. To some, it simply means balancing economic growth with environmental-protection goals, a relatively uncontroversial position. But to others, it means something different: dramatic reductions in economic growth in the industrialized countries coupled with massive international income redistribution.

According to advocates of the latter viewpoint, there are not enough resources left worldwide to sustain current economic growth rates, and these growth rates are also too damaging to the environment. Consequently, these advocates argue for government regulation of virtually all human behavior on a national and international scale and for governmental control of privately owned resources throughout the world. Such controls may be enforced by national governments or by international bureaucracies such as the United Nations. The "lesson" to be learned from the tragic failures of socialism, the sustainability advocates apparently believe, is that the world needs more socialism.

Such views would be dismissed as bizarre and irrational if they were not held by someone as influential as Norway's Prime Minister Gro Harlem Brundtland, who also chairs the United Nations World Commission on Environment and Development. This Commission published its views in a 1987 book, *Our Common Future*, which laid the groundwork for the June 1992 "Earth Summit" held in Rio de Janeiro.

But the policy proposals advocated by *Our Common Future* and the Earth Summit fail to recognize the many inherent flaws of governmental planning and regulation, and they ignore the important role of private-property rights, technology, and the market system in alleviating environmental problems.

What Role for Property Rights And Free Markets?

The final collapse of communism in 1989 revealed a dirty secret: that pollution in the communist world was far, far worse than virtually anywhere else on the planet. In theory, this should not have been the case, for it has long been held that the profit motive and the failure of unregulated markets to provide incentives to internalize external costs were the primary causes of pollution. Government regulation or ownership of resources was thought to be a necessary condition for environmental protection.

But that was just a theory. The reality is that, in those countries where profit seeking was outlawed for decades and where government claimed ownership of virtually all resources, pollution and other forms of environmental degradation were

devastating. According to the United Nations' Global Environment Monitoring Program, pollution in central and eastern Europe "is among the worst on the Earth's surface."

In Poland, for example, acid rain has so corroded railroad tracks that trains are not allowed to exceed 24 miles an hour. Ninety-five percent of the water is unfit for human consumption, and most of it is even unfit for industrial use, so toxic that it will dissolve heavy metals. Industrial dust rains down on towns, depositing cadmium, lead, zinc, and iron. Half of Poland's cities do not even treat their wastes, and life expectancy for males is lower than it was 20 years ago.

The landscape is similar in other parts of central and eastern Europe, in the former Soviet Union, and in China. Eighty percent of the surface waters in former East Germany are classified unsuitable for fishing, sports, or drinking. One out of three lakes has been declared biologically dead because of decades of dumping untreated chemical waste. Some cities are so polluted that cars must use their headlights during the day. Bulgaria, Hungary, Romania, the former Yugoslavia, and the former

Czechoslovakia suffered similar environmental damage during the decades of communism.

These sad facts teach important lessons that the sustainable development theorists have not learned. The root cause of pollution in the former communist world, and worldwide, is not the profit motive and unregulated markets, but the absence of property rights and sound liability laws that hold polluters responsible for their actions. The environmental degradation of the former communist world is an example of one massive "tragedy of the commons," to borrow the phrase coined by biologist Garrett Hardin. Where property is communally or governmentally owned and treated as a free resource, resources will inevitably be overused with little regard for future consequences.

But when people have ownership rights in resources, there is a stronger incentive to protect the value of those resources. Furthermore, when individuals are not held liable for damages inflicted on others—including environmental damages—then there is little hope that responsible behavior will result. Needless to say, the state did not

Industrial effluents pour into river near Bitterfeld in eastern Germany. Decades of dumping untreated chemical wastes in the former East Germany have left one out of three lakes biologically dead and 80% of all surface waters unsuitable for human use.

hold itself responsible for the environmental damage it was causing in the former communist countries. Thus, far from being the answer to environmental problems, pervasive governmental control of natural resources was the cause.

Our Common Future's Misinterpretations

Our Common Future neglects the role of property rights, and, consequently, it grants entirely too much credence to the efficacy of greater governmental controls and regulations as solutions to environmental problems. Several examples stand out.

• **Deforestation.** International economic relationships "pose a particular problem for poor countries trying to manage their environments," says *Our Common Future*. For example, "the trade in tropical timber . . . is one factor underlying

Factory in Espenhain, eastern Germany, emits foul pollutants, darkening the sky. Some cities in the formerly communist half of the country are so polluted that cars must drive with their headlights on during the daytime, says author DiLorenzo.

COURTESY OF GERMAN INFORMATION CENTER

tropical deforestation. Needs for foreign exchange encourage many developing countries to cut timber faster than forests can be regenerated."

But the need for foreign exchange is not unique to people in developing countries. All individuals prefer more to less, but they do not all cut down and sell all the trees in sight for economic gain. Deforestation was also a massive problem in the former communist countries, but the main reason was that the forests were communally owned. Consequently, anyone could cut them down, and there were virtually no incentives to replant because of the absence of property rights.

Deforestation has also taken place in democratic countries, primarily on government-owned land that is leased to timber companies who, since they do not own the land, have weak incentives to replant and protect its future value. Some of these same timber companies are very careful indeed not to overharvest or neglect replanting their own private forest preserves. They do so not so much out of a desire to protect the environment as to protect the value of their assets. Well-enforced property rights and the existence of a market for forest products will assure that forests are likely to be used wisely, not exploited.

• **Desertification.** The sustainable development theorists also misdiagnose the problem of desertification—

the process whereby "productive arid and semiarid land is rendered economically unproductive," as *Our Common Future* defines it. They blame capitalism for desertification, particularly "the pressures of subsistence food production, commercial crops, and meat production in arid and semiarid areas." Their "solution" is greater governmental controls on agriculture.

Desertification is undoubtedly a problem throughout the world—including parts of the United States. The primary cause is not commercial agriculture, however, but the tragedy of the commons.

A particularly telling example of the importance of private property to desertification was reported in *Science* magazine in a 1974 article on desertification in the Sahel area of Africa. At the time, this area was suffering from a five-year drought. NASA satellite photographs showed a curiously shaped green pentagon that was in sharp contrast to the rest of the African desert. Upon investigation, scientists discovered that the green blotch was a 25,000-acre ranch, fenced in with barbed wire, divided into five sectors, with the cattle allowed to graze only one sector a year. The ranch was started at the same time the drought began, but the protection afforded the land was enough to make the difference between pasture and desert.

• **Wildlife management.** The

Earth Summit advocated a "biodiversity treaty" whereby national governments would establish policies aimed at slowing the loss of plant and animal species. The type of policies most preferred by sustainable development theorists include prohibition of commercial uses of various plants and animals, such as the ban on ivory from African elephants, and the listing of more "endangered species," which may then be "protected" by governments on game preserves or elsewhere.

There is growing evidence, however, that the best way to save truly endangered species is not to socialize them, but to allow people to own them. As conservationist Ike Sugg has written:

[W]here governments allow individuals to reap the economic benefits of conserving and protecting their wildlife stocks—wildlife flourish. Where individuals are denied the opportunity to profit from wildlife legally, they do so illegally and without the sense of responsibility that comes with stewardship.

African elephants are protected by bans on ivory trade and other government measures. But there is growing evidence that private-property rights over wildlife may be more effective than "socialization" in protecting endangered species, according to DiLorenzo.

One particularly telling example that illustrates Sugg's point is the African elephant. Kenya outlawed elephant hunting in the 1970s; its elephant population quickly *dropped* from 140,000 in 1970 to an estimated 16,000 today as illegal poaching proliferated.

In contrast, Zimbabwe had only 30,000 elephants in 1979 but has over 65,000 today. The main reason for these differences, according to Sugg, is that in 1984 the government of Zimbabwe granted citizens ownership rights over elephants on communal lands—a large step in the direction of defining property rights. As expressed by one tribal chief who implicitly understood the value of property rights and the commercialization of elephants:

For a long time the government told us that wildlife was their resource. But I see how live animals can be our resources. Our wealth. Our way to improve the standard of living without waiting for the government to decide things. A poacher is only stealing from us.

The preservation of endangered wildlife through private-property rights and free markets is also prevalent in parts of the United States in the form of game ranching, which typically involves "exotic" or non-native animals. Game rancher David Bamberger of Texas, for example, has preserved 29 of the 31 remaining bloodlines of the Scimitar-horned Oryx, a rare antelope that is virtually extinct in its native Africa. Despite such successes, several states have outlawed game ranching because the notion of privatizing wildlife is blasphemy to the "religion" of environmentalism (not to mention "animal rights"), which holds that markets and property rights decimate species.

The principle of using property rights and market incentives to protect global resources and the environment applies to a wide range of problems, including the exploitation of water resources in the American West, the mismanagement of government-owned forest lands, the overfishing of public lakes and streams, and even the ocean commons: The Law of the Sea Treaty, which the United States has thus far refused to sign, would establish the oceans as the largest government-owned and regulated commons on Earth—and, inevitably, the largest tragedy of the commons.

This elementary principle, however, is not even acknowledged by the United Nations' sustainable development theorists. In answering the question, "How are individuals . . . to be persuaded or made to act in the common interest?," the Brundtland Commission answered with "education," undefined "law enforcement," and eliminating "disparities in economic and political power." No mention was made of the role of property rights in shaping incentives.

Sustainable Delusions

Sustainable development—as it is defined by the Brundtland Commission and the planners of the Earth Summit—can best be understood as a euphemism for environmental socialism—granting governments more and more control over the allocation of resources in the name of environmental protection. But if any lesson can be learned from the collapse of socialism in the former communist countries, it is that government ownership and control of resources is a recipe for economic collapse and environmental degradation. Socialism is no more effective in protecting the environment than it is in creating wealth.

Government ownership of natural resources inevitably leads to the tragedy of the commons, but that is all too often the "solution" offered by the Brundtland Commission. The Commission recommends government control of everything from outer space to energy, which is supposedly "too important for its development to continue in such a manner" as the free market allows.

Perhaps the top priority of sustainable development theorists is to expand the international welfare state by agitating for wealth transfers from "the rich" countries to the developing. But the whole history of development aid is government-to-government—most of it is typically used to finance the expansion of governmental bureaucracies in the recipient countries—which can be adverse to economic development. Even if most of the aid did make it into the hands of the citizens of the recipient countries, sustainable development theorists do not explain how that will translate into savings, investment, capital formation, and entrepreneurial activity—the ingredients of economic development.

THE SOVIET ENVIRONMENTAL LEGACY

Before its demise, the Soviet Union had the world's strictest, most detailed environmental regulations. But they were never enforced, says RAND researcher D. J. Peterson.

In his new book, *Troubled Lands*, Peterson points out that, in the late 1950s, once the Soviet economy began to decline and large-scale environmental degradation increased, the Communist Party passed a number of laws aimed at protecting land, water, air, and wildlife. But the regulations and codes met insurmountable obstacles in the Soviet bureaucracy itself, says Peterson. For one thing, responsibility for carrying out the government's efforts at environmental protection was divided among several ministries and state committees, all with other priorities besides protecting the environment. The regime's development imperative measured ministries' performance by how many tons of cement were produced, for instance, or how many hectares of land were irrigated.

Since environmental degradation—in the Soviets' mind—was associated with bourgeois development in the capitalist world, it took the Chernobyl disaster to wake the system up to reality. Peterson quotes one physicist: "Before the Chernobyl explosion, many important specialists and political figures believed that a nuclear reactor could not explode."

Peterson believes that environmental concern in the Soviet Union contributed to the forces that brought about the collapse of Communist control. "For Soviet society, the state of the environment, in physical terms, epitomized the state of the Union," he says. "Environmental destruction, added to social, economic, and political stresses, compounded the people's anger and ultimately undermined the Soviet regime. The Soviet Union can be relegated to history, but its *dostizheniya* (achievements), manifest in the legacy of widespread environ-

mental destruction, cannot be easily erased. Will the emerging post-Soviet societies cope with the challenge?"

There is a danger that zealous new capitalists, both those within the former Soviet Union and those representing international investors, will overexploit the region's vast natural resources, creating a backlash of public sentiment away from cooperation with the global community, Peterson warns.

"The West bears responsibility, when possible, not to violate the good faith and hopes of the emerging societies in the region. . . . The scale of the post-Soviet environmental challenge, in terms of global interdependence, mandates cooperation to support the democratic alternative," he concludes.

Source: *Troubled Lands: The Legacy of Soviet Environmental Destruction* by D.J. Peterson. Westview Press, 5500 Central Avenue, Boulder, Colorado 80301. 1993. 276 pages. Paperback. $19.95.

Finally, the theory of sustainable development commits in grand fashion the mistake of what Nobel laureate Friedrich von Hayek called "the pretense of knowledge." The detailed and constantly changing "information of time and place" required to produce even the simplest of items efficiently is so immense and so widely dispersed that no one human mind or group of minds with the largest computer in existence could imitate to any degree the efficiency of a decentralized market system. This, after all, is the principal lesson to be learned from the worldwide collapse of socialism.

Moreover, the larger and more complex an economy becomes, the more remote the likelihood that governmental planning could be anything but guess work. As Hayek states in his 1988 book, *The Fatal Conceit*,

By following the spontaneously generated moral traditions underlying the competitive market order . . . we generate and garner greater knowledge and wealth than could ever be obtained or utilized in a centrally directed economy. . . . Thus socialist aims and programs are factually impossible to achieve or execute.

The theory of sustainable development calls for myriad varieties of *international* central planning of economic activity. If the "pretense of knowledge" is fatal to attempts at governmental planning at the national level, the belief that international or global planning could possibly succeed is untenable.

The Brundtland Commission's recommendation that every govern-

ment agency in the world engage in economic planning and regulation in the name of environmental protection would lead to a massive bureaucratization of society and, consequently, a sharp drop in living standards. The image of millions of "green" bureaucrats interfering in every aspect of our social and economic lives is frightening.

The irony of it all is that the wealthier economies are typically healthier and cleaner than the poorer ones. By impoverishing the world economy, "sustainable development" would, in fact, also be harmful to the environment. Private property, free markets, and sound liability laws—anathemas to the theory of sustainable development—are essential for a clean environment and for economic growth.

THE GATT:
MENACE OR ALLY?

The world's free-trade interests seem bent on expanding their commercial powers even if that means jeopardizing any conflicting environmental laws. Can these powers be turned to the Earth's advantage?

HILARY F. FRENCH

Hilary F. French is a senior researcher at the Worldwatch Institute, and co-author of State of the World 1993.

From Embassy Row to Capitol Hill in Washington, D.C., it suddenly seemed as though they were everywhere: in the fall of 1991, posters began popping up around the city showing a "GATTzilla" monster with a dolphin in one hand and a can of pesticides in the other, crushing the U.S. Capitol under its foot. The caption: "What you don't know *can* hurt you." The posters were soon followed by a series of full-page advertisements in major newspapers around the country signed by a coalition of environmental and consumer groups warning that the General Agreement on Tariffs and Trade (GATT), the international agreement that stipulates world trade rules and arbitrates disputes over its terms, posed little-known but grave environmental threats. The ads called for a grassroots campaign to turn back efforts to expand GATT's powers through the Uruguay Round of negotiations, which had been underway since 1986 and was thought at the time to be nearing completion. (More than two years later, the Uruguay Round is still going around, though predictions are once again rife that a deal is near.)

How could an arcane international agreement to reduce trade barriers among more than 100 countries harm the environment? In a number of ways, according to the advertisements. Most fundamentally, the anti-GATT activists worried that environmental laws would be found to violate world trade rules—and would be overturned. The fear was aroused by a GATT dispute panel ruling that provisions of the U.S. Marine Mammal Protection Act violated the GATT, and it has been further excited by a rash of recent environmental trade disputes. For instance, Austria was recently forced to abandon plans to introduce a 70 percent tax on tropical timber, as well as a requirement that tropical timber be labeled as such, when the Association of Southeast Asian Nations (ASEAN) complained that the law violated GATT. In two ongoing disputes, the United States is charging that a levy imposed by the Canadian province of Ontario on non-refillable alcoholic beverage containers is a disguised trade barrier, and the European Community has formally challenged two U.S. automobile taxes intended to promote fuel efficiency—the Corporate Average Fuel Economy Law and the gas-guzzler tax.

The GATT-alarm ads painted a global conspiracy theory, according to which opponents of U.S. laws on environmental, health, and consumer safety legislation who had tried and failed to roll back decades of progress through the democratic process were now aiming to achieve their goals through the back door of the secretive, corporate-controlled GATT proceedings.

The international trade community was taken aback by this "demonization" of the

GATT, which many viewed as a key to the relative prosperity enjoyed by nations in the post-war era—a triumph of efforts to protect the collective good over the selfish goals of "protectionist" special interests. Since its creation in 1947, the GATT has indeed been remarkably successful on its own terms. Over the course of seven different negotiating rounds, tariffs have been cut in industrial countries from an average of 40 percent in 1947 to 5 percent in 1990.

The characterization of GATT as an imposing monster bore a certain irony, since many countries look to the multilateral trading system embodied by GATT as a means of protecting their interests against efforts by economic powerhouses, especially the United States, to unilaterally impose *their* will on the world. Developing countries viewed the environmental campaign against the GATT with particular alarm, both as part of what they saw as an unfortunate tendency on the part of Northern Greens to care more about whales and dolphins than about people, and as a cover for more sinister efforts to keep Third World goods out of northern markets.

In the intervening years, some progress has been made in merging these clashing views.

The heart of the GATT ruling on environment says that countries should not be allowed to use trade to influence practices outside their borders.

Governments have committed themselves to making trade and the environment "mutually supportive," though they have a long way to go before determining exactly how. The GATT itself, however, remains very much a product of its times. When the original agreement was forged in 1947, protecting the environment was not yet on most national agendas, let alone a pressing international concern. The General Agreement on Tariffs and Trade urgently needs updating and clarification if it is to become an instrument for furthering, rather than undermining, the goal to which governments

pledged themselves at the June 1992 Rio "Earth Summit." That goal, the environmentalists like to remind the GATT, is to find a path to development that does not deplete the resource base upon which future economic well-being depends.

The Tuna-Dolphin Challenge

What brought the issue to a head in late 1991 was the outrage over a GATT panel's ruling that Mexico had a valid case in arguing that it should be allowed to import tuna to the United States regardless of how it was caught. Mexican fishers use dolphins as markers for tuna swimming below, before setting out purse-seine nets which then ensnare the dolphins as well as the targeted tuna. Though this practice was once also prevalent among U.S. fishers, the 1972 Marine Mammal Protection Act effectively outlawed it by mandating tight dolphin mortality quotas for domestic and imported tuna alike (see "The Tuna Test," *World Watch,* March-April 1992). The panel ruling sent shockwaves through the environmental community, as it called into question the GATT-compatibility of a gamut of trade measures used to achieve environmental ends.

Though there had been cases in the past in which health and safety laws had been challenged as trade barriers, the tuna-dolphin ruling provoked a far greater backlash. That Mexico had won its case meant that a U.S. law had been not only questioned but struck down—and that could have led to the law's repeal, as bucking GATT's authority would not stand the United States in good stead when its turn came to charge another country with being out of step with world trade rules. As it happened, the U.S. law remains in place because Mexico decided not to press the point, not wanting to antagonize the United States in the midst of negotiations over a North American Free Trade Agreement (NAFTA). Most fundamentally, however, the ruling provoked cries of alarm not because of any nationalistic pride on the part of environmentalists, but because of the reasoning employed by the GATT panelists. If similar logic were applied in future cases, provisions of a large number of national environmental laws and even international treaties could be overturned. The ruling thus focussed attention on shortcomings in the existing GATT text, and provoked an international discussion on what—if anything—should be done to change it.

At the heart of the GATT ruling was the notion that countries should not be allowed to use trade tools to influence practices outside their borders for environmental ends. This, according to the panel, would amount to foisting a country's own environmental laws and values on the rest of the world, thereby riding roughshod over the once-inviolable principle of national sovereignty. More specifically, the panel decreed that the GATT rules, which generally allow countries to apply national laws governing *products* (such as car emissions standards or pesticide residue limitations) to imported goods at the border, did not cover this case because it was the *process* by which the tuna was produced (the setting of purse-seine nets on dolphins), rather than the tuna itself, that was being rejected by the United States—and this *process* took place outside U.S. jurisdiction.

The flaw in this logic, in the eyes of U.S. environmentalists, at least, is that it makes no distinction between environmental issues of purely national concern and those designed to protect the global commons—the oceans and the atmosphere. By applying the "domestic borders" criterion, the judges determined, in effect, that there is virtually no way short of an international agreement for nations to protect the Pacific dolphin, whose habitat is not contained in any country's borders. Actions to reduce the use of harmful drift nets in fishing, protect tropical forests, or stave off ozone depletion or global warming would also be severely circumscribed. Ominously, even provisions of *international* agreements designed to protect the global commons could be found to be GATT-illegal based on this reasoning.

It was thus particularly exasperating to environmentalists when the GATT judges further argued that the lack of an international agreement on dolphin protection practices in tuna fishing made the U.S. action suspect. Though international agreements are widely supported in principle, the process of reaching consensus can take years and even decades—time the world cannot often afford as global ecological decline continues its steady course. Indeed, nations had been trying for some time to reach an agreement on dolphin-friendly fishing practices, through the Inter-American Tropical Tuna Commission. In fact, it is most often a unilateral action by one country, sometimes backed by trade measures against others, that eventually spurs the international community to act collectively. Any challenges to the rights of countries to pursue these kinds of policies thus poses great threats to prospects for successfully heading off the deterioration of the biosphere.

The Implications

Under the logic of the tuna-dolphin ruling, other laws—many of them highly effective at achieving their environmental goals—could well be found to violate GATT if they were challenged. For instance, under a law known as the Pelly Amendment, the United States can prohibit the import of products from countries undermining the effectiveness of international fishery or wildlife agreements. Though the sanctions have never been invoked, the threat that they might be has brought about some significant changes in national behavior. It helped secure the participation of Iceland and Norway in the 1982 international whaling ban and of Japan and Taiwan in the U.N.'s 1993 worldwide moratorium on destructive drift net fishing, and helped convince Japan to stop importing endangered sea turtles for use in jewelry and eyeglass frames.

Though the United States has been the staunchest defender of the right to use unilateral trade tools for environmental goals, it is not alone in the practice. Despite its criticism of unilateral actions by the United States, the European Community has imposed a ban to take effect after 1994 on imports of furs from countries where painful "leghold" traps are permitted.

According to the GATT, at least 17 international environmental treaties involve limitations on trade—and could be rendered toothless if the tuna-dolphin reasoning holds up. Yet, in agreements like the Basel Convention on hazardous waste export, or the CITES treaty on endangered species, restricting trade is the very *purpose* of the agreement. In other cases, such as the landmark Montreal Protocol on depletion of the ozone layer, restrictions are used to try to prevent countries that have not signed the treaty from undermining its effectiveness. In the future, they may be needed to enforce compliance by uncooperative signatories.

This presents the world's governments with a momentous legal problem. With GATT aimed at limiting most restrictions on trade, and most environmental treaties requiring them, two sets of international agreements are in head-on conflict. Which treaty should take precedence? International law is unclear on this question. If the treaties in conflict are on roughly the same subject matter, and both parties to the dispute are signatories to both agreements, the most recent treaty generally prevails—which would tend to protect most environmental treaties. But trade and environmental treaties might not be viewed as sufficiently similar for this formula to apply. Furthermore, problems could develop if a country not party to the environmental treaty were to argue that GATT should rule—though so far, no such cases have arisen. To thicken the plot still further, successful conclusion to the Uruguay Round might mean that GATT would supplant the environmental treaties as the most recent agreement.

The Uruguay Round Threat

It was not just the tuna-dolphin wake-up call that riveted environmentalists' attention on the GATT in late 1991, but the Uruguay Round, which raised some troubling new environmental questions above and beyond the vulnerabilities revealed by the tuna-dolphin ruling. Now, in late 1993, the Uruguay Round is once again keeping taxis busy in Geneva, with governments working toward a December deadline—though longtime GATT-watchers are skeptical. Without considerable revisions, the Uruguay Round will likely meet with considerable environmental opposition if and when a deal is struck.

If completed, the Uruguay Round will expand GATT controls in a number of areas, including agriculture, services, and intellectual property, many of which promise to have wide-ranging environmental implications. Unfortunately, exactly how a given reform would

---

What looks to one country like a non-tariff barrier to trade is often another country's hard-won environmental law.

---

affect the environment is often a complicated question that can cut many different ways—some positive and some negative. And governments seem to be heading toward committing themselves to these changes with little study—or understanding—of their implications. For this reason, three U.S. groups that were prime backers of the late 1991 ad campaign (Friends of the Earth, Public Citizen, and the Sierra Club) filed suit with the U.S. District Court, arguing that trade negotiations should be subject under U.S. law to environmental impact statement procedures. In an important victory,

Judge Charles Richey ruled in the groups' favor in June, finding that such an assessment is required for the North American Free Trade Agreement. The Clinton Administration has announced plans to challenge the ruling. If the ruling holds, it would likely also apply to the GATT negotiations.

Though the environmental implications of a vast agreement like the Uruguay Round are not well understood, a diverse array of interests has raised concerns about aspects of the agreement. Family farm groups worry that the reduction of agricultural subsidies envisioned under the pact will be about as helpful to them as a plague of locusts. They have enlisted some environmental support for their campaign, arguing that smaller farms are often more ecologically sustainable than large ones. (A good example of how these things can cut both ways, however, is that reductions in agricultural subsidies can also mean reductions in production, which means less use of inputs such as toxic pesticides and scarce water.) Another concern is that exports of unsustainably-produced commodities, including agricultural, timber, and mineral products, might be stepped up as a result of tariff reductions on these goods.

Developing countries fear that provisions sought by the North to strengthen intellectual property rights protection in developing countries might impede the transfer of environmentally advanced technologies such as solar photovoltaic cells and energy efficient furnaces. They also fear that this would make it easier for pharmaceutical and agribusiness interests to monopolize products made with biotechnology, while jeopardizing developing countries' rights to remuneration for biological resources extracted from their territories—the recognition of which was viewed by many as the linchpin of the treaty on biological diversity agreed to at the Earth Summit.

The greatest focus of concern, however, has been that with many quotas already eliminated and tariffs drastically reduced through previous negotiating rounds, the Uruguay Round now has its guns aimed at the reduction of so-called non-tariff barriers to trade. The problem is that what looks to one country like a non-tariff barrier to trade is often another's hard-won environmental law, as the recent string of environmental trade disputes makes clear. For instance, in the Ontario-U.S. "bottle battle," U.S. negotiators are convinced the tax on non-refillable bottles is really aimed at keeping out U.S. beer, which is mostly sold in cans, while Ontario environmentalists insist that the levy is critical to preserving the province's 99 percent rate of bottle refilling, one of the highest in the world.

The draft text of the Uruguay Round addresses two different categories of product "standards": those designed to protect food safety, such as pesticide residue limits ("phytosanitary standards"), and so-called "technical barriers to trade"—a broad class that could include just about any specification, including car emissions standards, environmental labeling programs, and recycled content requirements, among others. In both cases, the Uruguay Round promotes the "harmonization" of these laws as a way to prevent unnecessary trade barriers. Environmental and consumer advocates fear that the "harmonization" will be downward, creating a least common denominator effect that would jeopardize countless environmental protections at the national and local levels.

The text does allow for nations to exceed the agreed international norm under certain conditions—such as a demonstration of "scientific justification," or a proof that the "least trade-restrictive" approach possible was used to meet a given environmental goal. Some trade specialists argue that these conditions are necessary to ferret out cases in which countries are wrapping what is really protectionism in a green cloak. They point out with suspicion, for example, that the Ontario non-refillable tax applies only to alcoholic beverages, and not to soft drinks, which, unlike beer, Ontario companies sell in cans in abundance.

Though at first glance these tests seem innocuous enough, they may prove a major obstacle to environmental progress. For one thing, scientists hold widely divergent views on questions of major importance to environmental policymaking, and might thus disagree among themselves on the question of "justification." And the messy political fact of the matter is that laws are often passed because of unholy alliances among those who stand to gain. A requirement that a given measure be the "least trade-restrictive" could easily be enough to doom any action at all.

Environmentalists also charge that the "harmonized" standards are set through a secretive, undemocratic process dominated by industrial interests. And if a national law were challenged as a trade barrier, the case would be heard behind closed doors by a panel of professors and bureaucrats steeped in the intricacies of world trade law, but not in the exigencies of the planet. Judgment on whether or not a law was "scientifically justified"

would be handed down by an appointed GATT panel, rather than by an elected legislature. To make matters worse, the latest draft of the Uruguay Round would make it far more difficult for a country to block a panel report not to its liking. Under the current rules, *adopting* a panel report requires unanimous consent; Under the new ones, unanimity would be required to *reject* a report.

The Uruguay Round also includes plans to create a new Multilateral Trade Organization (MTO) to give institutional form to the GATT, modeled on the International Trade Organization that was originally envisioned in the 1940s, but that was never created, in part due to concerns in the U.S. Congress over the potential invasion of sovereignty. It was a measure of the gulf between the trade and environment communities that just as public concern over possible conflicts between trade and environmental goals was reaching new heights, plans were moving ahead to create a sweeping new institution with little if any attention given to the environmental implications.

No Time Like the Present

As governments work to wrap up the seemingly-endless talks, they are resistant to introducing too many changes that could unravel the delicate balance achieved in the previous years of arduous bargaining. However, in the face of an international campaign by citizen's groups, some limited changes are likely to be introduced.

The U.S. government is finding itself in a particularly delicate position, as it was able to incorporate into the North American Free Trade Agreement some of the changes that the environmental and consumer activists are also urging for the GATT. It is thus under pressure to include some of these changes in the Uruguay Round discussions. For instance, under the NAFTA, the "scientific justification" test for standards deviating from international norms was loosened to require only the demonstration of a "scientific basis" for the law. In addition, the NAFTA suggests that harmonization should be in an upward direction. In the area of dispute resolution, NAFTA may again pave the way for needed changes. Unlike the GATT process, the NAFTA requires environmental expert advice to be provided if one party to the dispute requests it, and usually places the burden of proof on the country challenging, rather than the one defending, a domestic environmental law.

More far-reaching reforms are also being

contemplated. For instance, governments could agree to open the doors of dispute resolution proceedings to non-governmental observers, mandate environmental expertise for certain cases rather than merely permit it, and make documentation freely available. Meanwhile, hundreds of environmental groups from around the world have joined forces to protest the inclusion of the Multilateral Trade Organization (MTO) in the Uruguay Round. They argue that time should be taken to design an MTO that takes the environment and sustainable development adequately into account.

There are a number of issues that governments still need to take up if the GATT is to be thoroughly greened. With governments resistent to overloading the Uruguay Round agenda, pressure is building for a more extensive environmental negotiation once the current talks are completed. Indeed, governments are already laying the groundwork for such talks in negotiations at the Organization for Economic Cooperation and Development and in a working group of the GATT itself. Former GATT Director-General Arthur Dunkel has called for the next GATT round to be an explicitly "green" one, as has U.S. Senator Max Baucus, chair of the Senate Environment Committee and the Finance Committee's trade subcommittee. The largest U.S.-based environmental group, the National Wildlife Federation, is making its support for the conclusion of the Uruguay Round conditional on a commitment by governments to undertake these more far-reaching negotiations—and on a timetable for doing so in the event the Uruguay Round drags on for some years hence.

Fixing Tuna-Dolphin

The obvious first priority in such a negotiation would be to address the shortcomings of the current GATT agreement that were exposed by the tuna-dolphin ruling—the ominous ambiguity as to whether or not it is (or should be) consistent with GATT for countries to use trade tools to protect the environment outside their borders. Rather than enter into a protracted negotiation to amend the GATT articles in question in the tuna-dolphin case, the GATT member governments could simply clarify their interpretation of the existing text. Contrary to conventional belief, Steve Charnovitz of the Washington, D.C.-based Competitiveness Policy Council argues that at the time the agreement was written, governments were well acquainted with the use of trade tools for conservation purposes, and in-

tended to protect the prerogative of countries to use them. Charnovitz maintains that the tuna-dolphin panel members misread the existing text.

Another panel is currently re-examining the logic behind the earlier ruling, as the European Community—the target of a secondary U.S. embargo imposed on countries that purchase tuna caught in dolphin-killing nets—has now lodged yet another complaint against the disputed provisions of the U.S. Marine Mammal Protection Act. If the members of the new panel were to concur with Charnovitz's interpretation of the existing GATT, there would be far less need to amend the agreement in this area.

Though the unilateral use of environmental trade measures has surprisingly few defenders beyond the United States, there is somewhat greater support for them when undertaken through international environmental agreements. The NAFTA addressed the problem by stipulating that where there are conflicts between the provisions of the NAFTA and those of three international environmental agreements (Montreal, Basel, and CITES), the international environmental agreement shall in most cases prevail. The provision is not as far-reaching as some environmentalists had hoped for (agreements other than these three are not protected), but the provision represents a considerable improvement over the GATT status quo, and could serve as a model. Unfortunately, recent indications are that many GATT members may wish to limit, rather than protect, the use of trade measures even in international agreements.

Green Subsidies

The second priority in an extended "green" round is to address GATT's position on the relationship between subsidies and environmental protection. The GATT generally frowns on subsidies as trade distortions, and in some instances allows countries to impose "countervailing duties" on imports to compensate for them. These rules could threaten some environmentally helpful government programs, such as subsidies for the development of pollution control technology. On the other hand, GATT could also provide a powerful *green* weapon with which to attack environmentally damaging subsidies such as the $36 billion paid in energy subsidies by U.S. taxpayers, according to a recent report by the Washington, D.C.-based Alliance to Save Energy, or the implicit subsidy provided by granting logging companies cut-rate access to federal lands.

Under current GATT rules, the scope for challenging such subsidies is limited, though in one example of the potential, the European Community has recently listed subsidized water sales in California as an unfair trade practice. A "Green Round" could make a significant environmental contribution by overhauling its subsidy rules so that the harmful ones would be at least as vulnerable to challenge as the beneficial ones—if not more so.

Another idea gaining support is to define lax environmental protection or enforcement as an unfair subsidy, making it possible to levy countervailing duties or take other compensating action. Lower standards do add up to a sizable hidden subsidy, according to the World Commission on Environment and Development, which estimated that developing countries exporting to the OECD countries in 1980 would have incurred pollution control costs of at least $5.5 billion if they had been required to meet the requirements then prevailing in the United States. Both Senator Baucus and Vice President Al Gore (before taking his current office) have endorsed this idea.

Such tariffs would ensure that a country is not penalized in international markets for internalizing environmental costs more than its trading partners do. Without such measures, there is a danger that industrial production might increasingly locate in so-called "pollution havens"—areas where regulation or enforcement is lax. Fears of losing out to foreign competition in the global marketplace might also deter countries from adopting strict domestic environmental laws—as in fact happened recently in the debate over energy taxes both in the United States and in the European Community. However, proposals for levying border tariffs for this purpose raise a number of difficulties, such as that of evaluating exactly how much trade advantage is being gained (how large the countervailing duty should be), and whether such duties might create an opening for hidden protectionism.

There are ways, however, for governments to make such levies more politically palatable—particularly to developing countries. For instance, proceeds from these duties might be channeled into upgrading the environmental standards of those countries that fail to measure up. Or, if international standards were established, the duty could be based on the extent of deviation from the international minimum.

The NAFTA is likely to create some important precedents in this area. In one of its more innovative provisions, it will be considered "inappropriate" under the agreement to relax a nation's own environmental laws or enforcement of them in order to gain a trade advantage. The North American Commission on the Environment (NACE) now being negotiated to accompany the NAFTA will likely be charged with ensuring that domestic environmental laws are enforced, and perhaps with gradually negotiating harmonized production standards, particularly for industries in which environmental costs are high, such as paper manufacturing.

One highly controversial question in the negotiations is whether consistently lax enforcement of domestic environmental laws should be punishable by economic sanctions—whether, for example, the United States could levy a tariff or impose a fine on imports of Canadian wood pulp if it is produced in violation of laws regulating dioxins dumped into rivers by bleaching mills. Under heavy pressure from non-governmental groups, the United States is maintaining that some form of sanction should be available, but Canada and Mexico—as well as business groups in the United States and their backers in Congress—are far from enthusiastic about the idea.

Finally, there is the possibility of granting trade concessions such as preferential tariff treatment for environmentally sound goods, rather than simply penalizing the bad. Already, the United States makes tariff reductions on developing country imports under the Generalized System of Preferences program contingent on respect for internationally recognized worker rights. In one example of how this idea could be applied to environmental concerns, the government of Colombia has requested exemption from E.C. import duties for oils produced from certain organically grown plants such as lemon grass. The GATT could endorse this sort of initiative, and encourage its wider application.

The Greening of GATT

A historian combing the record several centuries hence may find the tuna-dolphin episode to be a revealing symbol of the passing of an era when nations could provide for the needs of their citizens by acting alone.

With security increasingly defined in economic and environmental rather than military terms, governments are coming to recognize that protecting their citizens from threats as diverse as sea-level rise induced by global warming, and unemployment created by industries migrating in search of pollution havens, will require an unprecedented level of international coordination.

If governments can work together to devise minimum rules of environmental conduct, it will greatly reduce the potential for trade conflict to erupt over environmental matters. Just as the International Labor Organization has formulated hundreds of workplace rules covering matters like child labor and occupational exposure to toxic chemicals, so could a U.N. environment agency be given the mandate to begin enunciating minimal standards of environmental behavior and generating the funds required for poorer nations to meet them. This process is already well underway in the European Community, and beginning to be developed under the NAFTA. Internationally, the more than 170 international environmental treaties that governments have agreed to constitute a decisive move in this direction.

In the meantime, if the GATT wishes to restore its tarnished reputation, it will need to be updated to reflect today's environmental imperatives. The buying power of consumers and nations is a powerful force that can be harnessed to encourage economic production that protects rather than ravages the earth's natural resource base. GATT should be a leader in this effort, rather than an obstacle to it.

The Environment of Tomorrow

Martin W. Holdgate

Martin W. Holdgate is director general of The World Conservation Union (IUCN) in Gland, Switzerland. This article is adapted from a lecture given for the David Davies Memorial Institute of International Studies.

Earth is the scene of constant change. The summits of the Jura Mountains in Switzerland, more than 1,500 meters above present sea level, are formed of limestone laid down as soft sediments in the bed of a warm and shallow sea about 175 million years ago. There were once forests in Antarctica and dinosaurs and ice sheets in England. As Alfred, Lord Tennyson wrote, "There where the long street roars has been the stillness of the central sea."[1]

Such changes will always occur. Some of today's seas will be squeezed out of existence by the collision of continents to form new mountain ranges, as the Himalayas are being shaped by the collision of India and Asia. The Arctic ice may well expand again, providing the ultimate solution to the architectural problems of Europe and North America. Life forms will continue to evolve and drive their predecessors toward the extinction that is the ultimate fate of every species. On a longer time horizon, as astronomer Fred Hoyle has put it, "we shall certainly be roasted"[2] when the sun emerges from its present stable phase and expands to engulf and vaporize the Earth.

However, I am concerned with a more limited perspective. Throughout this article, I will use *tomorrow* to mean 40 years from now, in 2030, and the phrase *the day after tomorrow,* viewed through a haze of uncertainty and for that reason receiving less attention, as 40 years later, in 2070. I will address three simple questions:

• What will the world be like as a habitat for life if present trends continue?

• What are the implications of these changes for humanity and for the world of nature?

• What can we do about it if we would like tomorrow to be different?

Throughout history, people have adapted perforce to the cycles of the changing Earth and, where those changes have proceeded slowly, have probably hardly noticed them. Even natural catastrophes, such as massive floods and volcanic eruptions like that of Mount Vesuvius in 78 and 79 B.C., though they left scars on the body of civilization, had only a local impact on its progress. Such events doubtless caused our ancestors much suffering and social upheaval, but at least people were able to excuse themselves from responsibility, unless they chose to ascribe the disasters to the vengeful acts of a god irritated by human sin. In Java and Bali, the gods are still held responsible for volcanic eruptions. However, most of the major changes on the planet today are very much acts of humanity, and they result from the cumulative impact of two linked processes: the growth in human populations and the process that we call development.

Until the last 10,000 years, our species was a relatively uncommon animal, slowly increasing in numbers and extending its range to reach a total global population of about 500 million by 1000 A.D. The number of humans then doubled by around 1800, doubled again in the following century, and is now rising past 5.2 billion. This growth has been made possible by development, or the alteration of the Earth's environmental systems so that an increasing proportion of their nonliving resources and biological productivity serves human needs. As a result of development, some 40 percent of terrestrial plant productivity—the basic fixing of energy by the green mantle of the planet—is today used in one way or another by people. Changes in ecological systems have been an inevitable result of development. Because we eat cereals and our livestock are grazers, much of our land has been deforested and converted into pasture or that highly modified grassland we call cereal cropland. To expand croplands, people around the world have greatly altered patterns of water flow, sometimes successfully in stable irrigation systems but often unsuccessfully, leading to salt accumulation and soil sterilization—which has now damaged some 60 million hectares worldwide, an area equivalent to two-thirds of China's cropland.

As agricultural methods become more dependable, people not directly involved in the business of subsistence were able to create objects and ideas that enriched the total community. Such craftsmen applied increasing skill to the use of nonliving resources in

From *Environment*, Vol. 33, No. 6, July/August 1991, pp. 14-20, 40-42. First published as a lecture in 1990 by the David Davies Memorial Institute of International Studies. Reprinted by permission.

buildings, metal goods, and other artifacts and transformed areas of the physical environment with mines, quarries, and other structures. As people began to smelt metals, the movement of these elements through the biosphere inevitably increased, just as their agricultural and fuel-burning habits increased the fluxes of carbon, nitrogen, and sulfur. Development, by permitting the dominance of our species, has inevitably altered the world both by the nature and the scale of the transformations involved.

In recent years, some so-called environmentalists in developed countries have spoken of development as if the word were dirty. That is nonsense, as the World Conservation Strategy, prepared in 1980 by The World Conservation Union (IUCN) in partnership with the World Wildlife Fund and the United Nations Environment Programme, makes plain.[3] Development has been essential to the evolution and expansion of human civilizations, and more will be needed to help millions of people escape from today's poverty and squalor and to feed tomorrow's added billions. What we have to be concerned about is the nature and quality of development and the social and political structures needed to bring it about. It is clear from the destruction that has been the price of today's uneven and unsatisfactory development that the world cannot afford much more of the same. We need something different. To use today's catch-word, we need development that is sustainable—that is, it must not overcrop soil, pastures, forest, or fisheries or create products that spread from a beneficial activity, like industry, to blight other essential ones, like agriculture, the supply of drinking water, or the stability of the world's ecosystems.

When we demand that development should be sustainable, we must be clear about our meaning. We do not mean that the growth in a human activity, such as the cultivation of new land, must be capable of indefi-

Throughout history, people have adapted perforce to the cycles of the changing Earth and, where those changes have proceeded slowly, have probably hardly noticed them.

nite extrapolation—little such growth will be. Rather, we mean that the changes we make in our environment must not only improve the yield of a useful product to-

day but also go on supplying that product tomorrow, without side effects or unforeseen consequences that undermine other essential environmental functions. If overuse has such an impact, we may have to accept adjustment to a lower level of sustainable production. To quote the World Commission on Environment and Development, sustainable development means "meeting the needs of the present without jeopardizing the ability of future generations to meet their own needs."[4] For IUCN, conservation means preserving the world's natural resource base as the indispensable foundation for the future.

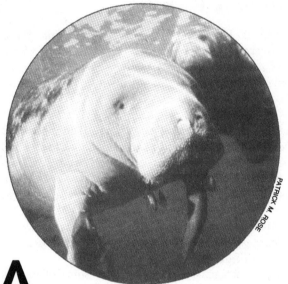

PATRICK M. ROSE

As we look to the possibilities of sustainable development, we would be wise to remind ourselves of three basic features of the Earth as a habitat. First, the Earth is a rather small planet with finite resources, and it receives a finite amount of energy from the sun. Its living and nonliving systems have interacted over time to create the habitats within which humanity evolved. Interactions between living and nonliving elements still operate in a fashion that regulates the overall environment in a manner analogous to the self-regulatory mechanism of a living being. James Lovelock pointed this out when he named the planetary organism Gaia, after the Greek goddess of the Earth. Hindu scholars had the same idea 2,000 years ago. There can be no sustainable development that does not preserve that essential planetary system on which all life depends.

Second, 70 percent of the Earth's surface is ocean, used only to a very limited extent by land animals. Although the plants in the sea fix about as much carbon annually as do those on land, humans only take from the sea about 100 to 120 million tonnes of assorted fishery products annually. Although I expect that our descendants will cultivate more marine plants and animals in shallow waters, most of the human future is going to depend on how we use the land.

Third, land will remain highly heterogeneous as a habitat. People living in different regions cannot expect to enjoy anything approaching environmental equality. It will always be easier to live comfortably off the produce of the temperate zones like western Europe than off the arid sands of the desert or the dry grasslands of the savannas. The broad pattern of life on Earth will prevail tomorrow as well as today, even though the detailed ecological pattern within those biomes will change. Biological diversity is and will remain concentrated in the tropics. Therefore, the nations of the world cannot be made environmentally equal. This would be true even if national frontiers had been drawn in an environmentally logical way. Instead, many boundaries bisect natural units like river basins and make it difficult for governments to manage in an integrated way the resources that support their populations. The implication is that, if humanity is to have a sustainable future on this diverse planet with many environmental inequalities, it can only be through a process of international cooperation that transcends anything we see today. To state that is immediately to pose an immense challenge if we are to seek success in as short a time as 40 years. And we have to contend with certain trends arising from the nonsustainable nature of the current development process that aggravate our problems and are likely to prove unstoppable on a 40-year time scale. I am particularly concerned with six of these trends.

The first is population growth. I have already mentioned the dramatic increase in the world's population from around 1 billion people in 1800, to 2 billion in 1900, to over 5.2 billion today. Although the rate of population growth has slowed somewhat from a peak of nearly 2 percent per year in 1970 to around 1.66 percent now, a cautious medium projection suggests that by 2030 there will be more than 8 billion people in the world and that our descendants will be lucky if, by 2070, stability has been achieved at around 10 billion. The trend is unstoppable because in many countries half the population is still under reproductive age. Even if these people only have two children per marriage, a near doubling is inevitable. Forty years from now, most of those people presently under 16 years of age will still be alive and their children will be between 10 and 35 years old—and some of them will themselves have children. Common sense tells us that not all people will limit their family size to two children per couple by 2030, and it will obviously require an immense change in attitude for this to happen by 2070. The total population achieved in a number of countries will depend on how many children survive, as well as on how many are born.

Today's population explosion is a tribute to the medical profession, which has greatly increased the average life expectancy in both developed and developing countries. Clearly, this advance could continue if the effectiveness of medicine is not undermined by malnutrition or social disorder and if new diseases do not appear suddenly to exploit the wonderful habitat represented by so much healthy human tissue. My own

guess is that new diseases will indeed appear because the ecological niche is there for rapidly evolving viruses, which are likely to exploit the particular means of dispersal provided by human behavior—as AIDS has exploited the unique human habit of transferring blood from person to person, a parasite's dream. However, I doubt that new diseases will limit the growth of human populations, because I have great faith in the resilience of medicine.

On the other hand, malnutrition will limit population growth in areas where water supplies are inadequate, the soil is overexploited, and there are no new reserves of fertile land. Agricultural science, along with medicine, has been a major supporter of population growth. But in large areas of Africa and parts of India, per-capita food production has been declining recently even though total food production has been increasing. Countries with high population growth, limited rainfall, and a hot climate will be at risk until population increase is curbed by measures more humane than famine. It is important to look now at places where people are on a collision course with their environment because

Interactions between living and nonliving elements still operate in a fashion that regulates the overall environment in a manner analogous to the self-regulatory mechanism of a living being.

these are the places that need the most help. We have learned the lesson of the recurrent tragedy of famine from Ethiopia and parts of the Sahel. We have also learned that food and medical aid cannot do more than stave off such misery.

Moreover, people are understandably reluctant to sit still and starve. Now, as in the past, if populations really deplete their environmental resources regularly and over significant areas, they are likely to migrate, and finding a frontier here or a different government there is unlikely to deter them. As an increasing number of people come up against environmental limits, real threats to the peace and stability of nations may arise. The only alternative is to accelerate programs of sustainable development that meet human needs and, at the same time, provide the means and the incentive for stabilizing human numbers, country by country, in the areas of greatest risk. Clearly, these programs will require levels of international assistance beyond anything available today, and politicians in the countries concerned must recognize that population pressure is a genuine and urgent problem—something that a number of governments are presently reluctant to accept.

The second trend is deforestation, which, I suggest, is also unstoppable because of the need to feed more people, especially in the developing world. Moreover, I do not think it is honest to present all tropical deforestation as an environmental disaster. There are parts of the tropics—including areas of the Amazon basin—where conversion of forests to well-managed agriculture, particularly agroforestry, could be a perfectly acceptable pattern of development, though the governments concerned are fearful about the damage caused by present methods. Further deforestation in the tropical regions of South America, central Africa, and Southeast Asia is virtually certain. The governments concerned should steer the process in the direction of sustainable agricultural systems, while halting destructive deforestation of areas that are especially valuable as reservoirs of biological diversity, that are essential to regulate local climate, or that protect the land from erosion. Looking beyond 2030 to 2070, my prediction is that the great forests of Earth will by then be concentrated in the boreal zone, in the more rugged and uncultivable mountains of the tropics, and in tracts of sparsely populated tropical lowlands, where the soils are poor. I also foresee significant but relatively small patches of natural forest in the temperate zones and in the densely

If humanity is to have a sustainable future on this diverse planet with many environmental inequalities, it can only be through a process of international cooperation that transcends anything we see today.

populated and heavily utilized tropical lowlands. We may not like that pattern, but it is the one that I suspect our descendants will have to live with.

Another trend that is unstoppable is desertification, or land degradation. I do not refer to the advance of the Saharan dunes but, rather, to erosive soil loss, salinization, and declining fertility as a result of poor or inadequate irrigation—which has, for example, affected some 60 percent of the cropland in Pakistan. However,

because of some good science in recent years, there are signs of a cure. In the Sahel, for example, food production has increased steadily over the past 20 years despite a lower average annual rainfall than in the preceding two decades. In Pakistan, measures to improve the quality of irrigation, to control salinization, and to plant crops that are resistant to salt are gaining ground. Accordingly, I do not expect the maps of the world to show more desert areas in 2030 than they do today, and by 2070 this phenomenon may have been brought under control or even reversed.

Fourth, I believe that the continuing loss of the planet's biological diversity is unavoidable. Before every conservationist starts jumping up and down and shouting, let me explain at once that I do not imply that destruction of biodiversity should continue at its present rate or that we should complacently accept this trend. We are, however, almost bound to lose a significant number of species from the Earth as a consequence of the development process impelled by the imperative of human need. Such loss is a logical necessity because, as I have said, it will be impossible to stop tropical deforestation. About 50 percent of the species believed to exist on Earth are insects and other small organisms living in the canopies of tropical rain forests. A significant number of these species are bound to disappear if the forests are destroyed. Islands, which are also significant reservoirs of unique species, are vulnerable to invasion by mainland life forms, which are increasingly spread around the world by today's unparalleled ease of transport.

The question we have to ask ourselves—and it is one that wildlife conservationists understandably shirk—is whether these losses will matter. My own answer is yes, they do matter, especially to those who can afford to care. But we must not oversell the disaster. The losses matter if you believe that species other than our own have a right to exist as a part of natural creation. They matter if you believe that the rich diversity of life maintains the equilibrium of Gaia, the planet, in ways we do not fully understand and that reducing diversity brings risks we do not comprehend. The losses also matter because there may be many species and genotypes in the wild that are or could be of considerable value to humanity and that will act as the genetic basis for evolution in response to future change. Despite all this, a sense of proportion is needed. The United Kingdom, for example, supports about 1,700 species of vascular plant, 25,000 invertebrates, 12 reptiles and amphibians, 54 freshwater fish, 50 mammals, and 200 birds. This is an extremely impoverished biota by world standards, yet I think we would regard it as one that is able to maintain essential ecological functions—even though people would soon grumble if their diet was based only on what they could grow or gather there. I am not arguing for impoverishing all of the world's ecosystems to this extent. It is always better to maintain a system with the widest possible range of ecological functions and the greatest practicable reservoir of genotypes. But I suggest that some loss of biological diversity does not inevitably bring collapse. The problem is that we do not know how much loss of what kinds of organism is tolerable. Figuring that out is one of the real challenges to science.

My final two unstoppable trends are quite different from the others. Trend five concerns pollution. Some kinds of damaging pollution are unstoppable over a 40-year time scale because they are already present in the environment and cannot be eliminated by 2030. For instance, polychlorinated biphenyls and other persistent chlorinated organic substances are widely dispersed in the ocean and in biological food chains, even though the most damaging kinds are no longer produced. Even if the production of chlorofluorocarbons (CFCs), which are responsible for depleting the ozone layer, ends by 2000, as many scientists wish, the persistence of these substances in the atmosphere is such that the ozone "hole" will not begin to fill in in less than 40 years, and the ozone layer will not be fully restored even by 2070.

The buildup of greenhouse gases is also most unlikely to be stopped by 2030; at best, the upward trend in carbon dioxide emissions will have been slowed, CFCs will no longer be released, and emissions of other greenhouse gases will be limited. Thus, the sixth and last unstoppable trend is climate change. Because of the many greenhouse gases already emitted, there is every reason to believe that by 2030 the world will be between 1° and 2° C warmer than it is now. Temperatures are likely to increase by more than this average in the higher latitudes of the Northern Hemisphere in winter, by slightly more than the global average in the temperate zones, and by slightly less than the global average in the tropics. Wet areas may get wetter, dry areas may get drier, and the force of storms may increase. By 2030, sea level may rise by 10 to 20 centimeters. As for the climate in 2070, much depends on how successful people are in curbing greenhouse-gas emissions. If they are not—if politicians drag their feet—then the world could be 2° to 4° C warmer, and sea level could rise 20 to 40 centimeters, posing severe risks to burgeoning human populations, especially those in the coastal zones of the tropics.

Because about half of the world's population lives in coastal zones, a 1-meter sea-level rise, which is greater than most experts consider likely, could spell disaster. Such a rise in Bangladesh would inundate nearly 15 percent of the country and displace some 10 million people.

In Guyana, 95 percent of the population and 90 percent of the agriculture are concentrated on a 3-mile-wide coastal strip that already stands near sea level and is protected only by slender barriers. The plight of fertile and densely populated deltas and low-lying coral atolls with most of their surface below the one-meter level is obvious.

Climate change also threatens the ways of life of many inland people. Each 1° C rise in temperature effectively displaces the limits of tolerance, or ranges, of species (including crops) by about 100 to 150 kilometers toward the poles or 150 meters vertically. Temperature changes may well prove less important than changes in

amounts of rainfall, but these figures do demonstrate the kinds of dislocation that could occur. The disruption of natural systems would be aggravated because the shifts in limits of tolerance could occur much faster than species of trees and other dominant plants would be capable of matching through natural dispersion. The spectacle of trees being "left behind" by their optimum habitats and, at the same time, ceasing to be able to maintain themselves by reproduction in considerable parts of their present range has been sketched out by many ecologists. The fact of the matter is, however, that scientists do not know how the world's living systems would respond to drastic change, and finding out is a major challenge that is only now being undertaken. There is little doubt that agriculturalists could help people grow new varieties of crops adaptable to the changed environment, but there would be considerable social stress nonetheless, especially if people had to change long-established dietary habits.

WORLD BANK—RAY WITLIN

Why are these trends unstoppable over such a long time scale as 40 years? All six trends have two features in common. First, all are caused by the innumerable small actions of a very large number of people, who do not see their actions as detrimental. Second, there is a substantial lag time between cause and effect for each trend. Let me prove my point by referring to one form of pollution I have not listed as unstoppable—acid rain. This phenomenon is stoppable because of two factors. First, the time lag between the emission of the sulfur as sulfur dioxide and its deposition as acid rain is quite short—only two or three days. Consequently, once emissions are reduced, acid depositions quickly decline, and the acidified ecosystems can begin to recover. Second, because the greatest proportion of the sulfur emissions come from a relatively small number of large point sources, specifically commercial power stations that burn coal, there is good reason to believe that by 2030 these sources can be cleaned up and acid rain will have been stopped.

However, the situation for greenhouse gases is quite different. CFCs enter the atmosphere from such products as leaky refrigerators, old-fashioned aerosol cans, industrial solvents, and the materials used to expand plastic insulating foam expand. The only solution here is to end CFC production and use in favor of less harmful substitutes. Even then, because of the 80- to 100-year residence time of CFCs in the atmosphere, the problems they cause will take decades to cure. Similarly, carbon dioxide, the principal greenhouse gas, is inevitably produced whenever forests are cleared or coal, oil, gas, or wood is burned in power stations, industrial heating plants, domestic stoves, or vehicles. Although carbon dioxide and other greenhouse gases, such as nitrous oxide

and methane, which come from livestock, cultivation, fertilizers, and fuel combustion, do not have as long a residence time in the air as do the CFCs, the process of their absorption by green plants or by the oceans and freshwaters is not rapid. The greenhouse-gas problem is intractable because of the multiplicity of sources, the fundamental importance of fuel burning and agriculture in human life, and the significant lead time between curative action and restorative effect.

Neither the causes nor the potential impacts of the trends are evenly spread throughout the world. In developed countries like those of Western Europe, population growth is not a problem (even if we might prefer less crowded conditions) and forest areas are actually increasing (although the species being planted are not those that naturalists or environmentalists would welcome most and may not be the most tolerant of climate change). Desertification is not a problem either, although soil loss and improper irrigation techniques can be problematic, and losses of biological diversity, though unwelcome, are not as heavy as in other parts of the world. However, all of these trends are grave concerns in the developing world. Rapid population growth is correlated with poverty and is, at the same time, an impediment to the economic growth that might end poverty and provide people with the means and the incentive to limit the size of their families. Moreover, the pollutants that threaten to perturb global climate, destroy stratospheric ozone, and accumulate in living systems come overwhelmingly from the developed countries of the Northern Hemisphere, and their spread imposes extra burdens on the nations experiencing the worst of the other trends. Put crudely, the poor are in a mess, and the rich are making that mess worse.

The solution to these problems, which are growing increasingly graver, does not lie in science or in technology, but in politics. Of course we need more and better science to understand the planetary systems more fully and to manage our use of them more wisely; of course we need a better understanding of the factors that determine human behavior; of course we need better medicine, better agriculture, better forestry, more efficient use of energy, technology that produces essential products without generating intractable pollution, and products that are environmentally benign and that use scarce resources more economically. All of these are necessary, and much more as well. But the lesson of the recent decades is that scientific understanding and technological capacity are not the limiting factors. The blockage lies in human perception and the willingness of people to change how they behave at the individual, group, and national levels.

Many world leaders would reject a great deal of my argument, and some would say that it is a typical manifestation of the obstructive and negative thinking that prevails in developed countries, where fear of overpopulation, arguments against deforestation, demands for the conservation of biological diversity, and calls to halt the production and use of global pollutants are deliberate devices to hold back the development of the Third World and to maintain the present inequalities among nations. These leaders would argue that the real problem of the global environment is chronic poverty, which has been imposed and aggravated by the economic dominance of developed countries that grew rich by exploiting the raw materials of poorer countries and that continue to consume more than their fair share of those resources, to control commodity markets in favor of their own terms of trade, and to depress the growth of the Third World by issuing loans that benefit the lender far more than the recipient.

These arguments can be put eloquently and, indeed, at one level, are true. I am less sure that the environmental speeches of leaders of developed nations are dishonest and actually designed to supply justification for keeping the rich rich, the poor poor, and the latter dependent on the former. However, it does not matter

VERNON SIGL

Each international agreement must be the culmination of a process thoroughly built from within each nation.

whether such a view is true or false. What matters is that it exists, and this argument will be heard a lot more as the United Nations prepares for the 1992 World Conference on Environment and Development. If we are seriously concerned about the future of the world as a habitat, with the quality of life for the people of tomorrow, with the peace of the world, and with the stability of the global economic system, we have to face this gap in perception and the suspicion that goes with it.

Perhaps the first thing to do is to let facts about the world environment speak for themselves. This means openly sharing all available scientific knowledge and letting the professional community in each country evaluate the significance of those facts for itself. This is

already happening to a considerable and increasing extent. There are internationally accepted compendia of environmental data, produced, notably, by the United Nations Environment Programme. There are many international bodies that bring together governments, nongovernmental organizations, and individual specialists from developing and industrialized countries. Some problems, such as climate change and ozone depletion, are accepted almost universally as serious and in need of concerted action. In both of these areas, the developing world wants and expects the developed countries to continue to share their knowledge; to help developing countries evaluate the unavoidable trends over the next 40 years; to take steps to reduce output of CFCs and greenhouse gases, which are already overtaxing the capacity of the environment and are leaving no room for development in the Third World; and to transfer alternative materials and technologies that will allow the Third World to develop without creating new environmental risks.

The costs of these four actions would all fall on the developed world. But is that unreasonable, when the causative agents of the ozone depletion and the greenhouse effect stem especially from the activities of developed nations? Unless the developed world shows itself willing to take these actions and shoulder a significant part of the costs, the credibility of its leaders as sincere advocates of action designed to benefit the whole world will be at stake. Any attempts by leaders of developed countries to sell the essential new technology at a profit or to transfer it on terms that add to the burden of debt that already hampers the capacity of the developing world to deal with other intense problems of sustainable development will be seen as proof of insincerity.

Such attempts will make it more difficult to persuade the leaders of the Third World to treat the issues of population growth, deforestation, and loss of biological diversity with equal seriousness. Similarly, if the developing countries are going to conserve their biological diversity, promote sustainable use of forests and rangelands, and experience positive trends in economic and social conditions that will favor population stability, they will need money, which today is largely concentrated in the developed north.

Convincing world leaders that they must talk and think as one community is essential, but I fear that it is the easiest of the tasks that lie ahead. Indeed, it is already half accomplished. For whatever suspicions and accusations they have, the leaders of the world do talk about the importance of the environment and have done so with increasing frequency since the first United Nations Conference on the Human Environment, held at Stockholm in 1972. Indeed, given the burgeoning number of international bodies, the almost incessant conferences, and the more than 100 international conventions, agreements, and action plans that now deal with the world environment, it might be deduced that action was well in hand. The problem is, as one of my colleagues put it recently, that "thunderstorms of rhetoric are followed by droughts of inaction." Words are cheap and action plans are easy to put on paper. Because such agreements create the cozy, inexpensive illusion of a problem solved, they are dangerous.

How do we turn the words into actions? Clearly, each international agreement must be the culmination of a process thoroughly built from within each nation so that, in signing the convention or plan, the leaders

Scientists do not know how the world's living systems would respond to drastic change, and finding out is a major challenge that is only now being undertaken.

concerned know that they express the will of their people and can commit themselves to putting the agreement into operation—either by acting within their own territory or by giving aid to other countries where the need is greatest. How can such a process be ensured? Within any particular country there are three components of certainty: a sound understanding of the problems and of the means to solve them; good organization

and administration; and the backing of a well-informed, committed public. None of these components can succeed without the others. Unfortunately, in many nations, environmental concerns have been grasped slowly and addressed half-heartedly through governmental measures that often lag behind public awareness and will. For example, most governments that have created ministries to address the environment have added them as new sectorial entities alongside the older establishments that deal with agriculture, energy, industry, defense, and local government. Many environmental ministries have been used primarily to govern national parks and nature reserves. However, if the environment is actually the fundamental resource base on which all development rests, then surely it is proper to see it as a country's "natural capital"—to use a term that is now widely used among economists. In other words, the environment is real national wealth, as

precious or more precious than the manmade wealth guarded by treasuries and analyzed by finance ministers in annual budgets.

Managing the environment as a national treasure would unite science, economics, and government in a new way. Science would be needed to understand both the potential of using the natural environmental capital for sustainable development and the sensitivity of the environment to various impacts; the techniques for doing this already exist. Economics is needed to place financial value on the resource and to convert the material income yielded by the environment into cash terms. Again, the techniques for more precise economic evaluation of environmental assets and for evaluating the true costs and benefits of alternative options for their development are now well advanced. Government is needed to define how the state's managerial system then proceeds to get the best value for its money out of those resources. Government must also ensure that individual sectorial departments do not undermine the whole system by the pursuit of narrow and often traditional policies.

However successful such national plans may be and however enlightened government administrations may become (and I would add that I know of no government in the world that has yet structured itself to deal properly with environmental issues), nothing will be accomplished without the support of the governed. This fact seems obvious in terms of the six unstoppable trends. For example, experience in China suggests that it is difficult for even an authoritarian and determined state to bring population growth rapidly under control. The power of governments to control pollution, desertification, deforestation and human-induced climate change is obviously limited, and it is expressed only by altering the behavior of individuals through education, training, incentives, help, and deterence of those individuals who nonetheless insist on seeking their own ends, regardless of wider interests. In the end, the problem and the solution are matters of perception.

Our ancestors viewed environmental disasters as "acts of God" and either tolerated them or sought to prevent them by appeasing affronted deities. Today, we are prone to regard such disasters as something done by someone else, which another "they," such as the authorities, must put right. Meanwhile, the chemicals we use, the gasoline we waste, the fumes we generate,

Real changes in direction are needed if we are going to build a world in which humanity lives in enduring harmony with nature.

the heat we let escape through poorly insulated walls and roofs, the litter we create, and the environmentally unfriendly pesticides we buy because they are cheap seem so trivial in proportion to the size of the problem that altering our behavior does not seem worth the effort—even if we can find alternatives to buy and places to send items for recycling. The perceived benefit is all too often of less value than the perceived cost. Yet, unless millions of people are helped, guided, and advised

to alter their behavior, the unstoppable trends will not stop and the day after tomorrow will be dark.

There is some indication that the 1992 United Nations Conference on Environment and Development could be a scene of conflict between the leaders of de-

veloped and developing countries, fueled by self-interest and greed for bigger slices of a finite cake. To avoid this disaster, all countries need to review and be prepared to adjust their strategies and policies so that the six unstoppable trends—and other apocalyptic horsemen—do not bring the roof down over our heads. We need to devote great effort to slowing or reversing the damaging trends, but real changes in direction are also needed if we are going to build a world in which humanity lives in enduring harmony with nature. Willingly, or perforce, many people face life-style changes, and many industries may face a smaller profit margin. But unless we begin to travel with hope we shall certainly never arrive, and unless we start quickly, we shall face regression rather than progress. Let us, therefore, begin to develop practical measures that will enlist the aid and inventiveness of all the world's peoples to make their contribution to tomorrow's world.

NOTES

1. *In Memoriam*, section 123, stanza 1.

2. F. Hoyle, *The Nature of the Universe* (New York: Harper, 1960).

3. International Union for Conservation of Nature and National Resources (IUCN), *World Conservation Strategy* (Gland, Switzerland: IUCN, 1990).

4. The World Commission on Environment and Development, *Our Common Future* (New York: Oxford University Press, 1987).

The World's Population: People and Hunger

One of the greatest setbacks on the road to the development of more stable and sensible population policies came about as a result of well-meaning and well-intentioned, but incorrect, population growth projections made in the late 1960s and early 1970s. The world was in for a population explosion, the experts told us back then, and they predicted that one-quarter of the world's population would starve to death between 1973 and 1983. The population projections were wrong for the same reasons that similar predictions made by an English clergyman named Thomas Malthus in the 1790s were wrong—they were based on extrapolating the future from the trends of the past and such extrapolations are rarely correct. Shortly after the publication of heralded works like *The Population Bomb* (Paul Ehrlich, 1975) and *Limits to Growth* (D. H. Meadows et al., 1974), the growth rate of the world's population began to decline slightly. There was no cause and effect relationship at work here; the decline in growth was simply the process of the demographic transition at work, a process in which declining population growth tends to accompany increasing levels of economic development. Unfortunately, since the alarming predictions of massive famine and starvation did not come to pass, the world began to relax a little. Population growth was viewed by some as good rather than bad, and human ingenuity, it was said, could cope with increasing populations that meant, after all, increasing markets for manufactured goods and a larger labor force to produce those goods. Indeed, a counterargument to the population explosion thesis began to gain favor by the late 1970s, and some theorists actually began to set forth the notion that gradual population growth not only does not harm the environment but accelerates human progress. This cornucopian thesis notwithstanding, two facts remain—population growth in biological systems must be limited by available resources, and the availability of Earth's resources is not infinite.

Consider the following: in Third World countries, high and growing rural population densities have forced the use of increasingly marginal farmland once considered to be too steep, too dry or too wet, too sterile, or too far from market for efficient agricultural use. Farming this land damages soil and watershed systems, creates deforestation problems, and adds relatively little to total food production. In the more developed world, farmers also have been driven—usually by market forces—to farm more marginal lands and to rely more on environmentally harmful farming methods, which use high levels of agricultural chemicals such as pesticides and artificial fertilizers. These chemicals create hazards for all life and rob the soil of its natural ability to renew itself. The increased demand for food production has also created an increase in the use of precious groundwater reserves for irrigation purposes, depleting those reserves beyond their natural capacity to recharge and creating the potential for once-fertile farmland and grazing land to be transformed into desert. The continued demand for larger production levels also contributes to a soil erosion problem that has reached alarming proportions in all agricultural areas of the world, be they high or low on the scale of economic development. The need to increase the food supply and its consequent effects on the agricultural environment is not the only result of continued population growth. For industrialists, the larger market creates an almost irresistible temptation to accelerate production, requiring the use of more marginal resources and resulting in the destruction of more fragile ecological systems, particularly in the tropics. For consumers, the increased demand for products means increased competition for scarce resources, driving up the cost of those resources until only the wealthiest can afford what our grandfathers would have viewed as an adequate standard of living.

The articles selected for this second unit all relate, in one way or another, to the theory and reality of population growth, to the conflicting viewpoints of economists and biologists regarding the benefits and risks of population growth, and to the consequences of increased human populations for environmental systems. The first article of this unit examines the debate between the proponents of unrestricted population growth as an economic stimulus and the neo-Malthusians for whom each additional mouth to feed brings Earth that much closer to ultimate carrying capacity. In "How Many Is Too Many?" both the economists' and the biologists' perspectives in layman's terms are explained with the intriguing conclusion that both points of view ignore what may be a more immediate question: How do we handle population growth in the short term with political systems and institutions that are simply not adequate to the task? Whether the economists' models for the outcome of population growth are correct or whether the biologists are right is less important than whether we can handle the short-term problems that arise.

In the next article, David Norse, a research associate in

the Environmental Change Unit at the University of Oxford, assesses the food supply question. Even though the original population estimates by Malthus failed to take into account the growth in technology and, hence, increasing ability to produce food, most recent estimates of population growth suggest that, in most geographic or regional instances, human systems are presently incapable of supporting themselves using existing environmental and food production systems. Norse advances a new strategy for planning both population control and food production; in this new strategy, the concept of carrying capacity is widened to include economic as well as environmental considerations.

In the next selection, Sharon Camp, of Population Action International, also focuses on the short-term scenario. Like Mann, Camp sees the problem of overpopulation as more immediate and real-world than long-term and theoretical. She suggests, in "Population: The Critical Decade," that the environmental and population problems are such that if solutions are not found by the end of this decade, global tragedy, both economic and environmental, will occur. Following the lead of estimates by biologists, Camp assumes that—at the levels of technology and political structure we can assume to be operative within the next half century—a population of twice the 1993–94 level puts Earth at maximum carrying capacity.

In "The Landscape of Hunger," Bruce Stutz addresses the notion that increased population leads to environmental deterioration which, in turn, leads to lessened food production and a population-food imbalance that becomes increasingly large. The clearest examples of this may be found in the African savanna lands where recent droughts have made it clear that the environmental costs of traditional economic development models have been far too high.

It is clear from most of the selections in this unit that the global environment is being stressed by population growth and that more people means more stress and more poverty. It should also be clear that we can no longer afford to permit the unplanned and unchecked growth of the planet's dominant species. No closed environmental system can long sustain the kind of pressures that a population of more than five billion persons places upon it. Continuation of those pressures will wipe out the tremendous strides in human well-being that have been made over the last few centuries and will ensure that the environmental problems of the present will pale in significance beside those of the future.

Looking Ahead: Challenge Questions

Explain the essential difference between the traditional economic model of population growth as an economic stimulus and the traditional biological model of population growth as a threat to regional and/or global carrying capacities.

What is the relationship between the biological concept of carrying capacity and human overpopulation? Has the past failure of population projections based on carrying capacity made all such projections suspect?

Explain why some social scientists believe that carrying capacity is an inadequate concept for the understanding of human population growth. Are there relationships between development or population control plans based on the carrying capacity concept and the relative inadequacy of those plans?

Explain how environmental deterioration and overpopulation are related. Does the decrease in the ability of the environment to produce food naturally result in a decrease in the rate of population growth? Why or why not?

Why are the world's tropical savannas in such poor environmental and social shape? Is there a relationship between economic development efforts and the current plight of savanna populations—particularly in Africa?

HOW MANY IS TOO MANY?

Biologists have argued for a century that an ever-growing population will bring the apocalypse. Economists argue that man and markets will cope—so far none of the predicted apocalypses have arrived. The near-term questions, though, are political, and they are overlooked in the fierce battles

CHARLES C. MANN

Charles C. Mann is a contributing editor of The Atlantic. *Mann has recently completed a book on endangered species, a preview of which appeared as* The Atlantic's *January, 1992, cover story.*

In 1980, when I was living in New York City, it came to my attention that the federal government was trying to count every inhabitant of the United States. In my building—subject, like many in New York, to incredibly complicated rent-control laws—a surprising number of apartments were occupied by illegal subtenants. Many went to elaborate lengths to conceal the fact of their existence. They put the legal tenant's name on the doorbell. They received their mail at a post-office box. They had unlisted telephone numbers. The most paranoid refused to reveal their names to strangers. How, I wondered, was the Census Bureau going to count these people?

I decided to find out. I answered an advertisement and attended a course. In a surprisingly short time I became an official enumerator. My job was to visit apartments that had not mailed back their census forms. As identification, I was given a black plastic briefcase with a big red, white, and blue sticker that said U.S. CENSUS. I am a gangling, six-foot-four-inch Caucasian; the government sent me to Chinatown.

Strangely enough, I was a failure. Some people took one look and slammed the door. One elderly woman burst into tears at the sight of me. I was twice offered money to go away. Few residents had time to fill out a long form.

Eventually I met an old census hand. "Why don't you just curbstone it?" he asked. "Curbstoning," I learned, was enumerator jargon for sitting on the curbstone and filling out forms with made-up information. I felt qualms about taking taxpayers' money to cheat. Instead, I asked to be assigned to another area.

Wall Street is not customarily thought of as residential, but people live there anyway. Some live in luxury, some in squalor. None were glad to see me, even though I had given away the damning U.S. CENSUS briefcase to my four-year-old stepson. The turning point came when I approached two small buildings. One was ruined and empty. The other, though scarcely in better condition, was obviously full of people, but not one of them would answer the bell. In a fit of zealotry I climbed through the ruin next door. Coated with grime and grit, I emerged on the roof and leaped onto the roof of my target. A man was living on it, in a big, dilapidated shack.

He flung open his door. Inside I dimly perceived several apparently naked people lying on gurneys. "Go away!" the man screamed. He was wearing a white coat. "I'm giving my wife a cancer treatment!"

My enthusiasm waned. I jumped back to the other roof. On the street I sat on the curbstone and filled out a dozen forms. When I was through, fifty men, women, and children had been added to the populace of New York City.

PROFESSIONAL DEMOGRAPHERS ARE NOT AMUSED BY this sort of story. This is not because they are stuffy but because they've heard it all before. Finding out how many people live in any particular place is strikingly difficult, no matter what the place is. In the countryside people are scattered through miles of real estate; in the city they occupy nooks and crannies often missed by official scrutiny. No accurate census has ever been taken in some parts of Africa, but even in the United States, the director of the Census Bureau has said, the last official count, in 1990, missed more than five million people— enough to fill Chicago twice over. If my experience means anything, that number is low.

It's too bad, because How many are we? is an interesting question. Indeed, to many people it is an alarming question. For them, thinking about population means thinking about *over*population—which is to say, thinking about poverty, hunger, despair, and social unrest. For me, the subject evokes the vague unease I felt toting around *The Population Bomb*, which I read in school. ("It's Still Not Too Late to Make the Choice," the cover proclaimed. "Population Control or Race to Oblivion.") In other people it evokes the desire to put fences on our borders and stop the most wretched from breeding.

The Population Bomb appeared twenty-five years ago, in 1968. Written by the biologist Paul Ehrlich, of Stanford University, it was a gloomy book for a gloomy time. India was still undergoing a dreadful famine, Latin American exports of grain and meat had dropped to pre-war levels, and global food production was lagging behind births. More than half the world's people were malnourished. Nobel laureates were telling Congress that unless population growth stopped, a new Dark Age would cloud the world and "men will have to kill and eat one another." A well-regarded book, *Famine 1975!*, predicted that hunger would begin to wipe out the Third World that year. (Fortunately, the book pointed out, there was a bright side: the United States could increase its influence by playing triage among the victims.) In 1972 a group of researchers at MIT would issue *The Limits to Growth*, which used advanced computer models to project that the world would run out of gold in 1981, oil in 1992, and arable land in 2000. Civilization itself would collapse by 2070.

The projections failed to materialize. Birth rates dropped; food production soared; the real price of oil sank to a record low. Demographers were not surprised. Few had given much credence to the projections in the first place. Nonetheless, a certain disarray appeared in the work of what Ansley Coale, of Princeton's Office of Population Research, called the "scribbling classes." Doubts emerged about the wisdom and effectiveness of the billion-dollar population-control schemes established by the United Nations and others in the 1960s. Right-wingers attacked them as bureaucratic intrusions into private life. Critics on the left observed that once again rich whites were trying to order around poor people of color. Less ideological commentators pointed out that the intellectual justification for spending billions on international family-planning programs was shaky—it tacitly depended on the notion that couples in the Third World are

somehow too stupid to know that having lots of babies is bad. Ehrlich dismissed the carpers as "imbeciles."

Population has become the subject of a furious intellectual battle, complete with mutually contradictory charts, graphs, and statistics. The cloud of facts and factoids often seems impenetrable, but after peering through it for a time I came to suspect that the fighters had become distracted. Locked in conflict, they had barely begun to address the real nature of the challenge posed by population growth. *Homo sapiens* will keep growing in number, as everyone agrees, and that growth may have disagreeable consequences. But those consequences seem less likely to stem from the environmental collapse the apocalyptists predict than from the human race's perennial inability to run its political affairs wisely. The distinction is important, and dismaying.

Cassandras and Pollyannas

HOW MANY PEOPLE IS TOO MANY? OVER TIME, the debate has spread between two poles. On one side, according to Garrett Hardin, an ecologist at the University of California at Santa Barbara, are the Cassandras, who believe that continued population growth at the current rate will inevitably lead to catastrophe. On the other are the Pollyannas, who believe that humanity faces problems but has a good shot at coming out okay in the end. Cassandras, who tend to be biologists, look at each new birth as the arrival on the planet of another hungry mouth. Pollyannas, who tend to be economists, point out that along with each new mouth comes a pair of hands. Biologist or economist—is either one right? It is hard to think of a question more fundamental to our crowded world.

Cassandras and Pollyannas have spoken up throughout history. Philosophers in ancient China fretted about the need to shift the masses to underpopulated areas; meanwhile, in the Mideast, the Bible urged humanity to be fruitful and multiply. Plato said that cities with more than 5,040 landholders were too large; Martin Luther believed that it was impossible to breed too much, because God would always provide. And so on.

Early economists tended to be Pollyannas. People, they thought, are a resource—"the chiefest, most fundamental, and precious commodity" on earth, as William Petyt put it in 1680. Without a healthy population base, societies cannot afford to have their members specialize. In small villages almost everyone is involved with producing food; only as numbers grow can communities afford luxuries like surgeons, scientists, and stand-up comedians. The same increase lowers the cost of labor, and hence the cost of production—a notion that led at least one Enlightenment-era writer, J. F. Melon, to endorse slavery as an excellent source of a cheap work force.

As proof of their theory, seventeenth-century Pollyannas pointed to the Netherlands, which was strong, prosperous, and thickly settled, and claimed that only such a populous place could be so rich. In contrast, the poor, sparsely inhabited British colonies in the New World were begging immigrants to come and swell the work force. One of the chief duties of a ruler, these savants thought, was to ensure population growth. A high birth rate, the scholar Bernard Mandeville wrote in 1732, is "the never-failing Nursery of Fleets and Armies."

Mandeville wrote when the Industrial Revolution was beginning to foster widespread urban unemployment and European cities swarmed with beggars. Hit by one bad harvest after another, Britain tottered through a series of economic crises, which led to food shortages and poverty on a frightful scale. By 1803 local parishes were handing out relief to about one out of every seven people in England and Wales. In such a climate it is unsurprising that the most famous Cassandra of them all should appear: the Reverend Thomas Robert Malthus.

"Right from the publication of the *Essay on Population* to this day," the great economic historian Joseph Schumpeter wrote in 1954, "Malthus has had the good fortune—for this *is* good fortune—to be the subject of equally unreasonable, contradictory appraisals." John Maynard Keynes regarded Malthus as the "beginning of systematic economic thinking." Percy Bysshe Shelley, on the other hand, derided him as "a eunuch and a tyrant." John Stuart Mill viewed Malthus as a great thinker. To Karl Marx he was a "plagiarist" and a "shameless sycophant of the ruling classes." "He was a benefactor of humanity," Schumpeter wrote. "He was a fiend. He was a profound thinker. He was a dunce."

The subject of the controversy was a shy, kindly fellow with a slight harelip. He was also the first person to hold a university position in economics—that is, the first professional economist—in Britain, and probably the world. Married late, he had few children, and he was never overburdened with money. He was impelled to write his treatise on population by a disagreement with his father, a well-heeled eccentric in the English style. The argument was over whether the human race could transform the world into a paradise. Malthus thought not, and said so at length—55,000 words, published as an unsigned broadside in 1798. Several longer, signed versions followed, as Malthus became more confident.

"The power of population," Malthus proclaimed, "is indefinitely greater than the power in the earth to produce subsistence for man." In modern textbooks this notion is often explained with a graph. One line on the graph represents the land's capacity to produce food; it slowly rises from left to right as people clear more land and learn to farm more efficiently. Another line starts out low, quickly climbs to meet the first, and then soars above it; that line represents human population. Eventually the gap between the two lines cannot be bridged and the Horsemen of the Apocalypse pay a call. Others had anticipated this idea. Giovanni Botero, an Italian scholar, described the basic relationship of population and resources in 1589, two centuries before Malthus. But

few read Malthus's predecessors, and nobody today seems inclined to replace the term "Malthusian" with "Boterian."

The *Essay* was a jolt. Simple and remorselessly logical, blessed with a perverse emotional appeal, it seemed to overturn centuries of Pollyanna-dom at a stroke. Forget Utopia, Malthus said. Humanity is doomed to exist, now and forever, at the edge of starvation. Forget charity, too: helping the poor only leads to more babies, which in turn produces increased hardship down the road. Little wonder that the essayist Thomas Carlyle found this theory so gloomy that he coined the phrase "dismal science" to describe it. Others were more vituperative, especially those who thought that the *Essay* implied that God would not provide for His children. "Is there no law in this kingdom for punishing a man for publishing a libel against the Almighty himself?" demanded one anonymous feuilleton. In all the tumult hardly anyone took the trouble to note that logical counter-arguments were available.

The most important derived from the work of Marie-Jean-Antoine-Nicolas Caritat, Marquis de Condorcet, a French *philosophe* who is best known for his worship of Reason. Four years before Malthus, Condorcet observed that France was finite, the potential supply of French infinite. Unlike Malthus, though, Condorcet believed that technology could solve the problem. When hunger threatens, he wrote, "new instruments, machines, and looms" will continue to appear, and "a very small amount of ground will be able to produce a great quantity of supplies." Society changes so fast, in other words, that Malthusian scenarios are useless. Given the level of productivity of our distant ancestors, in other words, we should already have run out of food. But we know more than they, and are more prosperous, despite our greater numbers.

Malthus and Condorcet fixed the two extremes of a quarrel that endures today. The language has changed, to be sure. Modern Cassandras speak of "ecology," a concept that did not exist in Malthus's day, and worry about exceeding the world's "carrying capacity," the ecological ceiling beyond which the land cannot support life. Having seen the abrupt collapses that occur when populations of squirrels, gypsy moths, or Lapland reindeer exceed local carrying capacities, they foresee the same fate awaiting another species: *Homo sapiens.* Pollyannas note that no such collapse has occurred in recorded history. Evoking the "demographic transition"—the observed propensity for families in prosperous societies to have fewer children—they say that continued economic growth can both feed the world's billions and enrich the world enough to end the population boom. No! the Cassandras cry. Growth is the *problem.* We're growing by 100 million people every year! We can't keep doing that forever!

True, Pollyannas concede. If present-day trends continue for centuries, the earth will turn into a massive ball of human flesh. A few millennia more, Ansley Coale, of Princeton, calculates, and that ball of flesh will be expanding outward at the speed of light. But he sees little point in the exercise of projecting lines on a graph out to

"OVERPOPULATION"

IS HARD TO

DEFINE EXACTLY. PART OF

THE REASON

IS THAT ATTEMPTS TO

ISOLATE SPECIFIC

SOCIAL OR ENVIRONMENTAL

CONSEQUENCES

OF RAPID POPULATION

GROWTH TEND TO

SINK INTO IDEOLOGICAL

QUICKSAND.

their absurdly horrible conclusion. "If you had asked someone in 1890 about today's population," Coale explains, "he'd say, 'There's no way the United States can support two hundred and fifty million people. Where are they going to pasture all their horses?'"

Just as the doomsayers feared, the world's population has risen by more than half since Paul Ehrlich wrote *The Population Bomb*. Twenty-five years ago 3.4 billion people lived on earth. Now the United Nations estimates that 5.3 billion do—the biggest, fastest increase in history. But food production increased faster still. According to the Food and Agricultural Organization of the UN, not only did farmers keep pace but per capita global food production actually rose more than 10 percent from 1968 to 1990. The number of chronically malnourished people fell by more than 16 percent. (All figures on global agriculture and population in the 1990s, including those in this article, mix empirical data with projections, because not enough time has elapsed to get hard numbers.)

"Much of the world *is* better fed than it was in 1950," concedes Lester R. Brown, the president of the Worldwatch Institute, an environmental-research group in Washington, D.C. "But that period of improvement is ending rather abruptly." Since 1984, he says, world grain production per capita has fallen one percent a year. In 1990, eighty-six nations grew less food per head than they had a decade before. Improvements are unlikely, in Brown's view. Our past success has brought us alarmingly close to the ecological ceiling. "There's a growing sense in the scientific community that it will be difficult to restore the rapid rise in agricultural yields we saw between 1950 and 1984," he says. "In agriculturally advanced nations there just isn't much more that farmers can do." Meanwhile, the number of mouths keeps up its frantic rate of increase. "My sense," Brown says, "is that we're going to be in trouble on the food front before this decade is out."

Social scientists disagree. An FAO study published in 1982 concluded that by using modern agricultural methods the Third World could support more than 30 billion people. Other technophiles see genetic engineering as a route to growth that is almost without end. Biologists greet such pronouncements with loud scoffs. One widely touted analysis by Ehrlich and others maintains that humanity already uses, destroys, or "co-opts" almost 40 percent of the potential output from terrestrial photosynthesis. Doubling the world's population will reduce us to fighting with insects over the last scraps of grass.

Neither side seems willing to listen to the other; indeed, the two are barely on speaking terms. The economist Julian Simon, of the University of Maryland, asserts that there is no evidence that the increase in land use associated with rising population has led to any increase in extinction rates—despite hundreds of biological reports to the contrary. The biologist Edward O. Wilson, of Harvard University, argues that contemporary economics is "bankrupt" and does not accommodate environmental calculations—despite the existence of a literature on the subject dating back to the First World War. A National Academy of Sciences panel dominated by economists argues in 1986 that the problems of population growth have been exaggerated. Six years later the academy issues a statement, dominated by biologists, claiming that continued population growth will lead to a global environmental catastrophe that "science and technology may not be able to prevent." Told in an exchange of academic gossip that an eminent ecologist has had himself sterilized, an equally eminent demographer says, "That's the best news I've heard all week!" Asking himself what "deep insights" professional demographers have contributed, Garrett Hardin answers, "None."

The difference in the forecasts—prosperity or penury, boundless increase or zero-sum game, a triumphant world with 30 billion or a despairing one with 10—is so extreme that one is tempted to dismiss the whole contretemps as foolish. If the experts can't even discuss the matter civilly, why should the average citizen try to figure it out? Ignoring the fracas might be the right thing to do if it weren't about the future of the human race.

2. THE WORLD'S POPULATION: PEOPLE AND HUNGER

Two Nations

POPULATION QUESTIONS ARE FUZZY. EVEN AN APparently simple term like "overpopulation" is hard to define exactly. Part of the reason is that evaluating the consequences of rapid population growth falls in the odd academic space where ecology, economics, anthropology, and demography overlap. Another part of the reason is that attempts to isolate specific social or environmental consequences of rapid population growth tend to sink into ideological quicksand.

By way of example, consider two nations. One is about the size of Maryland and has a population of 7.2 million; the other is as big as Montana but has a population of 123.5 million. The first has a population density of 703 people per square mile, a lot by most standards; the second has a density of 860 per square mile, among the highest on the planet. Country No. 1 has tracts of untouched forest and reserves of gold, tin, tungsten, and natural gas. Country No. 2 has few natural resources and little arable land. Life there is so crowded that the subways hire special guards to mash people onto the trains. Is it, therefore, overpopulated?

Most economists would say no. Country No. 2 is Japan. Paul Demeny, a demographer at the Population Council, in New York City, notes that Japan is where the Malthusian nightmare has come true. Population has long since overtaken agricultural capacity. "Japan would be in great trouble if it had to feed itself," Demeny says. "They can't eat VCRs. But they don't worry, because they can exchange them for food." Demeny is less sanguine about Country No. 1—Rwanda, the place with the highest fertility rate in the world. There, too, the production of food lags behind the production of people. But Rwanda, alas, has little to trade. "If something goes wrong," Demeny says, "they will have to beg."

Some economists might therefore attach to this crowded land the label "overpopulated." Others, though, might say that Rwanda has not yet reached the kind of critical mass necessary to develop its rich natural endowment. Fewer than 200,000 souls inhabit Kigali, its capital and biggest city, hardly enough to be the hub of a modern nation. In this case, a cure for having too many children to feed might be to have more children—the approach embraced by the Rwandan government until 1983.

Rwanda's leaders may well have been bowing to the popular will. By and large, people in the developing world have big families because they want them. "The notion that people desperately want to have fewer children but can't quite figure out how to do it is a bit simple," Demeny says. "If you picture an Indian who sees his children as capital because at the age of nine they can be sent to work in a carpet factory, his interest in family planning will not be keen." If the hypothetical impoverished Indian father does not today desperately need the money that his children can earn, he will need it in his dotage. Offspring are the Social Security of traditional cultures everywhere, a form of savings that few can afford to forgo. In such cases the costs of big families (mass illiteracy, crowded hiring halls, overused public services) are spread across society, whereas the benefits (income, old-age insurance) are felt at home. Economists call such phenomena "market failure." The outcome, entirely predictable, is a rapidly growing population.

Equally predictable is the proposed solution: bringing home the cost to those who experience the benefits. Enforcing child-labor and truancy laws, for example, drives up the price of raising children, and may improve their lives as well. Reducing price controls on grain raises farmers' incomes, allowing them to hire adults rather than put their children to work. Increasing opportunities for women lets them choose between earning income and having children. In the short term such modifications can hurt. In the long term, Demeny believes, they are "a piece of social engineering that any modern society should aspire to." Rwanda, like many poor countries, now has a population-control program. But pills and propaganda will be ineffective if having many children continues to be the rational choice of parents.

To ecologists, this seems like madness. Rational, indeed! More people in Rwanda would mean ransacking its remaining tropical forest—an abhorrent thought. The real problem is that Rwandans receive an insufficient share of the world's feast. The West should help them rise as they are, by forgiving their debts, investing in their industries, providing technology, increasing foreign aid—and insisting that they cut birth rates. As for the claim that Japan is not overpopulated, the Japanese are shipping out their polluting industries to neighboring countries—the same countries, environmentalists charge, that they are denuding with rapacious logging. "If all nations held the same number of people per square kilometer," Edward O. Wilson has written about Japan, "they would converge in quality of life to Bangladesh. . . ." To argue that Tokyo is a model of populousness with prosperity is, Wilson thinks, "sophistic."

Wait, one hears the economists cry, that's not predation, that's trade! Insisting on total self-sufficiency veers toward autarky. Japan logs other people's forests because its own abundant forests are too mountainous to sustain a full program of—*and wait a minute*, haven't we been here before? The competing statistics, the endless back-and-forth argument? Isn't there some better way to think about this?

Good News, Bad News

IN 1968, WHEN *THE POPULATION BOMB* WAS FIRST PUBlished, the United Nations Population Division surveyed the world's demographic prospects. Its researchers projected future trends in the world's total fertility rate, a figure so common in demographic circles that it is often referred to, without definition, as the TFR. The TFR is the answer to the question "If women keep having babies at the present rate, how many will each have, on average, in her lifetime?" If a nation's women have two children apiece, exactly replacing themselves and the fathers of their children, the TFR will be 2.0 and the population will eventually stop growing. (Actually, replacement level is around 2.1, because some children die.) In the United States the present TFR is about 2.0, which means that, not counting immigration, the number of Americans will ultimately hit a plateau. (Immigration, of course, may alter this picture.) But the researchers in the division were not principally concerned with the United States. They were looking at poorer countries, and they didn't like what they saw.

As is customary, the division published three sets of population projections: high, medium, and low, reflecting different assumptions behind them. The medium projection, usually what the demographer regards as the most likely alternative, was that the TFR for developing nations would fall 15 percent from 1965–1970 to 1980–1985. At the time, Ronald Freedman recalls, this view was regarded as optimistic. "There was a lot of skepticism that anything could happen," he says. He was working on family-planning programs in Asia, and he received letters from colleagues telling him how hopeless the whole endeavor was.

Now a professor emeritus of sociology at the University of Michigan, Freedman is on the scientific advisory committee of Demographic and Health Surveys, a private organization in Columbia, Maryland, which is funded by the U.S. government to assess births and deaths in Third World nations. Its data, painstakingly gathered from surveys, are among the best available. From 1965–1970 to 1980–1985 fertility in poor countries dropped 30 percent, from a TFR of 6.0 to one of 4.2. In that period, Freedman and his colleague Ann K. Blanc have pointed out, the poor countries of the world moved almost halfway to a TFR of 2.1: replacement level. (By 1995, Blanc says, they might be two thirds of the way there.) If the decrease continues, it will surely be the most astonishing demographic shift in history. (The second most astonishing will be the rise that preceded it.) The world went halfway to replacement level in the twenty years from 1965 to 1985; arithmetic suggests that if this trend continues we will arrive at replacement level in the subsequent twenty years—that is, by 2005.

That's the good news. The bad news is that since the late 1960s, 1.9 billion more people have arrived on the planet than have left. Even if future rates of fertility are the lowest in history, as is likely, the children of today's children, and their children's children, will keep replacing themselves, and the population will increase vastly. Nothing will stop that increase, not even AIDS. Pessimists estimate that by the end of the decade another 100 million people will be infected by HIV. Almost ten times that number will have been born. Barring unprecedented catastrophe, the year 2100 will see 10 to 12 billion people on the planet.

THE DIFFERENCE IN THE FORECASTS— BOUNDLESS INCREASE OR ZERO-SUM GAME, A TRIUMPHANT WORLD WITH 30 BILLION OR A DESPAIRING ONE WITH 10—IS SO EXTREME THAT ONE IS TEMPTED TO DISMISS THE WHOLE CONTRETEMPS AS FOOLISH.

Nobody will have to wait that long to feel the consequences. In a few years today's children will be clamoring to take their place in the adult world. Jobs, homes, cars, a few occasional treats—these are things they will want. And though economists are surely right when they say the lesson of history is that the great majority of these men and women will make their way, it is hard not to be awed by the magnitude of the task facing the global economy. A billion jobs. A billion homes. A billion cars. Billions and billions of occasional treats.

Trees and Soil

Few places in the United States better illustrate the unforeseen consequences of population growth than the lower valley of the Hudson River, which runs south from Albany to New York City. The river is wide and placid in appearance, and its banks sparkle with towns that were young when the nation was young. To the west rise the Catskills, blue at sunset and blanketed by trees. Interstate 87 makes a ribbon between the water and the mountains. A while ago I spent some time driving on that road, and down long miles of its length the forest stretched out so far and so dark and so empty that I imagined I was looking at the America of a hundred years ago, before there were millions of people like me around. How wonderful, I thought, that so close to Manhattan is a huge piece of real estate that we never trashed.

I was wrong. If I had traveled through the Hudson Valley at the end of the last century, I would have passed through an utterly different landscape. I would have been surrounded by small hardscrabble farms, fields of wheat and corn, and pastures ringed by stone walls. It might have looked picturesque—certainly guidebook writers of the day thought so. But I wouldn't have seen many trees, because long before, almost every scrap of land that wasn't vertical had been clear-cut or burned.

The forest was stripped to make way for agriculture and to supply New York's army of charcoal-burners (who needed lumber to make charcoal), tanners (who extracted tannin from bark), and salt-makers (who used wood fuel to boil down seawater). Loggers played a role too: Albany, the northernmost deepwater port on the Hudson, was the biggest timber town in the nation and possibly the world. When the first Europeans came to New York, the region was almost entirely covered by trees; by the end of the nineteenth century less than a quarter of the state was wooded, and most of what was left had been picked through, or was inaccessible, or was being kept by farmers as private fuel reserves. During the epoch that I, swooping along the tarmac in my minivan, was nostalgically picturing as a paradise, newspaper editorials were warning that deforestation would drive the valley toward ecological disaster.

Since then the collapse of small farming has allowed millions of acres to return to nature. When New York State surveyed itself in 1875, the six counties that make up the lower Hudson Valley—Columbia, Dutchess, Greene, Orange, Putnam, and Ulster—contained 573,003 acres of timberland, covering about 21 percent of their total area. In 1980, the date of the most recent survey, trees covered almost 1.8 million acres, more than three times as much. (This is no scrubby growth, mind you. Michael Greason, an associate forester in the state's Department of Environmental Conservation, calls today's Hudson and Catskill forest a "beautiful, diverse ecosystem.")

I was driving around the Hudson Valley partly because I was looking for a house in the country. My method of looking, insofar as one could call it a method, was to hunt in the counties with the lowest populations, figuring that they would be the least spoiled. I was trying to get away from people, and from the unpleasantness I associated with urban life. The more crowded an area, I thought, the more degraded its environment. I wanted natural beauty, and that meant "uninhabited."

Learning some local history gave me pause. Back in 1875 my six counties had a collective population of 345,679. The U.S. Census says the figure for 1990 was 924,075. In other words, the number of people living there almost tripled in the same period that the local ecosystem climbed out of its sickbed and threw away its crutches. This wasn't just some odd thing that happened

in New York. As a whole, American forests are bigger and healthier than they were at the turn of the century, when the country's population was below 100 million. Massachusetts and some other states have as many trees as they had in the days of Paul Revere. Nor was this growth restricted to North America: Europe's forest resources increased by 25 or 30 percent from 1970 to 1990, a time in which its population grew from 462 million to 502 million. Presumably the forest figure would have been yet higher without the continent's damaging acid rain.

People pollute. But more people does not always mean more pollution. Eco-critics can claim, with some justification, that the forests of the Hudson Valley recuperated because farmers abandoned them in order to wipe out the native grasslands of the Great Plains. But they can't explain away all the good news. Salmon are reintroduced to the Thames. White-tailed deer, almost extinct in 1900, plague New England gardens. Air quality in Tokyo improves remarkably. Wild turkeys have a greater range than they did when they were first seen by white settlers. If all this occurred during the population boom, why the belief, now frequently voiced, that overpopulation will lead to an eco-catastrophe?

"You can look at Lake Erie or Detroit and see it's gotten better," says Dennis L. Meadows, the leader of the research group that produced *The Limits to Growth*. "But to leap from that to the conclusion that there has been overall improvement is to look at one person getting rich and say that everybody is better off." Now at the helm of the Institute for Policy and Social Science Research at the University of New Hampshire, Meadows, with two co-authors, has recently published a sequel, *Beyond the Limits,* which is even more pessimistic than the first book. "When a rich country becomes concerned about environmental problems," Meadows says, "then it can typically develop effective responses." Lead additives in gasoline became a subject of American worries. The nation forced petroleum companies to phase out leaded gas, and lead levels diminished. Similar fears have led twenty-three industrialized countries since 1987 to halve the rate of release of the worst ozone-eating compounds. In Meadows's view, rich countries have the wealth to buy their way out of many such difficulties. But other difficulties remain, and their number grows. Soil erodes; draining ruins wetlands; contaminants infiltrate groundwater; toxic wastes keep accumulating. The situation is much worse in poor countries that cannot pay to clean themselves up. Behind all the concerns is the specter of growing numbers, of insatiable human wants, of continually increasing demand. "We're trying to sustain physical growth on a finite planet," he says. "Growth will stop in our lifetime." Meadows, who is fifty, confidently expects to see the end of population growth in his lifetime—probably through ecological breakdown.

Cassandras like Meadows, Ehrlich, and Lester Brown, of Worldwatch, regard land degradation as one of the worst and most obvious ecological consequences of crowding too many people into too little space. "Land degradation" is a catchall term covering such problems as wind and water erosion, soil pollution by urban wastes or pesticides, and the buildup of mineral salts caused by improper irrigation. The term may bring to mind newspaper photographs of faraway African husbandmen in terrain nibbled to exhaustion by cattle. But environmentalists say the problem is bigger than that—"virtually a worldwide epidemic," in the words of Anne Ehrlich, a research associate at Stanford University, a veteran Cassandra, and the wife of Paul Ehrlich. The International Soil Reference and Information Centre (ISRIC), in the Netherlands, estimates that since 1945 *Homo sapiens* has degraded 17 percent of the world's land, not counting wastelands like Antarctica and the Gobi Desert. Two thirds of the devastated area will require major restoration.

Every year, Brown warns, erosion steals 24 billion tons of soil from the world's farmers—a figure that the ecology-minded cite as plain evidence that humanity is exceeding the carrying capacity of the planet. The economics-minded see rhetoric and hand-waving. "Those figures Lester Brown is going on are based on really wild assumptions," says Pierre Crosson, a senior fellow at Resources for the Future, a nonprofit research group in Washington, D.C. "He's taking poorly understood sedimentation figures from big river deltas and extrapolating them to get a figure for productivity loss for the whole world." As for the ISRIC study, Crosson points out that it classifies most of Illinois, Iowa, Kansas, Nebraska, and the Dakotas as degraded enough to "greatly reduce" productivity. "The problem with this," he says, "is that in those six states yields have risen steadily over the last forty years."

The United States has the most carefully measured soil in the world. Every five years the U.S. Department of Agriculture evaluates the nation's land, county by county, in terms of something called "the universal soil loss equation," which assesses the soil movement in a given area. Because the equation measures movement rather than absolute disappearance, the results are hard to evaluate; they don't distinguish between soil that ends up on the bottom of the ocean and soil that is merely shifted to a neighbor's property, enriching yields there. (Sometimes the neighbors are far away: according to measurements by atmospheric scientists at the University of Virginia, more than 13 million tons of rich African soil are blown every year onto the Amazon forest floor.) In the 1980s three independent studies, one by Crosson, used the data to estimate actual soil loss. All concluded that the peril to U.S. agriculture from erosion is negligible. "The Soil Conservation Service is doing its job," Crosson says.

It is hard to be as sanguine about the developing world. Sub-Saharan Africa, for instance, is often described as a place where overpopulation has led to a terrible destruction of land. The region has high birth rates; per capita food production has consistently fallen in the past ten years. Rainless years in the same period led to

overgrazing, deforestation, wide-scale erosion, and the hunger shown in sad televised images. Civil strife and famine have driven at least two million people from their homes. The desert is said to be marching south at a rate of five to six miles a year, a scary prospect that Cassandras regard as dramatic evidence of Africa's population quandary.

No one doubts that Africa is in a dire position, but the Pollyannas think that the problem is bad luck, bad weather, and bad planning. Traditionally, African villagers often held land in common, with access regulated by unwritten cultural rules. "In those circumstances," Crosson says, "the people responsible for the management of the land take overuse into account, so they enforce rules of access that limit the use of the land." When modern crops and agricultural techniques appear, the system comes apart, because yields shoot high enough to give people a greater incentive to cheat. If they break the rules, they can make a bundle and skip town with the profit; those who play fair are left with the ruins. (Garrett Hardin calls this the "tragedy of the commons.") Population pressures exacerbate the problem, most economists concede, by shrinking everyone's share of the common land. Add drought or ethnic conflict and the result is disaster. But, they say, African nations without drought or conflict have done increasingly well. Given half a chance, people in Africa, like people anyplace else, seem to make their own way.

The Cassandras base their case "on intuitively powerful hypotheses that slide over the need for empirical verification," says Michael Mortimore, a senior research associate at Cambridge University, England. "Unfortunately, the only way to test them is to collect a large amount of empirical data." Mortimore has spent years gathering information on farms in Nigeria and Kenya. His comparison of contemporary and thirteen-year-old dirt samples in Nigeria, one of Africa's most heavily populated nations, shows that increasing population has, if anything, *raised* land productivity. In decades past, farmers could exhaust an area and then move on. Now rising numbers have made land more expensive, and people have greater incentives to take care of what they have. "You're not going to pass on a desert to your children," Mortimore says. Claiming that Africans are otherwise inclined is, to his mind, "straightforward cultural prejudice." The agricultural systems that he studies are "economically, socially, and ecologically sustainable." In Nigeria, he says, the 1992 harvest was the biggest in twenty years.

As for other parts of sub-Saharan Africa, no one denies the famine there. Yet recent independent studies have found no *long-term* environmental consequences of the recent and devastating drought; the southern border of the desert, one study shows, is in about the same place it was eighty years ago, suggesting that the desert expands and contracts with little regard for its human inhabitants. The drought may have led to temporary overuse of common property, economists concede, but the proper response would be to adjust land-use rules—change the zoning, so to speak—as societies did in other parts of Africa. That response's alleged failure to occur in sub-Saharan Africa says more about the pervasive corruption, inefficiency, and civil turmoil there than about the inherent evils of breeding.

"You always can blame any particular difficulty on something that is not overpopulation," Dennis Meadows says. "Nobody ever dies of overpopulation. They die of famine, pestilence, and war. There's always some proximate cause." Adjusting land-use rules is desirable in his view, but it does nothing to eliminate the fundamental problem. At bottom, more people means more resource use and more pollution and more loss of biodiversity, all of which must eventually stop. We can't go on forever, because the world is finite. No matter how smart we are, Cassandras say, we can't avoid being part of the web of life, governed by its laws. True enough, the Pollyannas say. But our part in that web is to be different. Other species die when they breed enough to wreck their environments, as they sometimes do. Human beings don't wait to be overwhelmed. When problems arise, we solve them—that's our nature. This argument makes Cassandras smite their collective brow and deride, as Ehrlich did recently, the "imbeciles running around today saying not to worry, that with the aid of science and technology we can take good care of many billions more."

Thinking Differently

IN NATHAN KEYFITZ'S VIEW, THE ARGUMENT AMOUNTS to a classic academic standoff. Keyfitz leads a group of demographers at the International Institute for Applied Systems Analysis (IIASA), a research group in Austria, and is one of the few people I heard praised by both sides in the dispute. Keyfitz recalls attending a meeting at Brandeis University convened with the noble intention of reconciling biologists and economists. For three days speakers from each side stated their points of view, completely ignoring everything the other side had said. Then, apparently satisfied, everyone went home. "Here," Keyfitz has written, "is a nightmare for democratic politics: what action to take on vital questions where the experts disagree violently."

A first step away from the impasse, Keyfitz suggests, might be for each side to accept the validity of the other's arguments, as long as they pertain to that side's discipline. Give the biologists their due, and agree that an increased human presence poses a huge threat to the environment, even if some of the biologists' claims are exaggerated. Then agree with economists that the problems are not due to population growth per se. Instead, population growth changes social and economic systems, which inevitably creates environmental problems: more food must be grown, so that once-adequate agricultural methods now overuse land.

But Keyfitz is hesitant to embrace the next step in the economists' logic. "I think we all know the idea," he says.

THE DISAGREEABLE CONSEQUENCES OF POPULATION GROWTH SEEM LESS LIKELY TO STEM FROM THE ENVIRONMENTAL COLLAPSE THE APOCALYPTISTS PREDICT THAN FROM THE HUMAN RACE'S PERENNIAL INABILITY TO RUN ITS POLITICAL AFFAIRS WISELY.

"As you catch up with these problems one by one, you turn the power of science on them and produce solutions." Sometimes the correction is expensive, Cassandras admit, but thoughtful government policies can minimize the pinch. Here we step outside the boundaries of economics into political science, into arguments nobody has promised to accept. The picture, Keyfitz says, has governments responding wisely—and, as everyone knows, they often don't. Indeed, an outside observer might find it curious that economists—disdainful of government in other matters—exhibit such touching faith in the power of their elected representatives to resolve the troubles exacerbated by rising population.

Even if governments try to respond, they are often unable to anticipate the consequences of their actions. As a means of fostering international communication, Keyfitz says, "the worldwide air-transport network does nothing but good. But it's responsible for the spread of AIDS. Otherwise it would be an unrecognized, unnamed disease in a corner of Africa." With the instant contact possible today, a local problem became a global catastrophe. "You'd like more chance to breathe," Keyfitz says. "I'm not saying that any of these problems are inherently unresolvable by themselves, but we wouldn't be running so fast if the world had half its present population."

As the human presence increasingly dominates the earth, new difficulties emerge at ever greater speed: The ozone layer. The exhaustion of fisheries. The greenhouse effect. The overuse of aquifers. The need to increase yields of tropical foodstuffs. Each must be evaluated, absorbed, treated, even as the next problem appears. The loss of biodiversity. The collapse of the infrastructure. The destruction of rain forests. And on and on. Maybe people can keep up; certainly Keyfitz's colleagues, the computer modelers at IIASA, think so. But in a world where every citizen is surrounded by examples of governmental incompetence, it is hard to imagine that nations will keep coming through, time after time. There'll be so much *juggling*. So many things will have to be fixed all at once. Societies will find themselves in the position of the antic ninjas in video arcades, hordes of enemies to every side. Whack! Bam! Pow! Eventually, as any adolescent knows, the ninja slips up, and the game is over.

My Madonna Problem

FOR MANY YEARS I HAD AN OFFICE ON THE FOURteenth floor of a building in Manhattan, overlooking a busy avenue. During that time I never had air-conditioning, because it does something to the air that makes my nose run. In hot weather I opened the window. A breeze came in, reducing the temperature to a range I found tolerable. The problem was that it was hard to speak on the phone with the window open. People drove around the city with their car stereo systems cranked up to medically unsafe levels. Abrupt blasts of noise filled the room.

A debacle occurred when I interviewed Mario Cuomo, the governor of New York. Arranging the call involved considerable negotiation with his office in Albany. Two minutes into the conversation Madonna arrived. I had to shout over her moans. The governor did not immediately believe my explanation of the sound's origin. The light changed; the noise went away. I asked him a question. Suddenly Madonna showed up again. I realized that she was circling the block, looking for a parking space. I slammed the window shut, raising the room's temperature by ten degrees. The song was still devastatingly loud. The third time the song came round, I draped my jacket over the window. The fourth time, the governor hung up.

Soon after, I began looking for a house in the country.

The noise, and my response to it, illustrate what social scientists might regard as typical population-related feedback. Congestion grows in a city; it becomes impossible to avoid Madonna indoors; people move; the city becomes less congested. Or, perhaps, the city enforces noise ordinances. In either case the problem is resolved after temporary conflict.

"Temporary"—note the hedge word. The window of my old office looked out over the remains of the West Side Highway, part of which was closed as unsafe in the

1970s. I began using the office after a lengthy squabble had broken out over the best way to repair the road. Some wanted to do the job cheaply, shifting the money to mass transit; others said that the city needed the highway to revitalize its waterfront. Politicians, environmentalists, neighborhood groups, and construction unions attacked one another. Reconstructing the West Side Highway became a legal imbroglio of dizzying complexity. Charges and countercharges, suits and countersuits—I couldn't begin to follow it. Meanwhile, huge rush-hour lines built up twice a day on the partially closed road. Not wanting to sit in traffic, people took the exit near my office and continued their journeys on my street. They've been doing that in one form or another for more than a decade—which is part of the reason there was so much congestion on the street, and part of the reason Madonna was in my office for so long.

Move to the country, buy extra-thick insulated windows—middle-class Americans like me are unlikely to face the extreme choices that will confront people in developing nations. As Meadows says, we will be able to buy our way out of some problems. But that does not mean that Americans will escape the repercussions of population growth. In 1990 the United States had 249 million inhabitants. Current Census Bureau projections foresee that births and immigration will drive that number to 345 million by 2030. Ninety-six million more Americans! Imagine everything that local, state, and federal governments will have to do to accommodate them. Think of all the thoughtful planning that will be required to make life pleasant. Whether it will happen "is an absolutely open question, in my judgment," says Allen Kelley, an economist at Duke University. "We're talking about political theory, and there just *is* no theory for this."

The future is not completely opaque. Demographers know roughly where those 96 million new Americans will live, who they will be, and what they will do. The Census Bureau projects that the extra people will not distribute themselves in an even blanket over the country. About half will choose to live in just three states: California, Texas, and Florida. Not one of these can absorb its share easily. California has too little water, Texas an unsteady economy, Florida a particularly troubled set of ecosystems. All three will also need many new roads, bridges, sewers, and schools—an expensive proposition.

Relatively few of the newcomers will be white. By 2030 about 40 percent of Americans will have ancestral roots in Africa, Asia, or Latin America. Few would disagree that people of color have long been at the bottom of the American pyramid. If that continues, many of the 96 million will be poor, and in the twenty-first century a bigger percentage of the U.S. population will be below the poverty level than is now. If that does not continue, white Americans will for the first time face nonwhite power centers. California, Texas, and Florida will have three of the four biggest voting blocs in the House of Representatives (the fourth will be New York's). All three could be dominated by Hispanics and African-Americans. If history is any guide, either alternative will put white Americans' moral grace sorely to the test.

The United States has a long border with Mexico, and thousands of miles of coastline. Migrants want to come in, and Americans seem increasingly diffident about putting out the welcome mat. (In 1991 Argentina's president publicly toyed with the idea of resettling skilled refugees at a price of about $50,000 a head.) Meanwhile, rising numbers will thrust together the blacks, whites, yellows, browns, and reds who are already here in ways they have not been before. Political turmoil is increasing in Europe—especially in France and Germany—as migrants from Africa, the Middle East, and the former Soviet bloc demand a piece of the better life. Population growth did not cause ethnic conflicts, but it could sharpen them.

At the same time, the United States will have to address its age shift. Census Bureau projections tell a sobering story. By 2030, one out of five Americans will be sixty-five or older. The ratio of elderly people to people of working age will have tripled since 1950, meaning that, on average, workers will have to support three times as many older people. (The figure may be offset somewhat by the smaller number of children there will be to support. But the problem will be amplified by the growing number of people in their sixties and seventies who will be taking care of still more elderly people, in their eighties and nineties.) Workers, meanwhile, will be facing gloomier prospects. Nowadays fewer older people are alive than younger people, which means that there is an approximate match between the small number of older workers and the small number of important jobs. As the population ages, this will change. "In a stationary population," Ansley Coale, the Princeton demographer, has written, "there would no longer be a reasonable expectation of advancement as a person moves through life." More and more people will find themselves trapped in dead-end jobs. In some ways, Coale suggests, we may end up looking back with envy at the days of rapid population growth.

People will have to deal with the frustration—at the same time that they seek politicians able to resolve all the other problems. An optimist can say that the American people are kind, clever, and adaptable. An optimist can say that Americans can resolve the difficulties described, and the others these will inspire. But even an optimist must conclude that the government will have its hands full, and might well wonder what it will be like to live in a future America that must balance so much so artfully. My guess is that it will be something like living in New York City today.

Because I had nothing else to do, I walked along the last remaining section of the West Side Highway one recent Friday afternoon—a masochistic act. The road is officially under construction, and so its shoulders have not been cleaned up. Heaps of trash lie in windblown aggre-

gations. The walls are gray and peeling. Cars move slowly and impatiently, in billows of exhaust. It is like an evocation of an unpleasant future. New York City took a hit in the recession, and the familiar assortment of urban evils got worse. Unemployment and crack use went up; real estate values, tax revenues, and bond ratings went down. The city fabric sprang a hundred big leaks. Like so many Dutch boys, the city fathers raced to plug them. Meanwhile, a thousand interest groups scrabbled to protect themselves. Understandably, fixing the highway seems not to have assumed pressing importance.

That day on the West Side Highway, I heard many radios played at high volume. I could see the point. License plates came from Connecticut, New Jersey, and Pennsylvania; drivers were commuting for hours to sit in traffic. Only part of their daily grind was due to the sheer size of New York; traffic jams afflict much smaller cities. But a three-hour commute is what some architects call a "disamenity." The waiting, the ugly walls, the blaring horns, the inability to get away from it all: these are the disamenities of overpopulation in rich societies. None is directly attributable to population growth. Each is an emblem of bureaucratic overload—overload caused by the hundred predicaments that population growth aggravates. The city could have fixed the West Side Highway long ago. So could the state, or the federal government. But they didn't. There are too many other, bigger concerns, and there will be for a long time. My Madonna problem is not going to vanish.

Population growth forces nations to confront problems that they could finesse in less crowded times. The inefficiencies of New York's highways surely mattered less when the metropolitan area held eight million people rather than fifteen. Governments may, as economists hope, avoid the major, life-threatening disasters. "Even Congress isn't stupid enough to avoid dealing with the ozone layer," one economist told me. Americans may create a society that can hold an older, darker-colored population. But while they are doing that, the disamenities accrue, making the future a grayer, grimier place. Per capita GNP may go up and up, but life will be less pleasant. How much less pleasant? The answer depends on the strength of the citizenry's desire to have a government that works much more flexibly and wisely than government does now. New Yorkers to date have shown little such inclination.

Down by Wall Street the West Side Highway has been demolished. The reconstructed roadway passes a few blocks to the west of the building where I saw those people on the gurneys. My guess is that most of its inhabitants—the real ones, I mean—have moved on. More have moved in, though. We're marching into the future together. Curtains are visible from the highway. There's some lawn furniture on the roof. It makes me wonder who owns it, and what they do, and whether, like me, they have children.

A New Strategy for
Feeding a Crowded Planet

David Norse

DAVID NORSE is a research associate for the Overseas Development Institute in London and a research fellow at the Environmental Change Unit at the University of Oxford.

How many is too many? Is planet Earth really in danger of collapsing because of overpopulation? How valid is the recent joint statement by the U.S. National Academy of Sciences and the U.K. Royal Society that, "If current predictions of population growth prove accurate and patterns of human activity on the planet remain unchanged, science and technology may not be able to prevent either irreversible degradation of the environment or continued poverty for much of the world"?[1]

Similar warnings were given by the British economist Thomas Malthus in 1798, when Earth's population was only about 1 billion;[2] yet 5.3 billion people are now supported, albeit inadequately in many cases, by Earth's bounty. Malthus's warning of collapse was wrong because he had a static view of science and technology and of how they could compensate for the lack of land by raising agricultural productivity, for example, through irrigation and fertilization.

More recent assessments have taken account of technology to a greater or lesser extent. Some of them, however, surpassed Malthus's perception of the problem to introduce technology itself as an additional factor that

could lead to collapse.[3] For instance, such production inputs as mineral fertilizers and pesticides have been cited for causing environmental damage when they are applied incorrectly or excessively.[4]

Earlier estimates of the maximum human population that Earth could support were methodologically weak and primarily concerned with the ratio of humans to land. More recent estimation techniques are more sophisticated and commonly use the concept of Earth's human-carrying capacity (or potential population-supporting capacity) to introduce agroecological and other factors into the analysis. The carrying capacity of a particular region is the maximum population of a given species that can be supported indefinitely, allowing for seasonal and. random changes, without any degradation of the natural resource base that would diminish this maximum population in the future.[5]

The concept was first used by ecologists and zoologists to describe the ability of natural ecosystems to support wildlife populations. If wildlife populations exceed the natural food supply available locally or through migration, population collapse is inevitable. But, whereas the carrying capacity of natural ecosystems is essentially static, that of managed agroecosystems is dynamic. A region's human-carrying capacity can be increased by raising land and labor productivity or through trade with better endowed regions.[6]

Prior to about 1850, the world's human population could be fed, clothed, and otherwise supported by relatively limited modifications to natural ecosystems. Since then, however, and at an accelerating pace since about 1945, the food supply—and hence the carrying capacity—has become increasingly dependent on science-based agriculture and on external inputs of nonrenewable resources (notably fossil fuels and rock phosphate). Thus has arisen the critical question of sustainable agricultural development and the role of the United Nations Conference on Environment and Development (UNCED), to be held in Rio de Janeiro this June, in effecting the necessary changes to achieve it.

The preparatory process for UNCED has highlighted a number of critical uncertainties regarding sustainable development and the long-term carrying capacity of the globe. The first uncertainty concerns the human population growth rate and whether Earth's natural resource base and technological prospects will be adequate to sustain future populations. In particular, is the planet's carrying capacity threatened, on the one hand, by climate change caused primarily by the developed countries and their wasteful consumption patterns[7] and, on the other, by irreversible land degradation in the developing countries?[8] The second uncertainty is the link between population pressure, poverty, and environmental degradation, which has both national and international dimensions and involves positive as

From *Environment*, Vol. 34, No. 5, June 1992, pp. 6-11, 32-39. Reprinted with permission of the Helen Dwight Reid Educational Foundation. Published by Heldref Publications. 1319 Eighteenth St., NW, Washington, DC 20036-1802.

Land reforms are needed to reduce poverty and malnutrition in Guatemala.

well as negative impacts on carrying capacity.[9] The third issue is the different attitudes of developed and developing countries toward the tradeoffs that must be made between the environment and development.[10] Developed countries tend to give priority to environmental conservation in spite of the high economic and social costs that may be associated with such measures. Developing countries, however, tend to emphasize the need to ensure that environmental measures do not have adverse economic effects on their development, in part because eliminating poverty is a prerequisite for limiting certain forms of land degradation that undermine carrying capacity.

Of course, a region's human-carrying capacity is not always defined in terms of food production potential; in some situations, the principal constraint on carrying capacity may be the lack of water or fuelwood.[11] In addition, the lack of food production potential can often be offset by other economic activities, but these factors are beyond the scope of this discussion. Nevertheless, even a general assessment of the major threats to Earth's long-term carrying capacity reveals the need for a new strategy for sustainable agriculture.

Before addressing these issues, however, it should be stated that asking what the human-carrying capacity is or asking how many people can be fed is dodging the issue in two important respects. First, it is posing the problem as a supply problem stemming from the lack of production potential, whereas the demand problem is more serious and its analysis should take precedence over any theoretical estimates of carrying capacity.[12] The 500 million to 1 billion people who go hungry each day do so because they are too poor to buy all of the food they need.[13] Thus, the central issues

should be the elimination of poverty, the resolution of the debt crisis, and the removal of developed countries' trade barriers that prevent developing countries and their people from raising their incomes.[14] Second, the momentum of human population growth is so great and so insensitive to minor changes in mortality, migration, and fertility rates that world population will grow by some 3 billion between now and 2025, the year chosen as UNCED's development horizon, unless the AIDS pandemic goes out of control or unless socially unacceptable forms of family size limitation are adopted.[15] Thus, the issue is not whether Earth can support a population of 8.6 billion or so in 2025—it will have to—but, rather, whether it can do so sustainably.

The Global Picture

Almost 200 years have passed since Malthus first raised doubts about the global carrying capacity, and the global population has increased fivefold. Still, the debate continues, with estimates of carrying capacity ranging from 2 billion to 30 billion people. The issue was widely discussed in the aftermath of World War II and came back into prominence in the late 1960s and early 1970s, when various

individuals and computer modeling groups, particularly those supported by the Club of Rome, argued that projected population growth, excessive demands on the natural resource base, and pollution would overburden the biosphere and cause its collapse.[16] This was the basic message of the club's *The Limits to Growth* report that shocked many people into examining the whole question of growth in a finite world.[17] Other analysts, however, while not denying the seriousness of present population and pollution trends, have concluded that collapse is avoidable with relatively minor and equally realistic changes in the assumptions about the rate of technological growth (that is, faster), and about how prices and demand respond to scarcity of natural resources.[18]

Dutch scientists belonging to one of the modeling groups set up by the Club of Rome produced the first comprehensive assessment of global carrying capacity.[19] The scientists disaggregated the world's land resources into 222 broad soil regions with known climatic, vegetative, and topographic characteristics and irrigation potential. Maximum production potential per hectare was calculated in units of "cereal equivalents" for each of these soil regions and then summed

to give a total maximum production of some 50 billion tonnes of cereal equivalent per year. However, this estimate is a theoretical maximum assuming the use of all of the land potentially suitable for farming (3.4 billion hectares compared with the 1.5 billion hectares that are currently farmed), no pests and diseases, no plant nutrient shortages, and no water constraints. Acknowledging that such assumptions are unrealistic, the scientists later applied a range of reduction factors for land availability, average attainable yields, seed requirements, storage losses, and so on to arrive at estimates of potential production of consumable grain.[20] A comparison was also made between potential production using the full package of modern inputs and that from labor-oriented agriculture without the use of tractors, modern machines, pesticides, or fertilizers. Table 1 on this page shows more than an eightfold difference between the production potentials of the two systems.

These output figures were roughly transformed into carrying capacity terms by assuming an average grain consumption of 800 kilograms per capita per year. The resulting global carrying capacities were 6.7 billion people for the modern system and 2.7 billion for the labor-oriented production system.[21] It is interesting to compare these results with some other, back-of-the-envelope estimates. F. Baade, for example, assuming that the arable land area was between 2 billion and 3 billion hectares and that average yields could rise to 5 tonnes of cereal equivalent per hectare, estimated the potential carrying capacity at some 30 billion.[22] No consideration was given to whether production could be sustained at this level; to whether all arable land would be used for agriculture; or to the increase in indirect demands for cereals as people switch to higher value but lower calorie-content foods when their income grows.

Roger Revelle and Bernard Gilland made assumptions similar to Baade's regarding potential crop yields, that is, about 5 tonnes of cereal equivalent per hectare.[23] Revelle also assumed

that much of the potentially arable land would be used for food production, whereas Gilland doubted whether the area available for sustainable agriculture would be much greater than the present one. At an average per-capita grain consumption of 800 kilograms per year, Baade's carrying capacity estimate becomes more than 18 billion, Revelle's becomes 14 billion, and Gilland's is 8.8 billion. These estimates exclude the output of rangelands, pastures, and fisheries, which could raise carrying capacity by more than 10 percent at the global level and by considerably more in certain regions.[24]

Critical Countries?

Methodological sophistication was taken a step further in a study completed in 1982 by the UN Food and Agriculture Organization (FAO) and the International Institute for Applied Systems Analysis (IIASA).[25] The study, on agroecological zones and

their population-supporting capacity, moved away from global and regional estimates to identify not just critical countries but also critical zones within countries. The analysis was completed on 115 developing countries, excluding China, North and South Korea, and some very small nations because of the lack of data. The land area within countries was subdivided into zones according to the length of the growing season, which was determined by the number of days in each year with temperature, rainfall, and soil moisture conditions suitable for crop growth. Each of these zones was further subdivided on the basis of soil characteristics, slope, and other factors influencing the suitability of land for crop growth. And, unlike most previous studies that estimated production potential only in terms of cereal equivalents, production was determined for 15 major crops: rice, wheat, maize, millet, sorghum, barley, cassava, sweet potato, white potato, banana and plantain, soybean, *Phaseolus* bean, peanuts, oil palm, and sugar cane.

Potential yields under rain-fed conditions were constrained according to the length of the growing season and were estimated for three different assumptions regarding the level of management and input use: a low level,

TABLE 1
PRODUCTION POTENTIALS OF DIFFERENT AGRICULTURAL SYSTEMS

	GRAIN PRODUCTION						
Agricultural system	South America	Australia	Africa	Asia (millions of tonnes)	North America	Europe	World[a]
Modern agriculture on all potential agricultural land[b]	2,932	623	2,863	3,770	1,870	1,100	13,156
Modern agriculture on present agricultural land[b]	379	100	636	2,929	712	582	5,338
Labor-oriented agriculture on maximum agricultural land[c]	241	57	236	426	391	256	1,606

[a]World totals do not reflect row totals because of rounding.
[b]Allows for seed requirements and storage losses amounting to 20 percent of total production.
[c]Assuming that only two-thirds of the land is cropped with grains.
SOURCE: P. Buringh and H. D. J. van Heemst, *An Estimation of World Food Production Based on Labour-Oriented Agriculture* (Wageningen, the Netherlands: Agricultural Press, 1977).

with traditional varieties, no mineral fertilizers or pesticides, and no specific soil conservation measures; an intermediate level, with a basic package of inputs and limited soil conservation measures; and a high level, involving optimum use of mineral fertilizers, pesticides, and improved crop varieties, plus long-term soil conservation measures regardless of the costliness of these inputs. Soil conservation measures were assumed to be totally effective at the high level, but, at the other levels, land and productivity loss from soil degradation were taken into consideration. Irrigation was not addressed in detail, nor were nonfood cash crops, although land was allocated to them. Allowance was also made for livestock production from rangelands unsuitable for crop production.

A computer program determined the optimum cropping pattern to maximize calorie production and then calculated the number of people that could be fed at the minimum calorie intake level for each country. Estimates were made, using the UN "medium variant" population projections, for each of the input levels and for the years 1975, 2000, and 2025. The resulting projections suggested that global carrying capacity is adequate for the year 2000 even at low input levels, provided that much of the potentially arable land can be used for food crop production instead of livestock or forest.[26] At the regional level, however, the picture is bleak, with only South America's developing countries having adequate carrying capacity under the assumption of minimal input use. It is not, of course, a realistic assumption; with the exception of sub-Saharan Africa, present input use is generally at the intermediate level or higher.

The crunch comes at the country level. At the low input level, 64 of the 117 countries studied would be unable to feed their projected populations in 2000 even if they use all of their arable land for food production. With greater input use at the intermediate level, the number of critical countries falls to 36, but 18 remain critical even at the high input level (see Table 2 on this page). Such estimates, however, assume that the countries concerned wish or have to be self-sufficient in food, whereas this is seldom the case. Most countries wish to exploit any comparative advantage they have regarding other natural resource endowments, like mineral deposits and fossil fuels, or in the production of manufactured goods. For example, the 18 critical countries include a number of oil producers in the Middle East, which, with little but desert for food production, can import their needs much more cheaply than they can grow them, and, if they use their resources wisely, could establish sustainable economies in spite of their limited carrying capacity in terms of food production. There are some countries, however, such as Burundi, Niger, and Rwanda, that have few natural resource endowments to exploit and,

TABLE 2 ▰▰▰▰▰▰▰▰▰▰
DEVELOPING COUNTRIES ABLE TO SUPPORT LESS THAN HALF OF THEIR PROJECTED POPULATION IN 2000

Level of agricultural inputs	Level of economic dependence on agriculture	Medium risk	Higher risk	Highest risk
High	High	Netherlands Antilles Barbados Cape Verde		Rwanda Democratic Yemen Yemen
	Significant exports of manufactured goods	Lebanon Israel Mauritius Singapore	Jordan	
	Significant fossil fuel resources	Kuwait Qatar United Arab Emirates Oman Saudi Arabia Bahrain		Afghanistan
Intermediate	High	Réunion Martinique Antigua	Lesotho Mauritania	Haiti Burundi Somalia
	Significant exports of manufactured goods		El Salvador Kenya	
	Significant fossil fuel resources		Iraq Niger Algeria	
Low	High	Guadeloupe Windward Islands	Namibia	Comoros Ethiopia Uganda
	Significant exports of manufactured goods			
	Significant fossil fuel resources			Nigeria

SOURCE: UN Food and Agriculture Organization, *Land, Food and People* (Rome: FAO, 1984).

Although Kenya exports some manufactured goods, the country may not be able to feed even half of its population by 2000.

WORLD VISION—DAVID WARD

because of their land-locked situation, have no comparative advantage in the production of low-cost, labor-intensive goods. These countries seem destined to remain on the critical list.

Limitations to the Concept

Refinements of the type introduced by FAO allow policy planners to go beyond global and regional generalities by helping identify where population pressures are most acute. Nevertheless, there are important limitations to the utility of the carrying capacity concept. As the saying goes, man does not live by bread alone, and people and countries are seldom entirely dependent on food production, agriculture, or their natural resource base for the means to ensure adequate access to basic needs. It is becoming increasingly clear, however, that at least part of this reduced dependence is a response to population pressures on the natural resource base. Poor farm families are diversifying their income sources to cope with population pressures and food production instability. Farmers may be able to meet their basic needs in a more sustainable way by producing export crops, for example, and buying their food requirements. Thus, in the Ivory Coast, oil palm production is four to five times more profitable than maize production. Moreover, oil palm cultivation is generally a more ecologically sound form of land use in tropical areas with high rainfall than is monocultural maize production.

Countries are seldom totally dependent on agriculture. Even in highly agrarian countries like Bangladesh, agriculture accounts for only 50 percent of the gross domestic product. At the other extreme, in countries like Switzerland and Singapore, the agricultural resource base no longer plays a significant role. As long as these countries use their capital bases well,

the carrying capacity of their own lands will be of secondary importance. Thus, many of the critical countries identified by the FAO/IIASA study are not in danger of collapse because they can buy their way out of trouble,[27] provided, of course, that the major food-exporting countries avoid serious land degradation and thus maintain their export potential.

Some of these limitations have been addressed in the second generation of natural resource models developed by FAO and IIASA. A case study of Kenya, for example, was developed to provide a district- and national-level planning tool rather than carrying capacity estimates.[28] Land resources were assessed at the district level, as was the potential for food production. Land requirements for and the potential productivity of export crops, primarily coffee, tea, cotton, sisal, pineapple, and pyrethrum, were estimated. Soil degradation was considered in detail, including erosion and loss of fertility through inadequate fallows. Livestock production and fuelwood supply were also assessed.

Such refinements allow a more precise assessment of local carrying capacity in physical terms, but several limitations remain. Risk is not taken into consideration, nor is the articulation between agroecological zones, migration, market infrastructure, and

the all-important economic dimension. Risk, for example, particularly that stemming from drought and unreliable rainfall, can be a major constraint to the adoption of yield-raising technologies. Purchased inputs such as mineral fertilizers may be profitable given the average rainfall assumed in carrying capacity studies but waste farmers' money in unpredictable drought years. Thus, the use of long-run averages for weather assumptions can result in overestimates of carrying capacity.

The importance of the economic dimension, migration, and market infrastructure is well illustrated by a recent study of the Machakos district of Kenya.[29] Large parts of the district were virtually uninhabited at the beginning of this century. Yet, by the 1930s, substantial areas were so degraded by agricultural activities that it was thought to be on the edge of ecological collapse.[30] Nonetheless, the population has increased more than fivefold during the past 60 years.[31] In terms of the per-capita food requirement assumed by Revelle and Baade (800 kilograms of cereal equivalents per year), the district's carrying capacity was exceeded in the 1940s. Using the lower food requirement assumed by the FAO/IIASA study (2,320 calories per capita per day), the carrying capacity was exceeded about 10

TABLE 3 ▰▰▰▰▰▰▰
SHARE OF HARVESTED LAND USED FOR EXPORT CROPS FROM 1987 TO 1989

Country	Importance of agricultural exports[a]	Percentage of harvested area used for export crops[b]
Burkina Faso	High	5
Ivory Coast	High	44
Ethiopia	High	4
Honduras	High	22
India	Low	1
Indonesia	Low	18
Kenya	High	8
Malawi	High	5
Rwanda	High	5
Uganda	High	7

[a]Defined in terms of share of primary agricultural commodity exports in total merchandise exports in 1989. High = 85–99 percent; Low = 19–21 percent.

[b]After an allowance for the area used to meet domestic requirements for those crops is subtracted.

SOURCE: Importance of agricultural exports estimated from data in World Bank, *World Development Report 1991: The Challenge of Development—World Development Indicators* (New York: Oxford University Press, 1991). Percentages derived from Supply Utilization Tables for 1987/89, FAO AGROSTAT Data Bank, FAO, Rome.

years later. Reality, however, has been quite different, and the district has continued to flourish.

A number of factors account for this inconsistency: emigration to urban areas and internal migration to more marginal lands in response to land shortages and rising land prices;[32] a reduction of the fallow period;[33] intensification of land use through multiple cropping and, particularly, through the use of manure (the supply of which is heavily dependent on intensified livestock production) and mineral fertilizers (in the case of export crops);[34] the adoption of soil conservation measures to rehabilitate land, notably conservation tillage, contour farming, and terracing (the terraced area rose from about 52 percent in 1948 to 96 percent in 1978);[35] and finally, the introduction of coffee, fruit, and other horticultural crops that provide higher incomes than do the basic staples and thus make soil conservation more profitable.[36] The introduction of

these crops was a response more to improvements in local processing facilities and market access than to population pressure per se. Thus, migration to marginal lands was probably less than that which would have occurred in the absence of agricultural intensification and export crops.

The Machakos experience and the example of palm oil production in the Ivory Coast are clear rebuttals to the argument that the competition for land between export crops and food crops is a threat to development, to food security, to the poor, and, implicitly, to carrying capacity.[37] Export crops seldom occupy more than 10 to 20 percent of a country's arable land, and, even then, it is not necessarily the best land (see Table 3 on this page). Most countries can compensate for the diversion of land to export crops by modestly increasing productivity on the remaining food-crop land and still achieve approximate self-sufficiency. Under careful management, both carrying capacity and food security will rise. Moreover, export crops can provide a range of economic benefits. Locally, their greater profitability means that the growers and their employees can both raise and diversify their food intake. The extra income also increases farmers' ability to pay for the education of their children. Education is a priority expenditure in much of the developing world and generally results in slower population growth. Nationally, the gains in foreign exchange earnings from export crops can eliminate or lower the need for food aid and can help to consolidate a country's political independence.

Environmentally, the threat to sustainability can be much less than with staple food crops.[38] The tree canopy of the oil palms in the Ivory Coast, for example, protects the soil surface from erosion almost as well as the original forest did. And the palms' deep roots raise mineral nutrients from the subsoil. Thus, oil palm production is often an ecologically sounder use of the land than is food crop production. Coffee trees in the Machakos district also protect the soil

from erosion better than do most annual food crops. With their extensive root systems, the citrus, mango, and other tropical fruit trees introduced into the district help to hold the soil in place on the terraces and provide greater economic incentives to farmers to construct and maintain the terraces. Other cash crops, such as cotton and pyrethrum, yield high returns that can make the use of mineral fertilizers profitable. These crops leave behind fertilizer residues that benefit the staple food crops that follow them in the production rotation and, hence, help to limit nutrient mining and soil degradation.[39] Finally, by providing greater employment and incomes, cash crops discourage migration into environmentally fragile marginal areas.

In light of these studies and assuming that recent growth rates in technology continue, it appears that most countries have adequate land or other resources to carry their projected populations for 2025. Land and resources may even be sufficient over the very long term, when Earth's population is expected to stabilize at around 11 billion after about 2100, provided that the wasteful consumption patterns of the developed countries are not taken as the universal model and that satisfactory answers can be found to the following questions:

• Can agricultural production be increased fast enough to keep ahead of population growth?

• Will some of the carrying capacity potential be lost because of land degradation and other factors?

• What changes are required to shift from the present unsustainable development path?

Production vs. Population

Most carrying capacity studies make three major assumptions regarding the first question—namely that crop production will win the competition for land against livestock and forestry, and losses from urban and industrial development will not be serious; that crop yields can increase fast enough; and that food prices will not rise beyond people's ability to pay.

Historically, crop production has almost always won the competition for land.[40] Increased desertification caused by crop production displaces livestock onto more and more marginal land and causes high rates of tropical deforestation, of which from one-half to two-thirds results from shifting cultivation. Competition for land from urban and industrial development, however, could become increasingly serious. Much of this development will involve the expansion of existing conurbations, which are often located on coastal plains and in river valleys with some of the best soils. Once these soils are built on, they are lost forever. Few of the developed countries had effective land-use policies to protect such soils before the 1960s and 1970s.[41] It is not surprising, therefore, that many developing countries are finding it difficult to introduce the necessary policies, and so, nearly 300 million hectares—one-quarter of all highly arable land—are at risk.

The expansion of crop production through 2025 seems likely to cause the loss of some 180 million hectares of rangeland and forests, whose soils are only marginally suitable for annual crop production and need very careful management if degradation is to be avoided.[42] Such losses are likely to have spill-over effects in terms of greater risks of overgrazing and intensification of the fuelwood crisis. Thus, it is vital for policymakers to focus on raising the productivity and employment requirements of the existing crop land so as to reduce the pressure for the cultivation of and migration onto these marginal lands.

Two facts cast doubt on whether it is possible for the rate of crop-yield increase to keep pace with population growth. First, staple crop yields in many sub-Saharan countries have been essentially static for the past 10 years or more.[43] The reasons are complex. In part, static yields stem from weak government policies that have failed to give adequate price incentives and infrastructural support to

producers. But they also reflect the widespread lack of appropriate technologies and the failure of research systems to develop sustainable technologies that match the perceptions and resources of small farmers.[44]

Second, although other regions have been more successful than sub-Saharan Africa at raising yields, there is growing concern that the future production potential may be insufficient. In parts of Asia, for example, experimental yields for irrigated wheat and rice fields seem to have reached a plateau.[45] Maximum yields have been static for some 10 to 20 years. Meanwhile, average yields still fall well short of experimental yields and are unlikely to match them entirely, but there is still a yield gap to be closed. Thus, it seems possible that recent trends in yield growth will continue for another one or two decades. Thereafter, however, the evidence suggests that it will become increasingly difficult to prevent a slow decline in yield growth unless research achieves another shift on the production frontier.[46] Whether advances in biotechnology will provide the solution is still an open question.

There remains the assumption regarding affordability. Current estimates of the number of malnourished people range from 500 million to 1 billion.[47] The numbers are not expected to decline appreciably, at least in the mid term.[48] By and large, people are malnourished because they are too poor to buy all of their needs. Purchasing power, however, has not been considered by a carrying capacity study conducted to date.

Projections suggest that future food prices will remain more or less constant in real terms for the next decade or so, as long as market mechanisms operate efficiently and the apparent crop yield ceiling is breached. If these projections are correct, the strains on carrying capacity will stem primarily from income growth. In the middle-income countries where people are consuming increasing amounts of livestock products, carrying capacity ceilings are already being breached because of the lack of grazing land and

the greater cost-efficiency of raising dairy cattle, pigs, and poultry on feed grains. Many countries are currently or will soon be unable to produce these feed grains in adequate quantities. They will become increasingly dependent on feed grain imports, and part of the environmental costs of production will be transferred to exporting countries like the United States, Canada, Argentina, and Thailand.[49]

The Effects of Soil Degradation

After *The Limits to Growth* was published and the 1973 oil crisis, many countries became concerned about the constraints on the raw materials for fertilizer and the energy supply. Neither concern withstands close analysis, however. The reserves of the main raw materials for fertilizer are adequate for well more than 200 years at projected levels of use, but prices will rise. If the global population is more or less stable in 100 years, agricultural production systems should be able to evolve so that a sustainable balance is reached between soil nutrient removal by crops and replacement by humans.[50] Fossil fuel energy resources will eventually run out, but, with greater policy and research emphasis on renewable energy systems and greater energy efficiency, agriculture's carrying capacity should be maintainable. Agriculture's share of primary energy use is only 3 to 5 percent of total use. Therefore, safeguarding its supply of energy is a question more of sociopolitical priorities than of resource scarcity per se.

There are other threats, however, that cannot be dismissed, particularly those from land degradation and climate change. Over the past millennium, humanity has degraded about 2 billion hectares of land, though only a small proportion of this area is too degraded to remain in agricultural use.[51] More recent estimates suggest that the present situation is much worse. For example, a recent study for the UN Environment Programme called the Global Assessment of Soil Degradation (GLASOD) concludes that people have degraded about 25 percent of the occupied land, and much of this damage has been caused in the past 50 years (see Table 4 on this page).[52] The most serious causes of degradation are water erosion through deforestation (43 percent), overgrazing (29 percent), and poor farming practices (24 percent). Population pressure on marginal lands plus mismanagement of the better soils may have acclerated land degradation over the past 20 years to reach some 5 million to 6 million hectares annually.[53]

Much of this degradation has stripped the land of soil. The quantities involved are immense. The current rate of soil erosion in excess of new soil production has been estimated at some 23 billion tonnes per year.[54] If the global soil reservoir is declining at about 0.7 percent annually, between 20 and 25 percent of the total could be lost by 2025.

These soil losses sound disastrous, but scientists and economists are unsure of their long-term implications. Estimating the consequences of this degradation for soil productivity and, hence, for human-carrying capacity is very difficult. There have been few well-conducted experiments on the relationship between soil degradation and crop yield losses, and most of them apply only to conditions in developed counties. Consequently, such estimations rely heavily on subjective judgements. Nonetheless, the GLASOD study concludes that some 295 million

hectares are so degraded that restoration of the land to full productivity is beyond the normal means of a farmer but that these hectares could be restored with major investments. (There is, of course, substantial uncertainty about the cost of land restoration. The GLASOD estimates are based largely on formal project costs, but the Machakos example given earlier and a number of NGO and community-based actions show that quite extreme degradation can be corrected profitably and at modest costs.)[55] An even greater area—910 million hectares—is moderately degraded to the point that the original biotic functions are partly destroyed and agricultural productivity is greatly reduced. If these estimates are correct, human-carrying capacity has already been seriously undermined, and the present trend is for even greater damage in the future unless corrective actions are given much greater priority.

Substantiating evidence for this view comes from two other directions. Studies of soil nutrient balance in which rates of nutrient removal by crops, soil erosion, leaching, and other factors are set against nutrient inputs from natural soil processes, dust, rain, manures, and mineral fertilizers indicate that widespread soil nutrient mining is taking place, particularly in sub-Saharan Africa, where food production has failed to keep up with population growth for the past 20 years.[56] This conclusion is supported by a number of other ex-

periments that have compared declining yields and fertilizer response ratios over time. It appears that one of the consequences of nutrient mining has been unfertilized base yields. In areas of Ghana, Malawi, Kenya, and Java, for example, the base yields of cereals have been falling by 2 to 10 percent per year for the past 10 to 20 years. Such losses are clearly unsustainable over the long term. The reasons are complex. Contributing factors include nutrient mining, as well as physical and chemical damage to the soil through erosion, loss of organic matter and soil moisture-holding capacity, and the buildup of soil pests and diseases caused by reduced fallows. Mineral fertilizers can compensate for such damage to a substantial degree, but they are too expensive for poor farmers or often simply unprofitable. And sometimes, mineral fertilizers are insufficient to restore fertility. Inputs of organic manures are required to achieve fertility, but the supply is commonly inadequate.

Attempts have been made to place these physical losses from land degradation in their national economic contexts.[57] Recent estimates suggest that, if remedial action is not taken, Ghana could lose 7 percent of its gross national product because of land degradation and that Nigeria could lose more than 17 percent of its gross national product because of deforestation, soil degradation, water contamination, and other environmental problems over the long term. Again the message is clear: Land degradation at current rates will reduce carrying capacity both by lowering agricultural production potential and by lowering the ability of countries to import food.

These losses seem destined to be compounded by climate change. A scientific concensus has emerged during the past five years or so that the threat to food production is real, though the timing and regional pattern of climate change's impacts is still uncertain.[58] Climate change could seriously threaten carrying capacity and sustainable agriculture in certain regions, though there may be both win-

TABLE 4
AREAS OF MODERATE TO EXCESSIVE SOIL DEGRADATION

Region	Water erosion	Wind erosion	Chemical degradation (millions of hectares)	Physical degradation	Total
Africa	170	98	36	17	321
Asia	315	90	41	6	452
South America	77	16	44	1	138
North and Central America	90	37	7	5	139
Europe	93	39	18	8	158
Australasia	3		1	2	6
Total	748	280	147	39	1,214

SOURCE: L. R. Oldeman, R. T. A. Hakkeling, and W. G. Sambroek, *Global Assessment of Soil Degradation* (Wageningen, the Netherlands: ISRIC/UNEP, 1990).

ners and losers from climate change. Some regions and countries may benefit from greater agricultural productivity because of temperature and rainfall increases and because of carbon dioxide's enhancement of plant growth.

Two major concerns are that some of the countries most at risk are already very vulnerable to food shortages and close to their human-carrying capacity if crop yields cannot be raised and that population pressure's negative impacts on the environment may increase both the intensity of and the land's vulnerability to climate change.[59] In the case of some Sahelian countries, for example, the analog approach to impact analysis suggests that the potential human-carrying capacity from domestic agriculture could fall by some 30 percent if climate change causes a decline in rainfall similar to that experienced in the 1965 to 1985 drought.[60]

A more recent analysis for Senegal that built on a carrying capacity model developed by the U.S. Geological Survey and on crop models from the Institut Senegalais de Recherche Agricole came to similar conclusions. Rain-fed production potential is projected to decrease 30 percent in spite of adaptive responses to climate change—a reduction equivalent to the food needs of 1 million to 2 million people.[61] Related impact analyses focusing on maize production in Zimbabwe, where conditions are not so arid as in Senegal, have projected a smaller impact.[62] Rain-fed yields should fall by less than 20 percent, in part because the positive effect of higher atmospheric carbon dioxide concentrations compensates partially for the negative impact of higher temperatures. What is more worrying, however, is the projected increase in the variability of rainfall and, hence, crop yield, which could seriously lower food security and increase the financial risk of using the mineral fertilizers essential for maintaining soil fertility.

A New Strategy

A new strategy is required that is the opposite in many respects to cur-rent policies. Present strategies tend to focus on four aspects of agricultural development—incentives, inputs, institutions, and infrastructures—and on the investments required by them and tend to give inadequate attention to sustainability. The new strategy should be more concerned with decentralized natural resource management by the farmers and communities that will ultimately decide the strategy's appropriateness and success. The strategy should therefore focus on local-level husbandry and development controlled by local community or user-based institutions, such as grazing associations and water-user groups, rather than on sophisticated institutional structures for national or regional land-use planning, for example, or for investment allocation. Of course, such structures do have an important role to play, provided that they are economically sustainable by the nations concerned and that they provide incentives rather than top-down directives for resource development and management. The issue is not top-down versus bottom-up; both approaches are needed, but they must be consistent with each other, and they must promote a convergence of national goals and local priorities.

The decentralization of decision-making can only lead to a socially desirable and sustainable allocation of productive and environmental resources if other requirements are met. For example, the responsibilities and entitlements of farmers and rural communities must also be decentralized through the clear allocation of property rights and environmental resources. User groups and similar bodies must have the power to raise revenues from resource users for operation, maintenance, and further development of those resources. Commodity prices must reflect as much as possible the full environmental costs of a given resource use, besides providing adequate incentives to producers.[63] Once governments have provided the infrastructure to ensure market access and have set appropriate environmental standards and taxes on discharges to protect public goods, government intervention should be kept to a minimum so that markets can function efficiently.

The strategy should be built around four critical components of resource management—soil fertility management; integrated pest management; water management; and integrated crop, livestock, and tree management—and around their greater integration. Past efforts by public and private development organizations, however, have often followed a relatively uniform approach centered on encouraging the use of one or a limited number of purchased production inputs.

Many past conservation measures that have failed were not profitable, except possibly in the long term. Priority should be given to resource conservation actions that quickly raise productivity and farmers' incomes. Some of the required conservation techniques are already available. In many cases, however, new technologies must be developed for, or existing ones adapted to, specific ecological conditions. This development and adaptation will take several years, substantial research funding, and a change in research and extension techniques.

More emphasis is required on biological approaches to resource management, including

• biological (as opposed to engineering) approaches to soil and moisture conservation, such as maintaining continuous ground cover with live mulches;

• biological inputs to integrated pest management systems to minimize pesticide inputs; and

• biological sources of nitrogen to replace or, in most instances, to complement mineral sources (phosphorus and other nutrients commonly have to come from mineral sources).

This emphasis on biological approaches would not contradict efforts to promote the use of mineral fertilizers and other off-farm inputs. These will continue to be a prime contributor to increased agricultural production for the foreseeable future, but there are situations where such inputs are too costly and are therefore not a viable solution. In these situations, biological approaches can complement or substitute for off-farm inputs to reduce production costs and maximize environmental sustainability. It is this balance that the proposed strategy seeks to achieve.

The new strategy differs from past ones in that it addresses sustainability problems through their social and cultural determinants or through the institutional constraints that have previously blocked such approaches, rather than treating them as environmental problems to be addressed by technical solutions developed in isolation from the ultimate users. Consequently, it stresses the evolutionary approach in which changes in farming practices build on indigenous knowledge, rather than the prevailing step-wise or single-component approach, and focuses on improvements in both the levels and the stability of yields instead of just maximum yields. This evolutionary approach concentrates on highly informed but low-risk and low-cost measures to minimize the need for credit-demanding technologies involving extensive use of off-farm inputs that are more suited to large-scale farmers, but it does not ignore the needs of the latter nor the fact that such technologies can be equally suited to small farmers in very fertile areas. Large farms will continue to play a critical role in producing food for urban populations and other net food buyers and as major suppliers of export crops. Most farmers need better support regarding water supply management and integrated pest management and help in minimizing or preventing some of the environmental problems commonly associated with large-scale, intensive farming systems.

Three main conclusions can be drawn from this discussion. First, there are some grounds for optimism regarding humanity's ability to raise and sustain agriculture's carrying ca-

pacity at the global and regional levels. This optimism, however, is contingent on major policy changes regarding international equity and the ecological soundness of technological growth. It is difficult, however, to see how some countries can sustain their projected populations through agriculture or other economic activities. The pressure for international migration may therefore be substantial.

Second, and further tempering optimism, there are a number of uncertainties about the future that support the increasing calls for the adoption of the precautionary principle to provide "a scientifically sound basis for policies relating to complex systems that are so poorly understood that the consequences of disturbances cannot yet be predicted. According to this principle, highest priority should be given to reducing the two greatest disturbances to Planet Earth: the growth of human population and overconsumption of resources."[64] These uncertainties include: the rate of deceleration in population growth, given that policy inaction could seriously delay the achievement of a stable population; the loss of arable land to urban development; the long-term consequences of soil degradation; the impacts of climate change; and the ability of science to continue to raise agricultural productivity.[65]

Third, if the concept of carrying capacity is to be introduced more centrally into the debate on sustainable development, it has to be widened to embrace economic as well as environmental considerations and complemented by a new strategy for sustainable agriculture.[66] Many poor farmers are forced to use unsustainable agricultural practices for a variety of institutional and economic reasons. In their struggle to satisfy current food needs, they have to place at risk the long-term carrying capacity of their land. Fine ecological words and appeals on behalf of future generations will not sway them unless the required changes in land management practices will also raise present-day household security.

ACKNOWLEDGMENTS

The author wishes to thank Tom Downing, Hartwig de Haen, and Mary Tiffen, as well as the reviewers, for their helpful comments on the first draft of this article.

NOTES

1. U.S. National Academy of Sciences and the Royal Society of London, "Population Growth, Resource Consumption, and a Sustainable World" (Joint statement released on 27 February 1992).

2. T. R. Malthus, *First Essay on Population 1798* (London: Macmillan, 1966).

3. E. Goldsmith et al., *Blueprint for Survival* (London: Penguin Press, 1972).

4. R. Carson, *Silent Spring* (Boston, Mass.: Houghton Mifflin, 1962).

5. G. Ledec, R. J. A. Goodland, J. W. Kirchner, and J. M. Drake, "Carrying Capacity, Population Growth and Sustainable Development," in D. J. Mahar, ed., *Rapid Population Growth and Human Carrying Capacity: Two Perspectives*, working paper no. 690 (Washington, D.C.: World Bank, 1985).

6. Mahar, ed., note 5 above.

7. Intergovernmental Panel on Climate Change, *Climate Change: The IPCC Impacts Assessment* (Canberra, Australia: Australian Government Publishing Service, 1990); and M. Parry, *Climate Change and World Agriculture* (London: Earthscan, 1990).

8. E. Eckholm, *Losing Ground: Environmental Stress and World Food Prospects* (New York: W. W. Norton, 1976); L. R. Brown, *The Changing World Food Prospect: The Nineties and Beyond*, Worldwatch Paper no. 85 (Washington, D.C.: Worldwatch Institute, 1988); and D. Norse, C. James, B. J. Skinner, and Q. Zhao, "Agriculture, Land Use and Degradation," in J. C. I. Dooge et al., eds., *An Agenda of Science for Environment and Development into the 21st Century* (Cambridge, England: Cambridge University Press, 1992).

9. R. W. Kates and V. Haarmann, "Where the Poor Live: Are the Assumptions Correct?" *Environment*, May 1992, 4.

10. D. Norse, "Trade-Offs Between the Environment and Agricultural Development," in *Proceedings of the European Association of Agricultural Economists Seminar on Environment and Agricultural Management* (Viterbo, Italy: European Association of Agricultural Economists, January 1991).

11. N. B. Ayibotele and M. Falkenmark, "Fresh Water Resources," in Dooge et al., note 8 above.

12. Even during famines, it is the collapse of household incomes, rather than the lack of food availability, that is the main cause of starvation. See A. Sen, *Poverty and Famines: An Essay in Entitlements and Deprivation* (Oxford, England: Clarendon Press, 1981).

13. R. W. Kates et al., *The Hunger Report: 1988* (Providence, R.I.: Brown University, Alan Shawn Feinstein World Hunger Program, 1988); and World Bank, *World Bank Development Report 1990* (Washington, D.C.: Oxford University Press, 1990).

14. Kates and Haarmann, note 9 above.

15. L. Arizpe, R. Costanza, and W. Lutz, "Primary Factors Affecting Population and Natural Resource Use," in Dooge et al., note 8 above; and R. M. Anderson, R. M. May, M. C. Boily, G. P. Garnett, and J. T. Rowley, "The Spread of HIV-1 in Africa: Sexual Contact Patterns and the Predicted Demographic Impact of AIDS," *Nature* 352 (15 August 1991):581.

16. F. Osborne, *Our Plundered Planet* (Boston, Mass.: Little, Brown & Co., 1948); W. Vogt, *Road to Survival* (Washington, D.C.: Population Reference

Bureau, 1948); H. Brown, *The Challenge of Man's Future: An Inquiry Concerning the Condition of Man During the Years That Lie Ahead* (New York: Viking Press, 1954); G. F. White, "Speculating on the Global Resource Future," in K. R. Smith, F. Fesharaki, and J. P. Holdren, eds., *Earth and the Human Future: Essays in Honour of Harrison Brown* (Boulder, Colo., and London: Westview Press, 1985); and D. H. Meadows, J. Richardson, and G. Bruckmann, *Groping in the Dark* (Chichester, England: John Wiley for the International Institute for Applied Systems Analysis, 1982).

17. D. H. Meadows, D. L. Meadows, J. Randers, and W. W. Behrens, *The Limits to Growth* (New York: Universe Books, 1972). The authors were not, as is commonly suggested, forecasters of doom proposing zero growth. They were considering alternative futures open to humanity, including sustainable futures. They returned to the issue in their recent sequel to the 1972 book. The dangers of rapid population growth and prolific resource use are again analyzed, as are the dangers of overshooting the limits imposed by finite resources and the ability of the environment to absorb emissions and waste. But in this new book, they consider in more detail the possible content of a sustainable future and the technological and other changes since 1972 that could facilitate the achievement of such a future. See D. H. Meadows, D. L. Meadows, and J. Randers, *Beyond the Limits: Global Collapse or a Sustainable Future* (London: Earthscan Publications, 1992).

18. H. S. D. Cole, C. Freeman, M. Jahoda, and K. L. R. Pavitt, *Models of Doom* (New York: Universe Books, 1973); H. Kahn, *The Coming Boom* (New York: Simon & Schuster, 1982); and J. Simon, *The Ultimate Resource* (Princeton, N.J.: Princeton University Press, 1981).

19. P. Buringh, H. D. J. van Heemst, and G. J. Staring, *Computation of the Absolute Maximum Food Production of the World* (Wageningen, the Netherlands: Agricultural Press, 1975). Their results were used in the development of the MOIRA model described in H. Linneman and M. A. Keyzer, eds., *MOIRA: A Model of International Relations in Agriculture* (Amsterdam: Economic and Social Institute, Free University, 1977).

20. P. Buringh and H. D. J. van Heemst, *An Estimation of World Food Production Based on Labour-Oriented Agriculture* (Wageningen, the Netherlands: Agricultural Press, 1977).

21. Carrying capacity estimates are highly sensitive to the assumptions regarding per-capita food requirements. There can be a twofold or greater difference in them, depending on whether food requirements are based on a primarily vegetarian diet or on one rich in grain-fed livestock products. See Kates et al., note 13 above.

22. F. Baade, *Der Wettlauf zum Jahre 2000* (Oldenburgh, Germany, 1960).

23. R. Revelle, "The Resources Available for Agriculture," *Scientific American* 235, no. 3 (1976):165; and B. Gilland, "Considerations on World Population and Food Supply," *Population and Development Review* 9, no. 2 (June 1983).

24. UN Food and Agriculture Organization, *The Fifth World Food Survey* (Rome: FAO, 1987).

25. UN Food and Agriculture Organization, United Nations Fund for Population Activities, and the International Institute for Applied Systems Analysis, *Po-*

tential Population Supporting Capacities of Lands in the Developing World (Rome: FAO, 1982); and P. A. Oram, "Building the Agroecological Framework," *Environment*, November 1988, 14.

26. UN Food and Agriculture Organization, *Land, Food and People* (Rome: FAO, 1984).

27. Ibid., 28.

28. UN Food and Agriculture Organization and the International Institute for Applied Systems Analysis, *Agroecological Land Resources Assessment for Agricultural Development Planning: A Case Study of Kenya* (Rome: FAO/IIASA, 1991).

29. Overseas Development Institute (ODI), *Environmental Change and Dryland Management in Machakos District, Kenya 1930–90* (London: ODI, 1991).

30. Ibid., 31; and M. Mortimore, ed., *Environmental Profile*, ODI working paper 53 (London: ODI, 1991).

31. See note 29 above, page 31; and M. Tiffen, *Population Profile*, ODI working paper 54 (London: ODI, 1991).

32. See note 29 above, page 33.

33. Overseas Development Institute, note 31 above; and R. S. Rostom and M. Mortimore, *Land Use Profile*, ODI working paper 58 (London: ODI, 1991).

34. Overseas Development Institute, note 31 above; and M. Mortimore and K. Wellard, *Profile of Technological Change*, ODI working paper 57 (London: ODI, 1991).

35. Overseas Development Institute, note 31 above; and F. N. Gichuki, *Conservation Profile*, ODI working paper 56 (London: ODI, 1991).

36. Overseas Development Institute, note 31 above; and M. Tiffen, *Farming and Income Systems*, ODI working paper 59 (London: ODI, 1991).

37. S. George, *How the Other Half Dies* (Totowa, N.J.: Rowman and Allanheld, 1977); and F. M. Lappe and J. Collins, *Food First: Beyond the Myth of Scarcity* (Boston, Mass.: Houghton Mifflin, 1977).

38. M. Tiffen and M. Mortimore, *Theory and Practice in Plantation Agriculture: An Economic Review* (London: ODI, 1990).

39. E. M. A. Smaling, "Two Scenarios for the Sub-Sahara: One Leads to Disaster," *Ceres: The FAO Review* 22, no. 2 (1991).

40. Norse et al., note 8 above.

41. Organization for Economic Cooperation and Development, *Land Use Policies and Agriculture* (Paris: OECD, 1976).

42. Norse et al., note 8 above.

43. UN Food and Agriculture Organization, *African Agriculture: The Next 25 Years* (Rome: FAO 1986); and World Bank, *The Population, Agriculture and Environment Nexus in Sub-Saharan Africa* (Washington, D.C.: World Bank, Africa Region, December 1991).

44. World Bank, page 45, note 43 above; S. J. Carr, *Technology for Small-Scale Farmers in Sub-Saharan Africa* (Washington, D.C.: World Bank, 1989); and D. Merril Sands, *The Technology Applications Gap: Overcoming Constraints to Small-Farm Development* (Rome: FAO, 1986).

45. P. L. Pingali, P. F. Moya, and L. E. Velasco, *The Post-Green Revolution Blues in Asian Rice Production: The Diminishing Gap Between Experiment Station and Farmer Yields* (Los Banos, Philippines: IRRI Social Sciences Division, 1990).

46. Y. Hayami and K. Otsuka, "Beyond the Green Revolution: Agricultural Development into the New

Century" (Paper prepared for the World Bank conference on Agricultural Technology: Current Policy Issues for the International Community and the World Bank, Virginia, October 1991).

47. Kates et al., note 13 above.

48. N. Alexandratos, ed., *World Agriculture: Toward 2000* (Rome and London: FAO and Belhaven Press, 1988).

49. M. P. Burton and T. Young, *Agricultural Trade and Sustainable Development* (Manchester, England: University of Manchester, 1990).

50. V. Smil, "Nitrogen and Phosphorus," in B. L. Turner II et al., eds., *The Earth As Transformed by Human Action* (Cambridge, England: Cambridge University Press, 1990); and Gilland, note 23 above.

51. Q. Zhao, "Land Degradation," *Acta Pedologica Sinica* 25, no. 4.

52. L. R. Oldeman, R. T. A. Hakkeling, and W. G. Sombroek, *Global Assessment of Soil Degradation* (Wageningen, the Netherlands: ISRIC/UNEP, 1990).

53. UN Food and Agriculture Organization, "Issues and Perspectives in Sustainable Agriculture and Rural Development" (Document prepared for the FAO/Netherlands Conference on Agriculture and the Environment,'s-Hertogenbosch, the Netherlands, April 1991).

54. L. R. Brown, "Conserving Soils," in L. R. Brown et al., eds., *State of the World 1984* (New York: W. W. Norton, 1984).

55. W. V. C. Reid, "Sustainable Development: Lessons from Success," *Environment*, May 1989, 6; and C. Reijk, *Soil and Water Conservation in Sub-Saharan Africa: Towards Sustainable Production by the Rural Poor* (Rome: International Fund for Agricultural Development, 1992).

56. Smaling, note 39 above; and Norse et al., note 8 above.

57. W. Magrath and P. Arens, *The Cost of Soil Erosion on Java: A Natural Resource Accounting Approach* (Washington, D.C.: World Bank, August 1989); J. Bishop, *The Cost of Soil Erosion in Malawi* (Washington, D.C.: World Bank, Malawi Country Operations Division, November 1990); and D. Pearce, "Conserving Global Natural Wealth: Economics and Politics," in J. Ball, ed., *The Creation of Wealth* (London: Edward Elgar, forthcoming).

58. See note 8 above.

59. D. Norse, "Population and Global Climate Change," in J. Jäger and H. L. Ferguson, eds., *Climate Change: Science, Impacts and Policy—Proceedings of the Second World Climate Conference* (Cambridge, England: Cambridge University Press, 1991).

60. The analog approach is described in M. H. Glantz, "The Use of Analogies in Forecasting Ecological and Societal Responses to Global Warming," *Environment*, June 1991, 10. The results of the Sahelian study were displayed at the Second World Climate Conference, see note 59 above.

61. T. E. Downing, "Climate Change and Vulnerable Societies: Case Studies in Zimbabwe, Kenya, Senegal and Chile," in *Climate Change and International Agriculture* (Oxford, England: University of Oxford, Environmental Change Unit, 1992).

62. Ibid., 26.

63. Burton and Young, note 49 above.

64. Dooge et al., note 8 above.

65. Arizpe et al., note 15 above.

66. Kates and Haarmann, note 9 above.

POPULATION: THE CRITICAL DECADE

Sharon L. Camp

Sharon L. Camp is senior vice-president of Population Action International (formerly the Population Crisis Committee), an independent, nonprofit research and advocacy organization based in Washington, D.C.

The 1990s mark a critical decade for many of the world's environmental problems. What the world community does or fails to do, led in part by the new Clinton administration, will determine whether or not we can stop the deterioration of natural life-support systems without irreversible damage to the Earth and to the quality of life for future generations.

The 1990s are clearly the decade of decision for one of the most basic environmental problems: the exponential growth in human populations. As Russell Peterson, former chairman of the Council on Environmental Quality, has put it, "The quality of all life on earth is increasingly threatened by a powerful and growing ecological force. We humans are that force, ever more of us using ever more materials, assaulting the environment with ever more machines, chemicals, weapons, and waste."

World population is now growing by 1 billion people every 11 years. This decade presents the last chance to stabilize human populations by the middle of the next century, through humane and voluntary measures, at something less than double the current world population of 5.4 billion. Meeting that goal would require a massive and immediate international effort to implement family planning and development programs known to reduce birthrates rapidly. While affordable, such actions require considerably more political will than has been demonstrated in recent years.

The absence of political commitment has been especially apparent in the United States. Over the last 12 years, a powerful anti-abortion lobby wrung a series of major concessions on population aid policy from the Reagan administration and then bullied the Bush administration into vetoing all congressional attempts to overturn them. Many readers will recall that at the 1984 United Nations International Conference on Population (held in Mexico City two weeks before the Republican party convention) the official delegation of the United States presented a startling White House–drafted statement that declared population growth "a neutral phenomenon" in international development and labeled government policies to deal with it an "overreaction." The same statement made opposition to abortion the new litmus test for international family planning organizations seeking U.S. assistance.

The U.S. policy reversal at Mexico City was followed later in 1984 by a decision to end 17 years of U.S. support for the London-based International Planned Parenthood Federation, and then in 1986 by the withdrawal of all support from the United Nations Population Fund. Twenty years of U.S. leadership on global population issues came, in effect, to an end. Those two organizations, with programs in almost every developing country, symbolize international cooperation on population. The Japanese and many European governments, for example, channel almost all their population assistance through them.

U.S. bilateral population assistance programs, many highly effective, did continue through the Reagan and Bush presidencies, owing in part to earmarked congressional appropriations. But in the absence of strong U.S. leadership, funding for family planning programs from foreign aid donors and developing country governments remained roughly static in real dollar terms. Expenditures on family planning have slipped well behind the growth in the number of couples of childbearing age—currently increasing by 18 million a year—not to mention the rising percentage of couples who want small families but cannot afford modern contraceptives. Over the last 12 years, while American policymakers politicized family planning, the world's population grew by a billion people.

The price of continued inaction during the 1990s will be high. If birthrates decline no faster in the 1990s than they did in the 1980s —that is, if the current demographic trajectory continues unabated—the world's population will triple, growing through the next century and beyond. So if President Bill Clinton serves a full two terms, his decisions will influence the course of world demography for a long time. How quickly the new administration puts American population aid policy back on track, and the quality of American leadership at the 1994 International Conference on Population and Development in Cairo, will be key to determining when and at what level world population can be stabilized.

Although high levels of per capita consumption make population growth in countries like the United States an important factor in global environmental problems, the populations of most industrialized countries are nearly stable or declining. More than 95 per cent of future growth will occur in the developing countries of Africa, Asia, and Latin America. Death rates in those countries have dropped steadily since about 1950, thanks to the wider availability of antibiotics, vaccines, and insecticides, as well as improved sanitation and nutrition. In virtually all developing countries birthrates initially declined more slowly than death rates, and actually increased in some African countries, creating the unprecedented population explosion of the last several decades.

Although birthrates are now finally declining faster than death rates in most parts of the world, the majority of developing countries outside of East Asia still have annual population growth rates of between 2.5 and 3.5 per cent. At those rates, populations double in fewer than 30 years. Africa is growing fastest and will have the largest relative increases in population, even with substantial increases in death rates from AIDS. But the biggest absolute numbers are still being added in Asia. India alone adds 18 million people to the world's population every year—more than all of sub-Saharan Africa. India and China together accounted for a third of the record 93 million–person population increase in 1992. For many developing countries, including most African countries, the tripling of world population that would result from current trends would mean a four- to six-fold increase in national populations. Such a

scenario has serious implications not only for global efforts to protect the environment, but also for efforts to alleviate widespread poverty and promote peaceful democratic change.

Pressure on Natural Resources

In the developing world, the problems of poverty, population growth, and environmental degradation are closely linked. Although global environmental problems such as climate change or ozone layer depletion are caused mainly by high consumption levels in the industrialized North, many environmental problems in developing countries, such as land degradation, deforestation, species extinction, and air and water pollution, are aggravated and accelerated by rapid population growth. Although the relationships can be complex, evidence is accumulating that impoverished families in much of Africa, Asia, the Middle East, and Latin America are becoming both the victims and the unwitting agents of ecological destruction: They are destroying their own natural resource base to survive. In their daily struggle for subsistence—for food, fodder, firewood, and water— they are changing local ecological systems, perhaps irreversibly, and contributing to global environmental change.

Human and livestock populations in some parts of the developing world have already exceeded local carrying capacity. Traditional patterns of land use, such as long fallow periods, have sustained smaller populations for generations, but they are being broken by growing population pressure on the land. Such trends could further undermine food production capacity. The U.N. Food and Agriculture Organization estimates that without major conservation efforts developing countries could experience an almost 30 per cent decline in agricultural productivity by the end of the next century, when their populations may have increased four- to six-fold. In sub-Saharan Africa, with food production growing at 2 per cent and population growing at 3 per cent, per capita food production has already dropped 15–20 per cent since 1970. Even without the problems of droughts and political turmoil, millions of African children still do not consume enough calories to become healthy adults. Although population growth is not the principal cause of hunger and malnutrition anywhere in the world, its pace in many developing countries worsens both the poverty and the environmental pressures that prevent many families from buying or growing enough food.

Increasing numbers of low-income families in

need of farmland or fuel for heating and cooking are cutting down tropical forests at a rate much faster than the trees can regenerate, contributing to deforestation that claims 40–50 million acres per year. Outside the Amazon region, 80 per cent of all the wood harvested in developing countries is used for firewood or charcoal. Among the nine countries accounting for the bulk of remaining tropical forest, population doubling times range from 35 years in Colombia to 22 years in Zaire.

Meanwhile, people and livestock are overcropping and overgrazing fragile soils, contributing to a process of desertification that threatens to put 15 million acres beyond hope of reclamation. Growing demand for water is depleting fresh-water supplies above and below ground. Swedish hydrologist Malin Falkenmark calculates that within the next 30–35 years a number of African countries will be entirely unable to meet their populations' needs for water (estimated at 1,370 liters daily per capita for all uses) from local aquifers and rivers. Falkenmark concludes that curbing population growth is vital to any strategy to meet Africa's water needs. In Northern China, rural water shortages already affect some 40 million people and 30 million livestock. Water tables in several regions are falling by 12–15 feet a year, and the aquifers supplying Beijing and Tianjin are already drained. In Mexico City, parts of the city are sinking as underground aquifers are pumped dry.

To escape deepening rural poverty, an estimated 10 million "environmental refugees" are on the move in Latin America, Africa, and parts of Asia, mostly from rural to urban areas. City services are collapsing under the weight of urban population growth, and unmanageable levels of pollution are creating a variety of threats to human health. Solid waste could quadruple (from 624 billion tons to 2,315 billion tons) in developing countries by the year 2025. In much of Southeast Asia, the rivers are virtual open sewers and many waterways flowing through metropolitan areas are biologically dead. Overall, at least four out of five residents of rural areas in developing countries lack sanitary facilities and an estimated 25,000 people, mainly children, die every day from waterborne diseases.

Potential solutions to those development and environmental problems do exist. The Earth's carrying capacity, and that of most developing countries, can be expanded through human ingenuity. The world community can and must develop alternative energy technologies and improve energy efficiency, launch major reforestation projects and develop faster-growing trees, create nitrogen-fixing grains and drought-resistant crops, invest in soil conservation and anti-erosion measures, institute land reform measures that give peasants a stake in conservation, put market-based pollution controls on industry and agriculture, and find new techniques for recycling.

Yet even if all those potentially costly efforts are successful, they will only allow us to accommodate, at some reasonable standard of living, the unavoidable addition of 4–5 billion people to global population. Such efforts will not absolve us of the need to slow the growth and eventually stabilize the size of human populations. As the U.S. National Academy of Sciences and its British counterpart, the Royal Society of London, said in a joint statement in 1991, "If current predictions of population growth prove accurate and patterns of human activity on the planet remain unchanged, science and technology may not be able to prevent either irreversible degradation of the environment or continued poverty for much of the world."

Tragically, the world population problem is bigger and more urgent than necessary. The U.S. government under Presidents Ronald Reagan and George Bush frittered away most of the 1980s on public debates about abortion and "supply-side demographics." That latter theory, developed by business professor Julian Simon, holds that since people create wealth, population growth can never long be a problem in a properly organized free-market economy. As Michael Teitelbaum, a resident scholar at the Alfred P. Sloan Foundation, points out in a recent issue of *Foreign Affairs*, the argument is the Siamese twin to traditional Marxist ideology, which maintained that overpopulation could not occur in a properly organized workers' paradise. Even China, the largest communist country, obviously no longer subscribes to that theory.

Supply-side demographics had considerable appeal to ideologues in the Reagan White House, becoming an important element in early attacks on what they referred to as the "globaloney" of President Jimmy Carter's *Global 2000* report. Among other things, the report recommended a huge increase in U.S. population assistance. Carter did propose a 1982 budget that would have increased population aid from $190 million to $345 million. Few of the wide-ranging environmental and population initiatives in the *Global 2000* report survived in

Reagan's 1982 budget revisions. In December 1981, conservative idealogues in the Office of Management and Budget and the State Department stripped the entire population aid budget from the 1983 presidential request. They were later reversed by administration moderates.

Supply-side demographics also proved a boon to groups of the religious Right. Opposed to all contraception, as well as to abortion, the supply-side theory allowed them to dismiss the population problem as a hoax, and along with it any need for organized family planning efforts. A large coalition of those groups met annually with Reagan on the January anniversary of the Supreme Court *Roe v. Wade* abortion decision and presented the administration with a laundry list of desired changes in federal policy. In 1984, a demand for the appointment of a "pro-life" delegation to the Mexico City Population Conference later that year was on the list, along with an end to U.S. support of the United Nations Population Fund and to almost any organization with "Planned Parenthood" or "population" in its name. Although Reagan and Bush failed to deliver on the most important of the right-to-life movement's demands—banning abortion in the United States—they did deliver on a number of the demands related to federal support for family planning, and in the process delivered several body blows to international family planning.

During the 1970s, with rapidly rising global expenditures on family planning programs and progress on several social and economic indicators, world contraceptive use had grown by 53 per cent (from 30 per cent of fertile-age couples in 1970 to 46 per cent in 1980). Average family size declined by 22 per cent (from 4.9 children to 3.8). In the 1980s, many governments adopted national population policies as the number of reproductive-age couples grew by about 150 million; yet assistance from the United States and other donor governments remained essentially flat. As progress on family planning and development efforts slowed, contraceptive use between 1980 and 1990 grew less than 20 per cent (to 55 per cent of couples) and family size declined a mere 8 per cent (to 3.5 children).

While one can argue about the extent to which U.S. reluctance affected birthrate declines in the developing world, it clearly helped delay a global commitment to early population stabilization and derailed improvements in a number of countries in reproductive rights for women. A senior Indian health official, S. K. Alok, put it this way in 1988:

The United States is no longer the only important source of technical or financial support. But U.S. political support remains important, and so it is enormously important that the United States stand up unequivocally for family planning.... [The United States] is undermining the international consensus on family planning.

That international consensus is important to many developing country leaders, who face indigenous opposition to family planning programs from both the Right and Left of the political spectrum.

The failure of U.S. leadership has rendered world population problems both more urgent and more expensive to solve. Halting world population at fewer than 9 billion people in the middle of the next century may no longer be possible. Investments in technology and conservation could, arguably, sustain a population of 9 billion at a reasonable worldwide standard of living without destroying the Earth's life-support systems. But there is little margin for error. Political leaders cannot be allowed to lose another decade. At the 1994 International Conference on Population and Development in Cairo, governments must agree to a plan for early population stabilization, one based on the principles of individual reproductive choice and responsible parenthood. And between now and then, Clinton must return the United States to the leadership ranks of world population control.

Population Policies

Measures like quality reproductive health care, greater educational and economic opportunities for women, and reductions in infant and child death rates can and will bring about rapid birthrate declines. If all developing countries were to emulate the most effective policies and programs and if donor governments such as the United States were to provide adequate levels of assistance, the population problem could be resolved in the lifetime of today's children.

The most cost- and time-effective measures are those that ensure a wider availability of birth control methods, along with the information and counseling needed to use them safely and correctly. Experience shows that even relatively poor women with little formal education can and want to take charge of their reproductive lives. Organized family planning programs are, not surprisingly, most successful where they operate in a political climate supportive of women's rights, so that women actually have the freedom to make decisions about their own fertility. In her 1979 book *Third World Women Speak Out*, noted feminist author Perdita Huston quotes a Tunisian woman who declares:

> Before, women were unhappy. They were always in poor health, nursing the children, having children, always weak. Now women are flourishing. They can take advantage of all that is offered. They can have family planning. They are clean and healthy and have freedom....That's all we women talk about: family planning and women's freedom....We are equal. The President says so— it's all thanks to him.

Recent surveys indicate that in many developing countries, a majority of young women and men now want only two or three children. Many of the women surveyed had not wanted to have their last child, and definitely did not want any more. Typically, half of those women were not using any form of family planning. That unmet need for fertility control affects somewhere between 90 and 160 million couples.

In the Asian and Latin American countries surveyed, the number of current and prospective contraceptive users is approaching an average 75 per cent of fertile-age couples. That statistic is highly significant because 75 per cent is the level of contraceptive use associated statistically with a two-child family average, or what demographers call "replacement level fertility." In other words, there may be sufficient demand for fertility control among women in many developing countries to produce a family average of two children early in the next century, and thereby begin the process of population stabilization.

Why then are an estimated 300 million couples, many of whom desire no additional children, not using a family planning method? One answer is lack of access to information and affordable services. The marketplace is not meeting the family planning needs of many low-income couples. In a number of African countries, the full commercial cost of a year's supply of condoms or birth control pills is equal to a third or more of per capita annual income, and supplies are available only in the major cities. Although more than 120 governments provide some budgetary support for family planning services, in many countries those services reach fewer than a third of all couples. In the United States, where federal support for family planning fell by two-thirds

in real dollar terms during the 1980s, many low-income women and sexually active adolescents lack services, and an estimated one-half of all pregnancies are unintended.

Quality is also a problem in family planning services. Often, contact with government programs is a dehumanizing, time-consuming activity, requiring a great deal of motivation. Organized usually for the convenience of the service-provider rather than the client, many programs offer only a limited range of contraceptive choices and not enough personal counseling to ensure that clients make a free and informed choice and understand how to use the method properly. One result is that for some contraceptive methods, like the pill, annual dropout rates can be 50 per cent or higher. A second is that service-providers in some countries show an overwhelming bias toward long-acting or permanent methods that they, not the client, control. In India, 70 per cent of couples using contraception rely on female sterilization; there are few temporary methods available for young couples who want to plan births. In China, 90 per cent of couples rely on sterilization or a fairly crude intra-uterine device, which has an estimated 20 per cent failure rate. Programs that rely on only one or two methods, especially poor-quality ones, are unlikely to generate widespread voluntary family planning.

Unfortunately, a number of developing countries have tried to solve their population problems on the cheap, relying on incentive and disincentive schemes and on community pressure to gain public compliance with demographic targets. Insensitive family planning services can discourage voluntary contraceptive use and reinforce misconceptions about contraceptive safety and side-effects. Moreover, the failure of family planning to address other important reproductive health care needs, including, for example, screening for infections and cancers as well as for sexually transmitted diseases, creates further problems. Particularly amid the spread of AIDS, family planning programs have an obligation to deal with broader issues of sexual health and to integrate concerns about fertility control and disease prevention.

Good quality, "user-friendly" family planning services are not necessarily more expensive if they result in effective long-term contraceptive use. Many community-based services make extensive use of local health workers, most of them women, and cost less than physician-dependent services confined to hospitals and clinics. In fact, most community programs have a record of safety and satisfaction that is as good or better than that of hospitals because the neighborhood workers have more contact with the client.

As a quarter century of experience shows, family planning efforts can be successful, even in relatively poor communities with low levels of education, if they embrace a few basic tenets:

•Make family planning and other preventive health services easily available to rural as well as urban populations by moving them close to the potential clients. Offer them in a manner and setting that is culturally sensitive.

•Allow people to choose freely from the broadest possible range of safe and effective reversible contraceptives, both temporary and long-acting, as well as contraceptive sterilization. Back up those methods with safe abortion.

•Allow people to choose freely among a variety of qualified family planning service-providers including private, commercial, and community programs. Enforce standards of care that protect clients' health and freedom of choice without creating unnecessary barriers to contraceptive availability.

•Link family planning with other health and community development initiatives, especially programs that help ensure that children survive and that help women make life choices about more than just childbearing.

If the world's governments fully employed public and private sector resources to meet the expressed need for fertility control, contraceptive use would increase substantially over current levels. In most settings, more extensive contraceptive use would lead to sharp declines in birthrates, given the number of young couples who prefer to have only two or three children. Field research also indicates that quality services create their own demand for fertility control, perhaps because satisfied clients act as role models for others. The most important challenge for world population efforts is not to create demand for family planning, but to do a better job of meeting the demand that exists. As the largest population assistance donor, the U.S. government can influence how quickly that happens.

Restoring U.S. Leadership

Many in the population community expect Clinton to reverse the policy disasters of the Reagan and Bush administrations rapidly. On his third day in office, Clinton overturned the infamous Mexico City policy—the gag rule for family planning programs abroad. The popula-

tion community also hopes that, in his first 100 days, Clinton will reestablish funding to the U.N. Population Fund, the International Planned Parenthood Federation, and other organizations targeted by the anti-abortion lobby. Those actions will help reestablish American leadership in global population efforts in time for the 1994 International Conference on Population and Development.

But there is less confidence that Clinton will treat population stabilization, or any other international development problem, as an immediate priority, given his stated desire to concentrate on major domestic issues. Unfortunately, global population problems cannot be put on hold while Americans reform their health care system, rebuild their inner cities, and reduce a record budget deficit. Avoiding another world population doubling or a tripling requires rapid action to accelerate family planning and development programs in the next few years. Stopping population growth by the middle of the next century at about 9 billion people will require, as a first step, the near universal availability of safe and effective birth control by the year 2000. Developing countries' expenditures on family planning would need to rise from an estimated $4.6 billion in 1990 (the last year for which complete figures are available) to $10.5 billion by the year 2000.

Initially, a large share of the increased expenditure, perhaps more than half, would need to come from foreign donors like the United States. A 20 per cent U.S. share of that expanded family planning assistance (the United States provides closer to 25 per cent in most ongoing multilateral efforts) would amount to $1.2 billion a year by the end of the decade. To attain that level over the remaining seven years, current U.S. population assistance should be $650–$750 million a year.

Another way to calculate funding is to start with current family planning needs. Taking the middle range of estimates, family planning programs require an additional $2 billion to provide for perhaps 125 million couples who today want to limit their families but lack access to services. If the United States were to contribute a modest 10 per cent of those additional funds, the total annual U.S. population assistance would rise to $635 million. The Clinton administration needs to put real resources on the table and challenge other governments to match them.

Expanded family planning programs would help move the world toward a two-child family average and eventual population stabilization. But they are not sufficient. Only a handful of countries, such as Bangladesh, China, India, and Indonesia, have achieved significant fertility declines without comparable progress on other social and economic indicators. For most developing countries, including major success stories such as Colombia, Mexico, South Korea, Thailand, and Tunisia, sophisticated regression analysis indicates that organized family planning programs have accounted for 40–50 per cent of the fertility decreases to date. Other social and economic factors, particularly those affecting the status of women, account for the remainder. Developing countries do not have to undergo a complete social and economic transformation in order to achieve low birth rates. But they will need to invest heavily in several measures that have a clear, if indirect, effect on birthrates. Donor governments like the United States will need to support those priorities. Most of the measures fall under the "basic human needs" approach that Congress used to reorganize U.S. development assistance in the early 1970s. That approach emphasized primary health care, basic education, and income generation for the very poor.

In virtually every country studied, rising levels of female education are strongly associated with declines in birthrates, and pursuing both goals simultaneously can produce dramatic results. The effect on birthrates is not consistent, however, in every setting or at every level of education. In some places, a few years of formal education is actually associated with higher birthrates, as women abandon traditional birth spacing methods, such as lengthy breastfeeding and sexual abstinence, without replacing such practices with contraception. The impact of female education on birthrates becomes very strong and consistent only at about eight years of formal schooling.

Unfortunately, in much of Africa, Asia, and the Middle East, many girls drop out before they reach secondary school, in part because parents see little value in educating girls. In conservative communities, girls are removed from school before they reach puberty. In both Africa and Latin America, rising rates of teenage pregnancy are in fact contributing to female dropout rates. Getting and keeping girls in school therefore often requires special initiatives, such as setting up scholarship funds exclusively for girls, building schools closer to the community, training more female teachers, and facing up to adolescent sexuality. The World Bank and other donors have begun to incorporate some of those measures in education loans and grants to developing countries where school enrollment for girls is very low. The Bank estimates that closing the gender gap in primary and secondary school education worldwide would cost an extra $3.2 billion a year. The United States, which recently has provided little support for basic education, ought to put its financial and political clout behind the effort.

Other women's development initiatives could influence fertility and family welfare, but the costs are not well documented. Women's paid employment outside the home appears to have a strong impact on fertility. Access to small amounts of credit, new labor-saving technology, or additional land or other productive resources can also further improve women's status and economic security and can change the opportunity costs associated with childbearing and childrearing.

Development programs, including many of those funded by the United States, have rarely created equal economic opportunities for women. Moreover, the recent shift to market-based economies appears to be leaving women even further behind men, as discriminatory wage, hiring, and credit practices go unchallenged. Such trends are most striking in the former Soviet republics. According to recent studies, structural adjustment policies promoted by the International Monetary Fund and the multilateral development banks have also harmed women disproportionately. Households headed by women, moreover, are over-represented among the very poor. Clinton's new management team at the Agency for International Development (AID), or its successor, will need to think hard and creatively about those issues in order to prevent a deterioration of women's economic status in many parts of the world.

A further reduction in infant and child mortality would reinforce the trend toward smaller families and may be easier to achieve quickly. In many developing countries, one in ten children does not live to age five. It is unreasonable to assume that most couples will stop at two children unless those two can be expected to survive into adulthood. Children remain an important source of old-age security, even though a growing number of parents recognize that their futures are best secured by having a few educated, successful children. Especially in areas where the status of women is low, women's only economic security may lie in having one or more sons. The United Nations Children's Fund estimates that 14 million children die every year from easily preventable causes, and that an additional $3 billion in annual expenditures could reduce infant and child mortality by a third by mid-decade, and maternal mortality by half by decade's end. Maternal and child health programs are highly popular with most Americans, as evidenced by charitable giving to groups like Save the Children.

Foreign aid that is clearly focused on a small number of international problems that most Americans consider important—slowing population growth, protecting the environment, eliminating hunger and endemic diseases, raising education levels, and saving children's lives—may be the *only* kind of foreign aid Americans can be convinced to support in the future. In surveys, 85–90 per cent of Americans say they are concerned about world population problems and support family planning as a way to reduce birthrates. In the fall of 1991, when Congress could not muster enough votes to pass a new foreign aid authorization, polls showed that 58 per cent of Americans wanted the U.S. government to resume support of the U.N. Population Fund. Only 17 per cent thought the American government was spending too much of their tax money on family planning aid. Those and other polls taken over the years indicate that most Americans oppose only foreign aid that is perceived as ineffectual or irrelevant to the interests of ordinary Americans. Clinton would do well to start from that premise as he undertakes foreign aid reform.

Such reform is badly overdue. Indeed, AID may not be capable, as presently constituted, of managing the kind of global cooperation that American taxpayers say they are prepared to support. Essentially leaderless since 1987, AID is increasingly bogged down by red tape, a confusion of more than 100 congressionally mandated priorities, and a preoccupation with country-level planning and macroeconomic policy. Many believe macroeconomic policy is best left to the World Bank and other multilateral agencies that have greater leverage over the governments of developing countries. Some observers believe the agency is no longer salvageable and needs to be scrapped—either folded into the State Department or transformed into a small grant-making foundation. Both those radical solutions are probably unworkable, but so is the status quo.

If Clinton wants a credible foreign aid program that can be sold to the American people, he must provide AID with dynamic new leadership and insist on a major organizational overhaul. The reorganization should give visibility and coherence to a half dozen international development initiatives where U.S. private and government expertise could make a difference. It should seek to insulate those long-term development initiatives from the vicissitudes of U.S. foreign policy, including shifting alliances, changes in governments, nuclear non-proliferation efforts, debt repayment, and other problems. Unrelated foreign policy considerations should not be allowed to distort the allocation of development assistance as they have in the past. In 1990, for example, the United States provided only $5.1 million in family planning assistance to India, with its record 26 million births a year, but provided $4.7 million to El Salvador, with 189,000 births a year.

The best way to protect global development initiatives is to make a clear distinction between development assistance and other foreign policy objectives requiring aid. That means transferring to the State Department whatever "walking around money" it needs to support allies or encourage democratic and market reforms, along with the staff AID now employs for such policy dialogues. A slimmed-down AID should then be organized, not along geographic lines (which merely invite political interference), but around specific development initiatives with clear goals, finite funding requirements, and measurable performance criteria. Such a structure would clarify what foreign aid programs are supposed to accomplish and provide a means for measuring progress.

Long-term population and development initiatives may appear costly, but not when considered in the context of the total U.S. budget for international affairs—about $21 billion in 1993. For the last two years, the Overseas Development Council has offered U.S. policymakers a revamped international affairs budget, which frees up $3–$5 billion a year for long-term sustainable development activities (including population and family planning), by shifting funds out of Cold War–era holdovers in military and related security assistance. Both Bush and the 102d Congress were slow to adjust foreign policy expenditures to reflect the new international reality, in which long-term economic and environmental problems have taken on greater significance among U.S. national security interests. Clinton, who may well be president through most of the critical 1990s, and the new Congress must face up to the new realities quickly.

THE LANDSCAPE OF HUNGER

We have seen the victims of mass starvation. We have shuddered at the images of millions of people arriving at remote camps for aid. Many of the world's famines result from wars or civil strife. But we rarely see that hunger also grows from environmental ruin.

BRUCE STUTZ

Bruce Stutz is features editor at Audubon. *He is the author of* Natural Lives, Modern Times.

The numbers alone are staggering: In 1990, 550 million people worldwide were hungry, 56 million more than in the early 1980s. During that time, the number of malnourished children in the developing world increased from 167 million to 188 million. And experts predict that the situation will only get worse as food production in the poorest countries continues to decline.

Deforestation, desertification, and soil erosion have devastating effects on food production. Where forests are cut down, the soils are washed or blown away. Where land is planted too often or grazed too long, it can no longer support crops or cattle.

Ironically, modern agriculture—the science of growing food—has had the greatest impact on the decline in the food supply. In the 1960s governments began encouraging nomads to settle in one place, to raise one cash crop instead of several, to herd only one kind of livestock, and the soils quickly became exhausted.

This "green revolution"—intensified farming of "improved crop varieties" with irrigation, chemicals, and pesticides—at first raised productivity, but the long-term results were just the opposite. According to Mostafa K. Tolba, former executive director of the United Nations' Environment Programme, the process "made agroecosystems increasingly artificial, unstable, and prone to rapid degradation."

Population growth and refugee migration add to the problems of environmental degradation. When the land becomes too crowded and the soil too exhausted to support life, farmers move into forests, slashing and burning new farms. As the land gives out, the people move on, again and again. Eventually, they find their way to the refugee camps.

At the 1972 United Nations Conference on the Human Environment in Stockholm, environmental issues were considered secondary to issues of economic development. But the 1968–74 drought in Ethiopia and the Sahel made it evident that the environmental costs of traditional economic development might be too high.

The Ethiopian government estimated that its highlands were then losing ap-

From *Audubon*, Vol. 95, No. 2, March/April 1993, pp. 54-63. Reprinted by permission.

proximately 1 billion tons of soil a year through water and wind erosion. So when development experts convened at the United Nations Conference on Environment and Development (UNCED) last June, the agenda had changed. The environment had to be protected, it was decided. The diplomats finally recognized the simplest truth: The land that produces the food must be preserved.

The food web inextricably connects plants, animals, and people with the water, soil, and atmosphere of the planet; the hunger web begins as those connections are severed. Air and water pollution contaminate and ruin food sources. The predictions are that global warming will also have an effect, changing planting times, growing seasons, even the ability of crops to survive in their present ranges.

On the following pages *Audubon* examines the environmental causes of mass hunger and some possible solutions. Most do not involve high-tech, grand-scale megaprojects; the best of them are low-tech, local, modest in scale. The solutions may not be as dramatic as scenes of armies massing to feed millions. But they are sustainable—which means that in the future hunger may be defeated without the help of armies.

DEFORESTATION

Slash, Burn, Plow, Plant, Abandon

Forests now cover 27.7 percent of all ice-free land in the world. In 1990 wood was the main energy source for 9 out of 10 Africans, providing more than half of their fuel. By the end of this decade, according to Mostafa Tolba, 2.4 billion people will be unable to satisfy their minimum energy requirements without consuming wood faster than it is being grown.

As human beings encroach on the world's remaining woodlands, deforestation will exacerbate the problems of hunger. For when hillsides are denuded, soil erosion sets in.

In Haiti, where forests once covered most of the land, 40 to 50 million trees are cut each year to supply firewood, cropland, and charcoal. At the current rate of deforestation, Haiti's forests will cease to exist within two or three years.

Already, loss of forests has caused massive soil erosion, and when drought strikes, the quality of the remaining soil will decline. When rain finally *does* fall, runoff will be too rapid and farmers will be forced to abandon cultivation. Since the 1970s, food aid to Haiti has risen sevenfold.

In Bangladesh and India, deforestation has caused another kind of problem, increasing the frequency and force of floods. Bangladesh used to suffer a catastrophic flood every 50 years or so; by the 1980s the country was being hit with major floods—which wash away farms and rice paddies—every four years. Between the late 1960s and late 1980s, India's flood-prone areas grew from approximately 25 million hectares to 59 million. (One hectare equals 2.4 acres.)

PROTECT AND PRESERVE

Although some countries, notably Brazil and Costa Rica, have preserved tracts of their forested land, less than 5 percent of the world's remaining tropical forests are protected as sanctuaries, parks, or reserves.

Regenerating woodlands by replanting them would provide some measure of relief. Over the past 10 years, China has reforested some 70 million hectares of endangered landscape. The U.N. Food and Agriculture Organization estimates that 1.1 million hectares of trees are successfully planted each year worldwide.

Modifying wood stoves to make them more efficient and increasing use of solar cooking would slow the decline of forests by decreasing reliance on wood for fuel.

Using the forests sustainably—by tapping trees for rubber, for example, or developing environmentally sound tourism—would provide more revenue than slash-and-burn agriculture.

DESERTIFICATION
The Spreading Barrens

Every year nearly 6 million hectares of previously productive land becomes desert, losing its capacity to produce food. The United Nations defines desertification as "land degradation in arid, semiarid, and dry subhumid areas [drylands] resulting mainly from adverse human impact." Translated into human suffering, that phrase means that by 1977, 57 million people had seen their lands dry up. By 1984 the number had risen to 135 million worldwide. Today, one-sixth of the total world population is threatened with desertification.

When drylands—which make up about 43 percent of the total land area of the world—revert to desert, hunger follows almost axiomatically. Most crops can't survive in the parched landscape, and harvests fail. Further, withered root systems can't hold the soil, and winds finally erode whatever topsoil remains.

Arid landscapes are so fragile that they break down quickly. A drought can mean catastrophe. In Mozambique, for instance, civil war combined with a worsening drought last year to leave 3.1 million people in need of food aid, 1.2 million more than in 1991.

But Africa is not the only place where food supplies are threatened by desertification. In Russia, annual desertification and sand encroachment northwest of the Caspian Sea were estimated to be as high as 10 percent. Around the drying Aral Sea, the desert has been growing at some 100,000 hectares per year for the last 25 years, an annual desertification rate of 4 percent.

STAVING OFF DISASTER

Agroforestry—planting trees as windbreaks and shade to protect pastureland—contributes to the maintenance of hard-used fields. In Kenya, for example, the Green Belt Movement has embarked on a large-scale tree-planting program.

Massive irrigation projects, such as those tried in Nigeria (see "Death of an Oasis," *Audubon* May-June 1992), are less practical, benefiting only a few at great cost to the environment. Small-scale projects, as low-tech and low-cost as collecting and managing rainwater, are often slow and laborious, but according to the Bread for the World Institute on Hunger and Development, they have succeeded in reclaiming hundreds of hectares of degraded land.

SOIL EROSION
A Worldwide Dust Bowl

Worldwide, erosion removes about 25.4 billion tons of soil each year. Deforestation and desertification both leave land open to erosion. In deforested areas, water washes down steep, naked slopes, taking the soil with it. In desertified regions, exposed soils, cleared for farming, building, or mining, or overgrazed by livestock, simply blow away. Wind erosion is most extensive in Africa and Asia. Blowing soil not only leaves a degraded area behind but can bury and kill vegetation where it settles. It will also fill drainage and irrigation ditches.

When high-tech farm practices are applied to poor lands, the result is often a combination of soil washing away and chemical pesticides and fertilizers polluting the runoff.

In Africa, soil erosion has reached critical levels, with farmers pushing farther onto deforested hillsides. In Ethiopia, for example, soil loss occurs at a rate of between 1.5 billion and 2 billion cubic meters a year, with some 4 million hectares of highlands considered "irreversibly degraded."

In Asia, in the eastern hills of Nepal, 38 percent of the land area consists of fields that have been abandoned because the topsoil has washed away. In the Western Hemisphere, Ecuador is losing soil at a rate 20 times what would be considered acceptable by the U.S. Soil and Conservation Service.

And even in the United States, 44 percent of cropland is affected by erosion.

DEFEATING THE ELEMENTS

According to the International Fund for Agricultural Development (IFAD), traditional labor-intensive, small-scale efforts at soil conservation—which combine maintenance of shrubs and trees with corp growing and cattle grazing—work best.

In the Barani area of Pakistan, a program begun by IFAD in 1980 to control rainfall runoff, erosion, and damage to rivers from siltation has resulted in a 20 to 30 percent increase in crop yields and livestock productivity.

POPULATION GROWTH

More Mouths to Feed

With a population growth rate of 1.7 percent, the world added almost 100 million people in 1992; an increase of some 3.7 billion is expected by 2030. Since 90 percent of the increase will occur in developing countries in Africa, Asia, and Latin America, the outlook is bleak: None of those countries can expect to produce enough food to feed a population increasing at such rates.

Population growth and environmental damage go hand in hand with poverty and hunger. In sub-Saharan Africa, for example, as colonial governments replaced pastoral lifestyles with sedentary farming, populations grew and farming and grazing intensified. Today, 80 percent of the region's pasture- and rangelands show signs of damage, and overall productivity is declining. Yet during the next 40 years the sub-Saharan population is expected to rise from 500 million to 1.5 billion.

Today only Bangladesh, South Korea, the Netherlands, and the island of Java have population densities greater than 400 people per square kilometer. (By comparison, the population density of the United States works out to 27 per square kilometer.)

By the middle of the next century, one-third of the world's population will probably live in overcrowded conditions. Bangladesh's population density could rise to 1,700 per square kilometer.

In Madagascar, population pressures have forced farmers to continuously clear new land. Virtually all the lowland forests in the country are gone. But cleared soil wears out quickly. Per capita calorie supply in Madagascar has fallen by 9 percent since the 1960s, probably the greatest decline anywhere in the world.

In Nepal, one of the world's poorest nations, increased population (700 people per square kilometer of cultivable land, the world's highest average) has forced villagers to expand their farm plots onto wooded hillsides. Marginal farmers rely on livestock, which they allow to graze in the remaining forests. Terraced soil once used for crop production has been abandoned for lack of nutrients, putting more pressure on ever-diminishing forest resources.

CHOOSING THE FUTURE

If a fertility rate of slightly more than two children per couple can be achieved by the year 2010, the world's population will stabilize at 7.7 billion by 2060. If that rate is not reached until 2065, world population will reach 14.2 billion by 2100.

According to a 1992 World Bank report, improving education for girls is an important long-term policy in the developing world. The more educated a woman is, the more likely she is to work outside the home and the smaller her family is likely to be. Choice also plays a role here: The United Nations' World Fertility Survey has found that women would have an average of 1.41 fewer children if they were able to choose the size of their family. Access to birth control methods could help lower the world's population by as many as 1.3 billion people over the next 35 years. During the Reagan-Bush years U.S. funding to programs offering such information was cut back; but President Bill Clinton has reversed that stand.

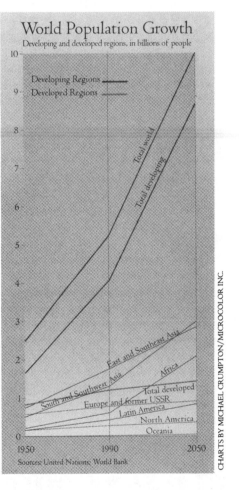

World Population Growth
Developing and developed regions, in billions of people

Developing Regions
Developed Regions

Total world

Total developing

East and Southeast Asia

Africa

South and Southwest Asia

Europe and former USSR

Total developed

Latin America

North America

Oceania

1950 1990 2050

Sources: United Nations; World Bank

CHARTS BY MICHAEL CRUMPTON/MICROCOLOR INC.

Somalia: An Ecopolitical Tragedy

Somalia's breakdown of law and order has created new waves of famine in recent months, but the groundwork for civil strife was laid by years of misuse of the nation's precious grazing lands and water resources.

Almost the entire country is categorized by the United Nations Environment Programme as susceptible to soil degradation, and most of the country is overgrazed.

Over the years the degradation was accelerated by the parceling out of communal grazing lands to private owners, which undermined traditional systems of land management. Private herds—which are generally larger than those owned communally—have stripped the hillsides bare, causing wind erosion during droughts and runoff during rains.

Building dams across valleys to halt water runoff in the north of the country has made matters worse by disrupting the natural drainage systems.

In the south the productivity of irrigated fields has been lessened by poor water management, which has created saline, waterlogged soils on the edge of the desert. There the salt will render the soil useless for food production in the decades to come.

—Fred Pearce

ENVIRONMENTAL REFUGEES
A Moveable Famine

The cycle of overpopulation, poverty, environmental ruin, and famine begins all over again when those trying to find a better life flee their ruined homelands. Land degradation is the largest cause of environmental-refugee movements.

According to Jodi L. Jacobson, a senior researcher at Worldwatch Institute, in Washington, D.C., 135 million people live in areas undergoing severe desertification.

But when those refugees move into areas that are already stressed by overpopulation or too intense agriculture, they place an added burden on the environment. Refugees need wood for fuel, water to drink, land on which to graze their livestock, and grain to eat—all of which are already scarce.

Jacobson estimates that some 10 million people worldwide are refugees from environmental ruin. "Competition for land and natural resources is driving more and more people to live in marginal, disaster-prone areas," she says, "leaving them more vulnerable to natural forces. Hence, millions of Bangladeshis live on *chars*, bars of silt and sand in the middle of the Bengal delta, some of which are washed away each year by ocean tides and monsoon floods."

CLIMATE CHANGE
Global Warning

Since 1800 atmospheric concentration of carbon dioxide (CO_2) has increased by about 25 percent and continues to rise each year. Over the same period atmospheric methane concentrations have doubled. Since the 1960s more than 100 separate studies have confirmed that a doubling of the CO_2 concentration would raise average surface temperatures by one to four degrees centigrade; three degrees is the figure used by the United Nations' Intergovernmental Panel on Climate Change.

Although a small number of scientists dispute these findings, weather and climate remain the biggest concern for farmers as warming begins to change growing seasons, irrigation needs, and land use. These changes will be especially serious in tropical regions, where farmland is already marginal and crops are growing near the limits of their temperature tolerance.

Those who live along shorelines will be hardest hit: The greenhouse effect could cause a global mean sea level rise of about six centimeters per decade. At that rate, many islands would become uninhabitable, and currently productive lowlands would be flooded. The developing countries that now experience the worst food shortages can expect to be hurt most by global warming. A study conducted by the University of Oxford and the Goddard Institute for Space Studies, funded by the U.S. Environmental Protection Agency, found that with an increase in average temperatures of three to four degrees centigrade, grain production in developing countries would decline by 9 to 11 percent by the year 2060, putting between 60 and 360 million people at risk from hunger—10 to 50 percent more than the currently predicted 640 million.

CLOSING THE GREENHOUSE

The use of renewable, or nonfossil, fuels—and more efficient use of all fuels—would go far to control the buildup of CO_2.

The World Bank recommends that governments remove energy subsidies and that they tax the use of carbon fuels. Maintenance of the world's large remaining forests would also help: Tropical deforestation accounts for 10 to 30 percent of the CO_2 released into the atmosphere.

The Oxford–Goddard Institute study found that slowing population increases could allow developing nations to cope more readily with the changing climate by changing land use and farming practices.

Hunger and the Ozone Hole

In early 1992 researchers from five U.S. marine-science institutes reported a drop of 6 to 12 percent in phytoplankton production under the Antarctic ozone hole. It appears that tiny marine organisms, which constitute more than half of all biomass on earth, may not be able to withstand harmful wavelengths of ultraviolet light—UV-B radiation—that penetrate the earth's thinning ozone layer.

When ozone holes form in the spring, most of the fish, shellfish, and crustaceans that humans harvest are in their larval, planktonic stages—floating in the topmost layer of the ocean. "Increasing intensities of UV-B radiation near the surface could negatively impact the reproductive potential of some of our most valuable marine resources, including tuna, pollock, cod, halibut, and flounder," wrote John Hardy, an associate professor at Huxley College, Western Washington University, in the November 1989 issue of *Oceanography*. Juvenile crabs, lobsters, shrimp, and anchovies are also vulnerable.

Just how such damage might move up the food chain is not known, but in Newport, Oregon, the Environmental Protection Agency's stratospheric ozone–depletion team found preliminary evidence of retarded growth in amphipods fed with phytoplankton that had been exposed to UV-B.

Although the leading industrial nations have agreed to halt production of the worst ozone-destroying compounds by 1996, ozone depletion is expected to continue for several decades as existing chemicals seep into the stratosphere. An ozone hole is likely to appear over the Northern Hemisphere, where ozone is dwindling at an estimated 1 percent per year. Each 1 percent decline in ozone is thought to increase exposure to biologically harmful ultraviolet light by at least 2 percent.

The United Nations Environment Programme warns that a 16 percent reduction in stratospheric ozone (which could occur in the next few years) would trigger a 6 to 9 percent drop in seafood production. Oceans now provide more than 30 percent of the animal protein eaten by humans.

—Brad Warren

Energy: Present and Future Problems

There has been a tendency, particularly in the developed nations of the world, to view the present high standards of living as exclusively the benefit of a high technology society. As recently as the 1960s, noted scientists described the technical-industrial civilization of the future as being limited only by a lack of enough trained engineers and scientists to build and maintain it. This euphoria reached its climax in July 1969 when American astronauts walked upon the surface of the Moon, an accomplishment brought about solely by American technology—or so it was supposed. It cannot be denied that technology has been important in raising standards of living and in permitting Moon landings. But how much of the growth in living standards and how many outstanding and dramatic feats of space exploration have been the result of technology alone? The answer is few—for in many of humankind's recent successes, the contributions of technology to growth have been less important than the availability of incredibly cheap energy resources, particularly petroleum.

As the world's supply of recoverable (inexpensive) petroleum dwindles and becomes more important as a tool (or, as events in Kuwait in 1990 and 1991 have shown, even a weapon) in international affairs, it becomes increasingly clear that the energy dilemma is the most serious economic and environmental threat facing the Western world and its high standard of living. With the exception of the population problem, the coming fossil fuel energy scarcity is probably the most serious threat facing the rest of the world as well. The economic dimensions of the energy problem are rooted in the instabilities of monetary systems produced by and dependent upon inexpensive energy. The environmental dimensions of the problem are even more complex—ranging from the hazards posed by the development of such alternative sources as coal and nuclear power to the inability of Third World farmers to purchase necessary fertilizers produced from petroleum that has suddenly become very costly to the enhanced greenhouse effect created by fossil-fuel consumption. The only answer to the problem of dwindling and geographically vulnerable inexpensive energy supplies is conservation. Conservation requires a massive readjustment of thinking away from the exuberant notion that technology can solve any problem.

The difficulty with conservation, of course, is a philosophical one that grows out of the still-prevailing optimism in high technology. Conservation is not as exciting as putting a man on the Moon. Its tactical applications— caulking windows and insulating attics—are dog-paddle technologies to people accustomed to the crawl stroke. Does a solution to this problem entail finding answers to energy conservation in the form of technological fixes with which people are so enamored? Conservation rather than technological fixes may well be the salvation of the world's energy consumers. In "What Would It Take to Revitalize Nuclear Power in the United States?" M. Granger Morgan, head of the Department of Engineering and Public Policy at Carnegie Mellon University, notes that while nearly everyone agrees on the need to reduce reliance on fossil fuels as a way to reduce the potential for the accumulation of greenhouse gases, there are no ready alternatives to fossil fuels. The development of the more traditionally understood alternate energy sources, such as wind and solar power, biofuels, and geothermal energy, all require massive infusions of capital and/or major readjustments in land-use policy. Waiting in the wings as an alternative to other alternatives is nuclear power, virtually a dead industry in the United States. Revival of this energy source is possible, Morgan concludes; but he cautions that it could be revived only with fundamental changes in design, operation, and waste removal strategies.

Profitability is at the heart of the argument for alternative energy advanced in the next selection. In "Here Comes the Sun," Christopher Flavin and Nicholas Lenssen, both of the Worldwatch Institute, conclude that scientific and engineering breakthroughs now exist that make it practical (meaning profitable) to begin producing our electricity, heating our homes, and fueling our cars with renewable energy. Because the conventional wisdom among government experts suggests that we are stuck with dependence on fossil fuels, Flavin and Lenssen contend, there has not been nearly as much government support for the continuing development of marketable alternate energy. A vigorous public commitment of research and development dollars could push renewable energy into the mainstream of energy use. The potential benefits of such a commitment to research are already being realized in one area of alternate energy development, according to a team of researchers from the Oak Ridge National Laboratory, authors of the selection "Energy Crops for Biofuels." Janet Cushman, Lynn Wright, and Kate Shaw describe research currently under way that aims to make cultivated trees and grasses an important source of fuel for transportation and electricity generation. While burning biomass (wood, for example) for fuel

In a discussion of still another alternative energy option, engineer Jon McGowan of the Renewable Energy Research Laboratory offers wind power as a solution. Wind power, long viewed as a quixotic energy source, is once again making strides, in spite of a long period of public and private abandonment of the idea. In "Tilting Toward Windmills," McGowan shatters many long-held myths as to the viability of wind power as a significant source of energy and concludes that it is not too late for the United States to again assume a leadership role in the development of this energy resource.

Finally, "All the Coal in China," by energy specialist Nicholas Lenssen of the Worldwatch Institute, examines the implications for the continued use of an energy source that does serve as an alternate to the use of other fossil fuels for the world's most populous nation. Lenssen points out that while the abundance of coal in China does provide that country with the necessary energy for industrial development, massive coal burning from full-scale industrialization of an economic system containing well over a billion people would probably overwhelm all existing efforts to control greenhouse warming. What petroleum has been for the world's currently industrialized nations, coal could prove to be for China; the implications for the improvement of the global environment are sobering.

is an ancient practice, the modern approach to biofuels is reliant on the conversion of biomass to fuel before combustion. Even more important, perhaps, are the potential developments in genetic engineering that make the prospects of producing plants, which can act as manufacturers of fuel, something more tangible than scientists of even a decade ago would have believed. This argument is carried another step forward in "The Great Energy Harvest" by Helena Li Chum and Ralph Overend, both energy scientists, and Julie Phillips, a communications consultant. They suggest that much of the world's energy problem could be solved within a relatively short period of time by the development of agriculture systems designed to produce crops that would provide much more of the world's transportation fuel and electricity than they do now. While the argument for biofuels has long been a staple in the alternative energy inventory, Li Chum, Overend, and Phillips argue that not only would such energy fuels take enormous pressure off dwindling fossil fuel reserves and contribute significantly to environmental cleanup, they also would provide a revival for moribund agricultural economies in many areas of the world, particularly in the lesser-developed tropical regions.

Looking Ahead: Challenge Questions

Why has the public perception of nuclear energy in the United States been so negative? What changes in the technology of the nuclear power industry are required to make nuclear power a viable alternative to fossil fuels?

What are the most important forms of alternate energy currently available? How might political barriers to the development of those energy sources be removed?

How can the development of biofuels and the application of biotechnology to crop raising contribute new sources of energy to countries like the United States? Are there environmental restrictions or drawbacks to the development of biofuels?

Why has wind power, once a favorite of alternative energy advocates, diminished in importance over the last few years? What conditions are necessary for the revival of this potentially important alternative energy source?

Why should the availability and use of cheap coal in the world's largest underdeveloped country be of such concern to countries in the developed regions of the world?

THE GREAT ENERGY HARVEST

Renewable energy crops could provide clean sources of fuel and industrial chemicals and at the same time revive rural economies.

Helena Li Chum, Ralph Overend, and Julie A. Phillips

Helena Li Chum is a physical chemist specializing in the chemical conversion of biomass into fuels, chemicals, and other materials. She is a fellow of the International Academy of Wood Science, and she is director of the Industrial Technologies Division of the National Renewable Energy Laboratory (NREL), 1617 Cole Boulevard, Golden, Colorado 80401-3393.

Ralph Overend is a principal research scientist of the Alternative Fuels Division of NREL. He is the editor of the journal *Biomass and Bioenergy* and a fellow of the Canadian Institute of Chemistry.

Julie A. Phillips is a communications consultant for the Industrial Technologies Division at NREL. Her address is J.A. Phillips & Associates, 2221 19th Street, Boulder, Colorado 80302.

In the twenty-first century, trees, grasses, and other renewable energy crops will provide much more of the world's transportation fuels and electricity than they do now. Green plants will also give us a growing variety of chemicals that industry can use to make automobile parts, household goods, and other products.

Environmental concerns will drive the acceptance of energy crops as a global energy resource. Many nations have already made a commitment to stabilizing or reducing carbon-dioxide emissions by 2005. Energy crops will help meet the need. They do not contribute to carbon buildup in the atmosphere as fossil fuels do, because energy crops remove as much carbon dioxide from the atmosphere when they are grown as the fuels emit when they are burned.

"Things are changing rapidly. The environmental push is very strong," says Donna Johnson, project director at Interchem Industries, a Leawood, Kansas, firm that makes diesel fuel from soybean oil and biocrude oil from wood wastes. "People are aware now that there may be no place in the world where their kids can grow up with the same quality of life they had. What's important is that we can change this. We are developing the technologies to do it."

Two decades of research and development (R&D) in biofuels and other renewable energy systems are bearing fruit. Energy crops—fast-growing trees and grasses grown specifically for energy production—are ready for large-scale experiments in Europe and the United States. Researchers at Oak Ridge National Laboratory are already studying energy crops' potential impact on wildlife, ecosystems, farmland, soil erosion, and water quality.

These researchers are working with universities, industry, and environmental groups to help energy farming avoid some of the drawbacks of the modern agricultural practices. For instance, industrial farming would shun single-crop farming (monoculture). Instead, islands of food crops such as corn, soybeans, or wheat would be surrounded by fields of perennial energy crops closely tailored to a specific region's climate and soil—a much more sustainable technique.

Alternatives to Imports

Biofuels offer nations an opportunity to replace their oil imports in the twenty-first century. The United States, for example, could derive anywhere from 4 to 14 quads of electricity or fuels from plants or plant-derived wastes by 2030. (A quad is 1,000 million million Btu.) In 1990, Americans used about 25.5 quads of transportation fuels, and the nation imported 15.3 quads of petroleum. Increasing average biological productivity fourfold over 1992 levels *and*, at the same time, boosting energy efficiency by a factor of four (including vehicle-fuel efficiency) would relegate these expensive oil imports to the past.

Sweden and Finland now have active R&D programs to support increased reliance on biofuels in the twenty-first century. These nations already obtain about 16% of their primary energy from trees and wastes—the highest percentages of all industrialized countries.

Since biofuels are renewable, they offer environmental benefits as well. They recycle carbon dioxide, greatly reduce sulfur emissions, and are biodegradable. With another 10 years of research, development, and demonstration, biofuels could easily be as affordable as fossil fuels at that time. Because they produce the same secondary energy forms that people are used to—electricity and liquid transportation fuels—biofuels will

The Farm of the Future

Ten to 50 years in the future, a traditional farm is converted into a fully integrated system for producing energy, chemicals, plastics, and other products, in addition to food.

The traditional farmhouse (A) and barn (B) have a high-tech twist, receiving power from a photovoltaic array (C) and advanced windmill machines (D). Beside the barn is a grain storage and transfer system (E), and in front is a greenhouse (F). Livestock (G) provide manure that is transferred to a biogas plant. In the foreground, a truck (H) sprays fertilizer made from a byproduct of the waste-management system. Beyond the farm is a plantation of trees (I) genetically created to grow to harvestable maturity in six to eight years.

Three sections of crops at the bottom left of the drawing (J) represent feedstocks such as switchgrass, sorghum, and corn that are grown for both food and fuel production. The central diagonal of the drawing illustrates several crop-to-product conversion facilities, including an ethanol fuels production plant (K), which transforms feedstocks into the alcohol-based fuel. One byproduct is a dry material made into pellets for feed. To the right (L) is a facility for converting logs into biocrude, a petroleumlike material.

Another facility (M) converts biocrude into plastics, and another (N) uses woodchips, recycled paper, and sawdust to make paper. The biogas facility (O) converts cow manure into biogas, which is then converted into electricity and sent to the city via powerlines.

The fountains at the upper left of the drawing (P) are sewage-treatment facilities, which draw out gas for electricity. To the right is the high-efficiency bioelectric plant (Q), which takes almost anything and burns it, producing steam-generated electricity. These facilities are also attached to the power grid (R) leading to the city on the horizon. All vehicles depicted operate on hybrid fuels and are highly efficient, running as much as 100 miles per gallon.

Artist and source: Raymond David, National Renewable Energy Laboratory, 1617 Cole Boulevard, Golden, Colorado 80401-3393.

RAYMOND DAVID / NREL

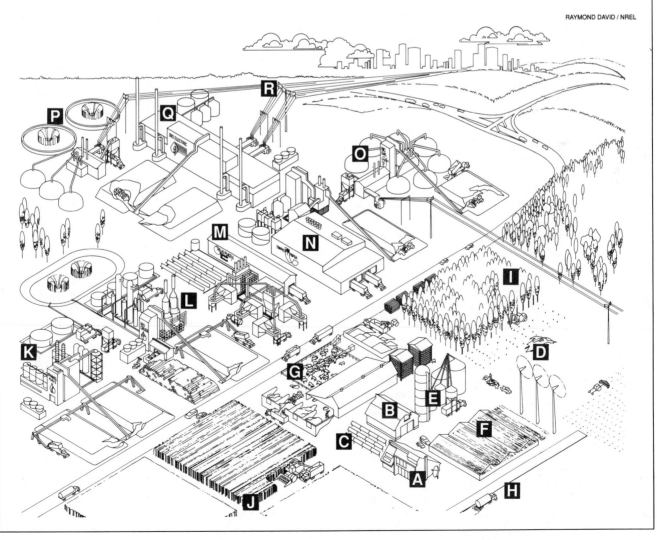

3. ENERGY: PRESENT AND FUTURE PROBLEMS

Hybrid poplars are harvested in Minnesota. Grown as energy crops, many trees and perennial grasses also retard soil erosion and build up new and productive soils.

integrate easily with existing energy infrastructures. Some waste-to-biofuels routes are already competitive.

Clean-burning, biodegradable diesel fuels made from vegetable oils are already required for use in environmentally sensitive areas of Europe, such as the Alps. New energy firms in Austria and Italy are leading efforts to increase the use of these low-emission fuels in heavily polluted areas.

In the United States, the Clean Air Act amendments of 1990 will soon prohibit the use of regular diesel fuel for public transportation in U.S. cities with serious air pollution problems. Biodiesel fuels made from soybeans, sunflower seeds, microalgae, or rapeseed are an excellent replacement.

Revitalizing Rural Economies

The need to adopt creative strategies to curb global carbon emissions and solve other environmental problems dovetails nicely with the need to implement new agricultural crops to revitalize faltering rural economies in the United States and the European Community. Modern agricul-

ture has been enormously successful in producing food. Less and less land and labor are required to produce record-breaking harvests. In recent years, these developments have occurred worldwide, creating intense global competition in foodstuffs along with low income for farmers.

The result has been a decline in the number of jobs in rural areas. Rural economies in the developed countries are languishing. The European Community, in particular, would like to find ways to redress these festering rural problems as a strategy for maintaining a population balance between rural and urban areas.

Growing crops for energy, fiber, and chemicals would create new jobs in rural areas. In addition, energy-conversion industries would spring up near biofuel farms to reduce the transportation costs that drive up the total costs of the new fuels.

As rural energy industries grow, they will attract new business and stimulate regional economic development. James Stricker, an extension agent for agriculture and natural resources at the University of Florida, says this is a key factor in his interest

in promoting energy farming. Stricker is part of a pilot project to grow sugar cane and elephant grass on what is now useless, unreclaimed phosphate mining land in central Florida.

By 1995, the Florida project sponsors hope to have a 1,000-acre energy farm integrated with a fuel plant producing 2 million gallons of ethanol a year. If this venture is successful, the project leaders plan to bring all 100,000 acres of the old mining land into energy production. Within five to 10 years, energy farming will revitalize the soil, attract new wildlife to the area, and redress the extensive environmental damage from strip mining, the project sponsors predict. In the process, the project will create at least 5,000 jobs. This is good news for a region that now suffers from an unemployment rate greater than 10%.

Demonstrating Home-Grown Energy

In the United States, a national program to demonstrate the benefits of home-grown energy is under way. During the mid- to late-1990s, the U.S. Department of Energy (DOE) would like to participate in joint ventures designed to link energy-crop production to energy-conversion facilities. The DOE, in collaboration with other government agencies and industries, may support as many as 10 different energy-crop/energy-conversion systems tailored to the resources and energy needs of specific regions of the country. Coalitions of farmers, foresters, and energy industries from 16 states have already expressed interest in participating in these ventures.

The projects will test energy crops as diverse as elephant grass, eucalyptus, bagasse (a byproduct of sugar-cane refining and ethanol production), black locust, poplars, switchgrass, sorghum, sugar cane, and sycamores. With the exception of sorghum, most energy crops will be trees or perennial grasses. Such crops not only retard soil erosion, but can also help build new and productive soils. Switchgrass, for example, was one of the tall prairie grasses originally responsible for building up the fertile soils of the Midwest.

Farmers will be able to grow tree crops or perennial grasses on partially eroded, marginal agricultural lands—a technique used in Brazil since the 1960s. In 1992, Brazil had about 5.5 million acres of high-yield eucalyptus and pine plantations, much of it on what was once eroded farmland. The plantations now supply raw materials for Brazil's paper, pulp, and charcoal/steel industries. Forestry experts have also developed profitable tree crops in Portugal, Sweden, Finland, China, Canada, and the United States to supply their respective wood and paper industries.

Research at Oak Ridge National Laboratory has found that appropriate selection leads to trees and grasses that take far less energy to farm than food crops like corn or sugar cane. On average, energy crops require less fertilizer and fewer pesticides than traditional food crops do. Irrigation is less often needed.

As a result, trees and perennial grasses yield about 10 times as much usable energy as farmers expend in growing them. With another decade of research and experience, farmers could raise the energy output/input ratio even higher. This translates into a golden opportunity to develop crops to produce electricity, fuels, and chemicals. That's exactly what the proposed U.S. energy-crop/energy-conversion projects hope to accomplish. And there's good reason to believe that utilities, along with fuel and chemical industries around the world—especially in the United States, Europe, and Brazil—will give these new technologies a warm reception.

Energy Crop Projects And Prospects

Many exciting projects are already under way to commercially develop biofuels and other agriculturally derived industrial products.

• **Electricity from biomass.** In 1992, wood and other plant-derived materials (known collectively as biomass) produced as much electricity in the United States as six nuclear power plants. Most biomass power plants are small—about 10% of the size of a typical coal-fired power plant. They generate electricity using relatively low-tech boilers and steam turbines. This is all about to change in response to environmental pressures.

"Utilities are taking the issue of global climate change seriously," says Jane Turnbull, project manager

Biomass power plant converts wood wastes (foreground) into electricity: the Wheelabrator Shasta Energy Company near Anderson, California. Biomass power plants, small and relatively low tech, use boilers and steam turbines to convert plant-derived materials into electricity.

Energy from Wind and Sun

Wind and solar farms could provide a significantly greater proportion of energy in the twenty-first century, according to *Vital Signs 1992*, a report of the Worldwatch Institute.

Wind farms are already well established in Denmark and in California, where independent power producers have installed hundreds of wind turbines and hooked them up to utility grids, reports Christopher Flavin, a senior researcher at Worldwatch.

"In tiny Denmark, a flat and windy country, wind turbines have been integrated into the agrarian economy during the past decade," says Flavin. "They are found throughout the country in ones or twos rather than clusters. Altogether, Denmark has about one-fifth as many wind turbines as California, but these provide a full 2% of the nation's electricity." Other large wind-farm projects are planned for Germany, Italy, the Netherlands, Spain, and the United Kingdom.

Solar-electric systems are also a high-growth alternative energy, according to Flavin. New materials are reducing the costs of photovoltaic cells, improving their competitiveness as a power source.

"The next step in the development of solar electricity is to deploy solar cells on rooftops and at desert-based power plants. Several suburban solar homes have already been built in the United States, and large grid-connected systems are being tested as well. By early in the next century, solar electricity may be ubiquitous," Flavin concludes.

Source: *Vital Signs 1992: The Trends That Are Shaping Our Future* by Lester R. Brown, Christopher Flavin, and Hal Kane. W.W. Norton and Co., Inc., 500 Fifth Avenue, New York, New York 10010. 1992. 131 pages. $10.95. Paperback.

The Carbohydrate Economy

Future products may be derived increasingly from plants rather than petroleum, according to the Institute for Local Self-Reliance, a Washington, D.C., based research organization. If so, the switch will represent a return to past patterns.

Vegetables, trees, and other crops were once the raw materials for the majority of manufactured goods—medicines, inks, dyes, paints, industrial materials, clothing, and fuels. But some 150 years ago, hydrocarbons—fossil fuels and petroleum-derived goods—took hold and by 1970 had captured more than 95% of the market once held by plant matter.

Now, carbohydrates are making a strong comeback against hydrocarbons. The reasons: Advances in

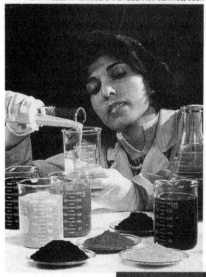

KEITH WELLER / AGRICULTURAL RESEARCH SERVICE, USDA

Newsprint inks made from 100% soybean oil are tested by U.S. Department of Agriculture chemist Sevim Erhan.

Soybeans produce a wide variety of industrial products, from printing inks to foam used in fighting fires.

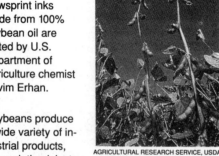

AGRICULTURAL RESEARCH SERVICE, USDA

biotechnology are reducing the costs of manufacturing plant-derived products, while environmental regulations are increasing the costs of their petroleum-derived competitors, the Institute reports in a new study, *The Carbohydrate Economy*. "Moreover, the growing environmental consciousness has prompted many consumers to pay willingly a 'green' premium for carbohydrate-derived, environmentally benign products," note authors David Morris and Irshad Ahmed.

As demand for cellulose, starches, vegetable and plant oils, and biofuels grows, prices will rise, boosting farmers' profits and spurring a renaissance for rural economies. Since plant matter is expensive to transport, processors will be encouraged to locate their operations closer to the farmers growing industrial crops, thus providing jobs and adding to local tax revenues, the report suggests.

One high-growth plant product is soy ink, which is increasingly being used in newspaper printing instead of petroleum-based inks. In 1987, when it was first introduced, only six newspapers in the United States used soy ink; today, 75% of the 1,700 daily newspapers in the country use soy ink, according to the report. Soy inks aren't just cheaper or more environmentally sound than petroleum inks—they do a better job. "Some printers report that they can print more copies with soy oil inks. Soy inks flow more smoothly, reducing paper waste. Since newsprint accounts for 40% of the operating costs of newspapers, the cost savings can be significant," note Morris and Ahmed.

Other plants and their potential industrial uses include:

• **Coconut:** resins, cosmetics, soaps, pharmaceuticals, plasticizers, and lubricants.

JACK DYKINGA / AGRICULTURAL RESEARCH SERVICE, USDA

Lipstick and other cosmetics can be produced from lesquerella rather than petroleum. This promising commercial oilseed crop could also be used for resins, waxes, nylons, plastics, and lubricants.

• **Corn:** ethanol, fermentation products, and resins.
• **Jojoba:** cosmetics, pharmaceuticals, plastics additives, varnishes, adhesives, and inks.
• **Linseed:** drying oils, paints and varnishes, inks, polymer resins, and plasticizers.
• **Palm oil:** fermentation products, soap, wax, fuel processing, and polymers.
• **Rapeseed:** plastics, lubricants, cosmetics, and adhesives.
• **Safflower:** paints, varnishes, and adhesives.
• **Soybean:** paint solvents, resins, plasticizers, adhesives, lubricants, and pharmaceuticals.
• **Sunflower:** plastic resins, plasticizers, fuel additives, cosmetics, agrichemicals.

Source: *The Carbohydrate Economy: Making Chemicals and Industrial Materials from Plant Matter* by David Morris and Irshad Ahmed. Institute for Local Self-Reliance, 2425 18th Street, N.W., Washington, D.C. 20009-2096. 1992. 66 pages. Paperback. $28.75 postpaid.

▲ Bus in Denver runs on methanol, a clean-burning alcohol derived from gasifier technologies.

▶ Truck's engine has been converted to run on 100% ethanol, a biofuel produced from corn.

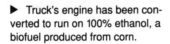

for the Electric Power Research Institute's storage and renewables department. Turnbull believes that biomass is the renewable resource with the greatest potential to make a significant contribution to power generation in the next 20 years.

New technologies are on the horizon for biomass power generation. Energy Performance Systems of Minneapolis, Minnesota, is developing a new process for burning whole trees to make electricity. Hawaii's Pacific International Center for High Technology Research (PICHTR) is building a small commercial-scale gasifier to convert sugar-cane-processing waste into gas for combustion in turbines to produce electricity.

• **Methanol.** A number of companies plan to develop gasifier technologies for producing methanol, a clean-burning, alcohol-based transportation fuel. By the mid-1990s, Hawaii's PICHTR plans to build an integrated gasifier system capable of producing electricity, methanol, or both, depending on market demand. By the end of the decade, the gasifier technology could be adapted for making a variety of industrial chemicals.

• **Ethanol.** In the United States, corn crops produced about a billion gallons of ethanol in 1992. Although this represents less than 1% of the transportation fuels Americans use each year, the ethanol industry is well established. Attractive tax incentives and growing respect for the environmental benefits of alcohol fuels support the current industry.

The corn ethanol industry provides an excellent launching pad for a much larger alcohol-fuels industry supported by energy-crop farming.

The same is true for Brazil's alcohol-fuel program, Proalcool, now the largest in the world. Brazil currently produces more than 4 billion gallons of ethanol a year from sugar cane. This is enough to meet about 20% of its transportation fuels requirements. Both the U.S. and Brazilian ethanol industries depend upon food crops containing sugar or starch, which yeast can ferment into alcohol. The basic conversion technology has been practiced for thousands of years.

The National Renewable Energy Laboratory (NREL) in Golden, Colorado, is developing new technology to make ethanol from cellulose-rich trees, grasses, crop residues, or wastes. The process transforms lignin (a glue-like structural material) into boiler fuel and cellulose (fiber) and other fermentable materials into ethanol. Since cellulose is the largest repository of biomass energy on the planet, this innovative technology is opening up a huge fuel resource.

• **Biocrude oil.** NREL has also developed a promising new technology that converts energy crops and cellulose-rich wastes to biocrude oil. This thick, gooey liquid, with the color and consistency of molasses, is the raw material for making a variety of high-value chemicals, including biodegradable plastics and adhesives, as well as gasoline oxygenates such as MTBE (methyl tertiary-butylether)

for reducing emissions of carbon monoxide and other pollutants. In the future, reformulated gasoline will essentially replace today's more-polluting gasoline-type fuels.

The Future of Fuel Farming

After nearly 20 years of R&D, biomass technologies for producing power, fuels, and chemicals are just now entering the commercial arena. The rate at which biomass technologies capture a share of the energy market will depend on a host of interconnected variables: the verification of climate change; the rate of industrialization of the Third World; energy and agriculture policy decisions in the United States, Japan, and Europe; international cooperation; and the commercial successes of new energy crops and conversion technologies.

Sometime between 2010 and 2020, global energy-use patterns will begin to change rapidly. Renewable energy technologies will begin to replace fossil fuels at a much faster rate. By then, it may even appear that this transition is happening overnight.

Throughout the 1990s, joint ventures between the public and private sectors will help both researchers and entrepreneurs learn more about using energy crops for making electricity, fuels, and chemicals. These projects will provide an opportunity to evaluate the impact of energy crops on biodiversity, soil erosion, and wildlife. And they will give farmers the opportunity to prove that energy-crop farming is sustainable.

3. ENERGY: PRESENT AND FUTURE PROBLEMS

As new energy businesses prove they can extract usable energy from cellulose-rich wastes, agribusinesses will learn to use manure from feedlots and treated sewage from cities to revitalize their soils. Sewage sludge isn't a good fertilizer for food crops because of the potential for disease. But people won't eat energy crops. And the soils these crops grow in will profit from the nutrients.

In some ways, the high-tech farm of the future will resemble the farms our great-grandparents knew. In the past, farmers grew crops that provided energy for horses and draft animals. Today, farmers depend on diesel fuel and gasoline. In the future, they'll grow crops to produce electricity, biodegradable diesel fuel, alcohol transportation fuels, reformulated gasoline oxygenates, plastics, lubricants, adhesives, and other chemicals.

In the United States, agriculture is likely to become one of the nation's fastest-growing businesses. With a major commitment to producing bio-fuels from energy crops, the $50 billion a year now spent on oil imports could be redirected into farms, tree plantations, regional energy firms, and rural infrastructure development. The integration of agriculture, forestry, and energy should bring enough new business to rural areas to reverse the migration to large cities that began early in the twentieth century.

The world isn't running out of fossil fuels: There are sufficient petroleum reserves to support the world's present needs for at least another two generations, and there are hundreds of years' worth of coal reserves. But the global environment can't wait for humans to deplete all fossil fuels before we decide to switch to more-sustainable sources of energy.

Within about 40 years, the world's farmers could grow all the food needed, with enough capacity left over to meet up to 30% of global energy requirements. And they will continue to deliver ample supplies of food, feed, and fiber as they get into the business of providing energy, liquid fuels, and chemicals in biomass refineries. Farming will become more lucrative—and sustainable.

In industrial nations such as the United States, money formerly flowing out of the country for oil imports will stay at home revitalizing rural economies, creating new jobs, attracting new business, and luring well-educated specialists back from the cities. In developing nations, energy farming will help improve the lives of the poorest citizens. Reliable, sustainable energy for cooking, heating, transportation, and electricity will help bring villagers in the world's most-remote regions into the twenty-first century.

The great "harvest" from energy crops will be an atmosphere that accumulates less carbon dioxide, cities that enjoy cleaner air, and a planet that breathes a little easier.

What would it take to revitalize

Nuclear Power

in the United States?

M. Granger Morgan

M. GRANGER MORGAN is head of the Department of Engineering and Public Policy at Carnegie Mellon University in Pittsburgh, Pennsylvania.

lthough the U.S. government has not yet responded to the threat of global warming with policy initiatives that involve more than expanded research, a growing number of people have concluded that the time has come to get serious about limiting emissions of greenhouse gases. One of the most popular means proposed is limiting emissions of carbon dioxide from fossil fuel-fired power plants. But easy alternatives to fossil fuel use are not readily available. Hydropower reserves are limited, for instance, and all new, large projects, such as the James Bay project in northern Quebec, face intense environmental opposition. A switch to bio-fuels, on the other hand, could consume enormous amounts of land and have serious ecological implications. Moreover, progress on other renewable energy sources and on conservation has been slower than enthusiasts have predicted. As a result, optimism has been stirred up within the U.S. nuclear power establishment, which has begun to argue that soon the United States and the world will have no choice but to return to their senses and embrace nuclear power for their energy salvation.

The issue is not so simple, however. A number of expensive but plausible scenarios could limit greenhouse-gas emissions without requiring a major rebirth of nuclear power. Nevertheless, now is a good time to ask what it would take to bring back nuclear power. The answer depends on who is asked. At least a few people have concluded that nuclear power is fundamentally evil; no amount of readjustment in the organizations that manage the technol-

ogy or in the nature and balance of the risks and benefits it brings will make nuclear power acceptable. But most people still view this issue as a balance of risks, costs, and benefits. Those who oppose nuclear power do so because, at least in a rough qualitative way, they have balanced the risks and benefits as they understand them and have concluded that nuclear power is a bad deal. Proponents of nuclear power argue that, with much re-education and a bit of fine tuning of organizations and technology, the public can be persuaded to rebalance the equation and welcome back nuclear power.

In the author's view, U.S. public acceptance of nuclear power will require more fundamental readjustments. Although nuclear power continues to play an active role in the energy planning of other nations, the U.S. nuclear power industry is dead. Its rebirth will take more than increasing energy supply pressures, public relations, and a little fine tuning. Five basic domestic problems plague the current U.S. nuclear power system: The nation has been building the wrong kind of reactors; has organized and managed reactor construction improperly; has taken the wrong approach to regulation; has taken the wrong approach to handling radioactive wastes; and has failed to resolve issues that can be solved only through high-level political will and leadership. A critical element that pervades much of the nuclear issue is a failure to treat the public with respect. With a change in philosophy and some bold new programs, these five problems could be resolved domestically. A sixth problem, involving the more effective management of nuclear weapons and their proliferation, will require collective international solutions.

A Shift in Design Philosophy

There are two reasons why a fundamentally new reactor-design philosophy is needed if nuclear power is to make a comeback. First, the design for today's nuclear reactors evolved incrementally from the design of reactors devel-

From *Environment*, Vol. 35, No. 2, March 1993, pp. 6-9, 30-32. Reprinted with permission of the Helen Dwight Reid Educational Foundation. Published by Heldref Publications, 1319 Eighteenth St., NW, Washington, DC 20036-1802.

oped for the propulsion of submarines and other naval ships. Military systems, however, often push technology to the edge of performance and are inherently risky. In the midst of the cold war, the risks faced in nuclear submarine warfare overshadowed the potential environmental impacts and the level of reactor safety. Moreover, such issues as inherent stability and ease of decommissioning did not figure very highly among naval design objectives. However, U.S. Admiral Hyman G. Rickover, who directed the development of the atomic submarine, did stress safety, in part because he recognized that a reactor accident could have caused a public outcry that might have stopped the nuclear submarine program.

Once the technology was transferred to civilian applications, however, concerns about reactor safety began to grow. The response has been to add to or elaborate on existing safety systems. Today's reactors are extraordinarily complex systems, so much so that the complexity itself has become a source of risk.

A number of quite different approaches to reactor design have been proposed by nuclear engineers. Some of these designs have been subjected to fairly substantial evaluation; others exist only as preliminary ideas. Before nuclear energy makes a major comeback, the design slate must be wiped clean, and a systematic review of alternative designs must be mounted. Although individual companies may have large investments in particular technologies, there is no good reason why a society should not opt for whatever best meets its overall needs. Practical engineering experience and costs are obviously both very important. It may be that, when all the factors are considered, incremental designs based on current practice will prove the most desirable—but that conclusion is not apparent today.

Issues that need to be considered in a comprehensive review of reactor design include safety (in both normal and abnormal operations), reliability, complexity, cost, design maturity, and ease of assembly and decommissioning. Particular attention should be given to how optimal size varies with different designs; to how risks, costs, and benefits vary with size; and to such issues as economies that may result from siting a number of reactors and their supporting infrastructure in one location.

Such a systematic review is unlikely to be accomplished in a few months. Indeed, the entire process could take years. Because there is little practical experience with a number of promising advanced reactor designs, an iterative approach probably will be necessary. One of the recommendations from the first stage of the review would likely be a call for several major new efforts in applied research.

To be persuasive, this systematic national rethinking of nuclear power must be done in a very open fashion. Probably the best institutional arrangement would be a presidential commission working with a number of analysis groups and resources to produce serious studies as they are needed. The review process must involve many credible independent parties, including faculty from a number of leading universities; nuclear engineers with substantial experience; and a significant number of responsible critics of the current system. Although they should be able to contribute much useful technical analysis and support, the U.S. Department of Energy's national laboratories should not be the focal point of such an effort; they carry too much old baggage.

The second reason for such a review of nuclear power reactor designs is that people will want clear and unambiguous evidence that the revitalization of the nuclear power business does not mean picking up with business as usual. Some proponents of advanced reactor designs have argued that a switch to inherently safe reactors would be sufficient to assure general public acceptance of and resurrection of nuclear power. However, although the adoption of a better design may be a necessary step, it is only part of a much broader set of philosophical and institutional changes that will be necessary to achieve a safe, cost-effective, and politically acceptable system of nuclear energy supply in the United States. A much more important requirement is a fundamental change in the philosophy of the nuclear power industry.

Turbine inspection at a nuclear power plant.

A Shift in Manufacturing Philosophy

U.S. nuclear power reactors have been custom-built one at a time, which affects their costs, safety, and reliability. An analogy can be made with the construction of commercial aircraft. Suppose that, rather than using standard designs and mass production, commercial aircraft were custom-built one at a time. Airplanes would take considerably longer to build, and, because good quality control in complex systems is to a large extent a cumulative process that depends on organizational learning, problems in quality would be rampant and the level of safety would be lower. Obtaining a certificate of airworthiness for each craft from the Federal Aviation Administration would probably take years. Many travelers would find the level of safety unacceptable, and air travel would be much more expensive. Pilots and mechanics would have to be specially trained to operate each aircraft and, as a result, could not readily be moved around the fleet. Many replacement parts would have to be custom-made, as well. Moreover, every time an aircraft experienced a problem, engineers and managers would be unsure how to extrapolate the lessons to other aircraft because each would be different.

If the phrase *nuclear power reactor* is substituted for the word *aircraft*, the preceding description provides an accurate picture of the way the United States built its current stock of nuclear power reactors, as well as a fairly accurate description of the origins of many of the problems. However, the manufacture of complex systems is not just a technical problem; it is also a problem of social organization. Although a few U.S. utilities, such as Duke Power in North and South Carolina, have understood this fact and done an excellent job of acquiring the skills needed to build nuclear power plants reliably and at modest cost, most utilities have not understood the enormous organizational difficulties that are involved. As a result, they have experienced large cost overruns and problems in operating reliability.

If nuclear power is to have a future in the United States, the industry must make use of standard designs and mass production. Construction must be managed by project teams that are experienced in managing complex social and technical processes. Also, these teams must be in a position to learn from repeated experience. Although some strong forces work against such developments, thoughtful regulatory strategy should substantially counter these pressures.

A New Regulatory Apparatus

Although the nuclear power system is complex, interconnected, and dynamic, regulatory problem-solving is typically adversarial, piecemeal, and incremental. John G. Kemeny, chairman of the presidential commission on Pennsylvania's beleaguered Three Mile Island power plant, accurately described the state of nuclear power regulation in 1980 when he wrote, "suppose Congress designed an airplane, with each committee designing one component of it and an eleventh hour conference committee deciding how the various pieces should be put together."[1] No rational person would fly in such an airplane. However, this is precisely the way the United States develops policies for regulating the complex nuclear power system. In reality, the system is worse, and not just Congress is at fault. Various regulatory commissioners, executive branch administrators, federal bureaucrats and bureaucracies, state and federal courts, state and local governments, and private interests also are key players.

If the mess is to be fixed, simpler and more capable organizations are needed that must have simpler and more realistic mandates and greater authority. Nuclear power in the United States is regulated by the Nuclear Regulatory Commission (NRC), which grants licenses and monitors utilities for safety violations, and by the Environmental Protection Agency (EPA), which sets standards for exposure and waste disposal. Also, the Department of Energy (DOE) implements NRC and EPA rules and regulations mostly at the beginning and end of the "nuclear fuel cycle."

The NRC/EPA/DOE regulatory system is a bureaucratic quagmire. Many of the officials are technically mediocre and have been in the nuclear regulatory establishment most of their professional lives; they have grown used to following daily procedures and not asking embarrassing questions; and their commitment often is to self-preservation rather than to the future of the industry of the country. A hard look at much of what goes on in these organizations would show that, despite grand justifications, the nuclear power regulatory system serves no purpose beyond providing work for functionaries in the industry, the agencies, and a group of parasitic consulting organizations.

Environmental monitoring at a cooling pond.

3. ENERGY: PRESENT AND FUTURE PROBLEMS

It is unlikely that the state of the current regulatory organizations can be fixed with a bit of institutional fine tuning. They carry too much old baggage—too many habitual ways of thinking, traditions, legal precedents, and other institutional encumbrances—ever to become the kind of smart, efficient, and responsive regulatory organizations that are needed.

Although several ways exist in which dramatic change might be achieved, the best strategy probably would be to establish a new regulatory organization that contains two very separate branches. One branch, staffed largely with new people, would oversee the new nuclear industry. This new operation should include a few of the best, most thoughtful, and experienced people from the current system because experience and a knowledge of previous mistakes are important. The other branch would continue to deal with the existing industry but should slowly shrink over time as existing nuclear plants live out their useful lifetimes. This separation into two branches would allow limited high-level coordination and would not burden the new industry and regulatory organization with 50 years of past mistakes. Given the substantial cutbacks in nuclear power-related university programs, staffing a new organization will pose challenges, but, with effort and some intensive training, an appropriate new staff could be organized.

A New Approach to Waste

One of the most difficult aspects of the regulatory quagmire is the problem of dealing with waste, especially spent nuclear fuel and contaminated material such as worn out reactor parts. Most people view the problem of waste handling and disposal and the closely related problem of reactor decommissioning as difficult technical problems that must be solved before nuclear power can become acceptable. Actually, these problems are mainly philosophical and institutional.

Today, the official policy objective for nuclear waste is to fix it so that society can walk away from it, forget all about it, and be confident that it will never do any appreciable damage to anyone, under any circumstances, for all time to come. The moment our society adopted this notion of the waste problem, we set ourselves up for failure. For some nuclear materials, "all time to come" is measured in many thousands of years. It is a simple fact of geology and engineering that people do not know how to build anything that will reliably last untended for thousands of years.

Although nuclear waste carries some hazard for a long time, as a result of radioactive decay, most of the hazard is gone after the first few hundred years. A well-designed waste storage system would probably not present significant risk even if it were left alone, but future geological processes and human activity are not perfectly predictable. If engineers occasionally check that erosion, groundwater, earthquakes, and other unpredictable natural processes (or human activities) are not beginning to erode the safety barriers, a waste storage system can be made as safe as one

wants for as long as is needed. If we do not include occasional inspections and insist that an untended system present no appreciable damage to anyone, no matter what the circumstances, for all time to come, there can be no solution. However, the U.S. political, legal, and regulatory systems have pretended that there is a solution. Until the statement of the problem is changed, things cannot improve.

The policy objective for nuclear waste should be to fix it so that it poses no significant risk of damage, as long as it is monitored occasionally. Because scientists cannot be sure about the future, the policy objective should also be to secure waste in a way that is unlikely to do significant damage even if people completely forget about it at some point in the future. With such a change in philosophy, geologists and engineers could produce workable designs to meet any reasonable desired level of safety. In short, our nation's regulatory approach needs to take a lesson from nature and opt for an adaptive strategy. In the words of Chris Whipple, a waste management system is needed "in which change is not an admission of error" but an indication that things are working as they should.[2]

If a storage system is designed so that, at some time in the future, spent fuel can be removed, it probably will not have to be tended forever. Technologies are now available that separate the longer-lived elements from spent fuel and burn them in reactors; however, these technologies pose economic and safety problems. In the future, these technologies are likely to be much cheaper, easier, and safer. Even if the separation of longer-lived constituents never becomes attractive, the need for monitoring will become less and less demanding as the radioactive materials decay and the waste becomes less and less hazardous.

Because future societies may become preoccupied with pressing events and forget to keep tabs on the storage system, some precautions should be taken. Scientists and engineers can certainly design and build storage facilities that could remain untended for decades or even centuries without presenting significant risk. The further into the future the system is forgotten, the lower the risks will be. It is also possible that society may regress to a lower state of technology and lose the ability to keep tabs. Such a future low-tech society would have had to go through some wrenching changes as it lost its technical capabilities. Like other societies with limited technical knowledge and resources, it would face low life expectancy and high death rates because of diseases and other natural causes. Under such circumstances, the small risk from a very old and largely decayed store of spent nuclear fuel would likely be a very minor concern in the broad scheme of things. To guard such a future society from risks posed by our current society's activities, many other things, such as chemical wastes, probably deserve greater attention than does the disposal of nuclear fuel.

In summary, if our society persists in adopting a regulatory approach that ignores the physical facts, we are sowing the seeds for long and unproductive social discord.

Pacific Gas and Electric Company's Diablo Canyon Nuclear Power Plant in Avila Beach, California.

Political Will and Leadership

Individual actors, such as nuclear fuel suppliers and electric utilities, may be able to implement some of the previously mentioned changes if they can get their act together. However, regulatory changes and changes in waste management philosophy will not happen unless there is high-level political leadership. To date, there has been no such leadership, and, for the near term, it is hard to see why it would emerge. After all, given all the problems faced by the nation, why should any U.S. president cash in the political chips that would be needed to put the nuclear waste problem on a rational and feasible footing? Politically, it is better to let the spent fuel pile up in swimming pools near reactors all around the country. There are just too many other problems that politicians see as more pressing. The risks to health and safety may be higher this way, but probably nobody will get hurt over the next four to eight years of a presidential term.

There are three things that might change the situation. First, an accident or some other "event" might occur that could focus high-level attention on the waste problem but leave most of the other problems unsolved. Indeed, the inevitable furor that would follow such an accident would probably worsen some of those problems because, in the highly charged political environment that would result, careful, deliberate analysis might become even more difficult. Second, the country might get into another serious energy crunch and decide to revisit the nuclear option, which could focus attention on all the issues discussed above. Finally, political pressures to reduce carbon dioxide emissions and greenhouse warming might lead to a systematic rethinking of nuclear power. However, growing public pressure to address the problem of climate change will not necessarily lead to this rethinking. Studies of public understanding of the climate issue suggest that relatively few people understand the link to energy use and to carbon dioxide. Indeed, a few people even see nuclear power as a source of climate change, although it is not. Therefore, politicians will have to exercise some leadership and not just follow the path of least political resistance. Without a major rethinking and reworking of the entire system, it is very unlikely that nuclear power can be resuscitated in the United States.

Proliferation

Solving the five domestic problems might be sufficient to resurrect nuclear power in the United States. However, if international events focus U.S. public attention on the growing problem of nuclear proliferation in the developing world, progress also will be required on managing nuclear weapons and their proliferation. Such progress can only be achieved through cooperative international effort.

Nuclear power was rushed into civilian use under the banner of "atoms for peace," in part as a public relations ploy to offset uneasiness about the nuclear weapons program. In the intervening years, the nuclear power establishment has worked very hard to separate in the public mind civilian nuclear power and military weapons. This effort has been only partly successful. The loose, negative associations that many people now make with nuclear weapons may be partly misinformed, but they also contain elements of reason. Proliferation of nuclear weapons poses grave threats to world peace and security. Widespread availability of civilian nuclear technology and materials

Control room of a nuclear power plant.

contributes to that threat, though the risk is modest if developed countries exercise appropriate safeguards.

If, however, international developments focus U.S. public attention on proliferation, acceptance of domestic nuclear power may not be possible without dramatic progress in reducing the large existing nuclear arsenals, most of which are now superfluous. Even greater progress would likely be required to build a collective international system of proliferation controls that is far more vigorous and effective than the International Atomic Energy Agency and the Non-Proliferation Treaty have been so far. Among the accomplishments that might be needed would be a comprehensive test ban. Such an agreement might still allow occasional tests under cooperative international supervision to assure safety.

Public Perception

Many members of the nuclear power old guard will object that the six key problems identified here do not address the most important problem of all—changing public perception. Clearly, public understanding and perception are important issues but, for the most part, involve an effect, not a cause. The public and even many otherwise well-informed opinion leaders have serious misunderstandings about nuclear power. But until the basic problems begin to be addressed, it seems unlikely that resolving public misunderstandings through better risk communication will cause many people to change their judgments about where the balance lies between risks, costs, and benefits. From time to time, people may do dumb or silly things, but, on average, the U.S. public is pretty smart. The focus on public relations has been a distraction and has kept the nuclear power establishment from looking hard at its own performance so that it is forever blaming someone else. If the nuclear power establishment can get past blaming others and get its technical, managerial, and regulatory houses in order, public understanding and perception will come around and support will grow. If, on the other hand, the nuclear power establishment does not get itself in order, the public will continue to sense that things are a mess, and, for better or worse, nuclear power will have no future in the United States.

NOTES

1.   J. G. Kemeny, "Saving American Democracy: The Lessons of Three Mile Island," *Technology Review* 83 (June/July 1980):64-75.

2.   C. Whipple, "Reinventing Radioactive Waste Management: Why 'Getting It Right the First Time' Won't Work," *Waste Management '89* (Phoenix, Ariz.: University of Arizona, 1989).

HERE COMES THE SUN

The technology exists today to produce most of our energy from the sun, wind, and heat from the earth. Tapping these sources, though, will require a vigorous public commitment to push renewable energy into the mainstream.

Christopher Flavin and Nicholas Lenssen

Christopher Flavin is vice president for research and Nicholas Lenssen is a research associate at the Worldwatch Institute.

Imagine an energy system that requires no oil, is immune to political events in the Middle East, produces virtually no air pollution, generates no nuclear waste, and yet is just as economical and versatile as today's.

Sound like a utopian dream? Hardly. Scientific and engineering breakthroughs now make it practical to begin producing our electricity, heating our homes, and fueling our cars with renewable energy—the energy of the sun, the winds, falling water, and the heat within the earth itself.

The conventional wisdom among government leaders, energy experts, and the public at large is that we are stuck with dependence on fossil fuels—whatever the cost in future oil crises, air pollution, or disrupted world climate. But, with continuing advances in technology and improvements in efficiency that make it possible to run the economy on reduced amounts of power, a renewables-based economy is achievable within a few decades.

In California, the future has already begun to emerge. The state that always seems to be a decade ahead of everyone else is once again ushering in a new era. The current revolution is subtle, yet momentous. It's evident in the spinning white wind turbines on the hills east of San Francisco and the glinting mirrored solar-thermal troughs set in rows in the Mojave Desert.

Since the early 1980s, California has built no coal or nuclear power plants and has been harnessing renewable energy and improving energy efficiency with a vengeance. The state gets 42 percent of its electricity from renewable resources, largely from hydropower, but also 12 percent from geothermal, biomass, wind, and solar energy—virtually all of it developed in the past decade.

But, for all its success, California's energy revolution is a bit one-dimensional. Its electricity system has been altered, but its cars and homes are still powered largely with fossil fuels. The next step is to find a way to run the whole economy on renewable energy sources.

The missing link is hydrogen—a clean-burning fuel easily produced using renewable power and conveyed by pipeline to cities and industries thousands of miles away. Hydrogen shows great promise as the new "currency" of a solar economy. It can be used to heat homes, cook food, power factories, and run automobiles. Moreover, the technologies to produce, move, and use hydrogen are already here in prototype form.

The challenge of creating a clean, efficient, solar-powered economy is essentially that of reducing the cost of the various constituents of a solar-hydrogen system—from the manufacturing costs of wind turbines to the efficiency of new automobiles.

Here, the pace of progress will be heavily influenced by government policies. The change will come slowly if governments continue to shower favors on fossil-fuel based energy sources. To encourage the adoption of alternative energy sources, policymakers will need to reduce subsidies, raise taxes on fossil fuels, increase research funding on new energy technologies, and provide incentives to private industry for renewable energy development.

Power from the Sun

Renewable resources now provide just 8 percent of the energy used in the United States, but government scientists estimate that renewables could supply the equivalent of 50 to 70 percent of current U.S. energy use by the year 2030 if the government got behind the effort. This estimate is based on the abundance of renewable resources and the technological advances made in tapping them since the mid-1970s.

Such improvements have reduced the cost of renewable energy technologies by 65 to 90 percent since 1980, a trend that is projected to continue through the 1990s [see Table 1]. Increasingly, as governments begin to consider the full costs of fossil fuels (including air pollution and threats to national security), renewables look like a bargain.

Renewables' ability to go head-to-head with coal, oil, and nuclear power has credence even among some utility executives. Greg Rueger, senior vice president of California's Pacific Gas & Electric Company, the nation's largest utility, says "many renewable-generation options are technically feasible today, and with encouragement can prove to be fully cost-competitive...within 10 years."

Solar energy probably will be the foundation of a sustainable energy economy, because sunlight is the most abundant renewable energy resource. Also, solar energy can be harnessed in an almost infinite variety of ways—from simple solar cookers now used in parts of India to gleaming solar collectors on rooftops in Beverly Hills.

Using sunlight to generate electricity has been a dream of scientists and energy planners since the early 1950s, when the first practical photovoltaic cell was invented. This device converts the sun's rays directly into electric current via a complex photo-electric process. Photovoltaic technology has advanced for four decades now, making it possible to convert a larger share of sunlight into electricity—as much as 14 percent in the most advanced prototype systems. Manufacturing costs also have fallen, making this technology a competitive energy source for some limited applications.

Photovoltaic solar cells are now widely used, for example, to power electronic calculators, remote telecommunications equipment, and electric lights and water pumps in Third World villages. These and dozens of other uses created a $500-million market for photovoltaics in 1990, with sales projected to double every five years. The 50 megawatts' worth of cells produced in 1990, though, is only sufficient to power about 15,000 European or Japanese homes.

Table 1: Costs of Renewable Electricity, 1980-2030[1]

Technology	1980	1988	2000	2030
		(¢ per kilowatt-hour)		
Wind	32[2]	8	5	3
Geothermal	4	4	4	3
Photovoltaic	339	30	10	4
Solar Thermal				
trough with gas assistance	24[3]	8[4]	6[5]	—[6]
parabolic/central receiver	85[7]	16	8	5
Biomass[8]	5	5	—	—

[1]All costs are averaged over the expected life of the technology and are rounded; projected costs assume return to high government R&D levels. [2]1981. [3]1984. [4]1989. [5]1994. [6]Estimates for 2030 have not been determined, primarily due to uncertainty in natural gas prices. [7]1982. [8]Future changes in biomass costs depend on feedstock cost.
Source: Worldwatch Institute, based on Idaho National Engineering Laboratory et al., *The Potential of Renewable Energy,* and various sources.

During the past two decades, the cost of photovoltaic power has fallen from $30 a kilowatt-hour to just 30 cents. (This figure is composed almost entirely of manufacturing costs, since solar power requires no fuel.) This is still four to six times the cost of power generation from fossil fuels, so further reductions are needed for solar power to be competitive with grid electricity.

In April, the Texas Instruments Company of Dallas announced plans to produce a new solar cell that costs half as much to manufacture as existing models. Several American, Japanese, and European companies expect advances in other photovoltaic designs, cutting manufacturing costs, and improving efficiency levels.

Photovoltaics are already the most economical way of delivering power to homes far from utility lines. This technology will soon become an economical way of providing supplementary utility power in rural areas, where the distance from power plants tends to cause a voltage reduction that is otherwise costly to remedy.

New applications will spur further cost reductions, which is likely to lead to widespread use of solar cells. As they become more compact and versatile, photovoltaic panels could be used as roofing material on individual homes, bringing about the ultimate decentralization of power generation. Around the same time, perhaps a decade from now, large solar power plants could begin to appear in the world's deserts—providing centralized power in the same way as do today's coal and nuclear plants.

Another source of centralized electricity is solar-thermal power, a technology already proving its viability in California's Mojave Desert. Luz International of Los Angeles has developed a solar-thermal system using large mirrored troughs to reflect the sun's rays onto an oil-filled tube, which in turn superheats water to produce the steam that drives an electricity-generating turbine.

Since the mid-1980s, Luz has installed 350 megawatts of solar systems across three square miles of southern California desert—enough to electrify 170,000 homes. The Luz systems turn 22 percent of incoming sunlight into electricity, which is higher than for any commercial photovoltaic system so far. And because they are of modular design, they can be built on a variety of scales.

Solar-thermal electricity is now produced for about 8 cents per kilowatt-hour, close to the cost of that from fossil fuels in California, where extensive pollution controls are required. However, because it relies on mirrored concentrators, solar-thermal power is only practical where there is intense, direct sunlight—conditions found only in arid regions.

Photovoltaics, which are much more effective in hazy or partly cloudy conditions and can be installed even on a very small scale on residential rooftops, are likely to become the more common power source in the long run. Still, both solar technologies will play important roles.

Contrary to what their critics charge, solar energy systems won't require unusually large areas of land to power the econ-

omies of tomorrow. In fact, they need less space to produce a megawatt of electricity than does coal-fired power when the land devoted to mining is factored in [Table 2]. One-quarter of U.S. electricity needs could be met by less than 6,000 square miles of solar "farms" according to the Electric Power Research Institute in Palo Alto, California, the research arm of the U.S. electric utility industry. That's about the area of Connecticut, or less than 8 percent of the land used by the U.S. military.

Table 2: Land Use of Selected Electricity-Generating Technologies, United States

Technology	Land Occupied
	(square meters per 1,000 megawatt-hours, for 30 years)
Coal[1]	3,642
Solar Thermal	3,561
Photovoltaics	3,237
Wind[2]	1,335
Geothermal	404

[1]Includes coal mining. [2]Land actually occupied by turbines and service roads.
Source: Worldwatch Institute, based on various sources.

More Renewable Options

Another form of renewable energy, wind power, can be captured by propeller-driven turbines mounted on towers in windy regions. Though wind power has a rich history in areas such as Holland and the American Great Plains, it has been taken seriously as a major energy source only since the late 1970s.

Technological advance in the design of wind turbines brought down the cost of wind electricity from more than $3.00 a kilowatt-hour in the early 1980s to the current average of just 80 cents. By the end of this decade, the cost of newer models is expected to be around 50 cents per kilowatt-hour, while the cost of coal-fired power will rise above 50 cents as a result of tightening pollution standards.

Most of the cost reductions for wind energy stem from experience gained in California, which has 15,000 wind machines producing about $200 million worth of electricity annually, enough to power all the homes in San Francisco. Denmark, the world's second-largest producer, received about 2 percent of its power from wind turbines in 1990—but still only about one-fifth of that produced in California.

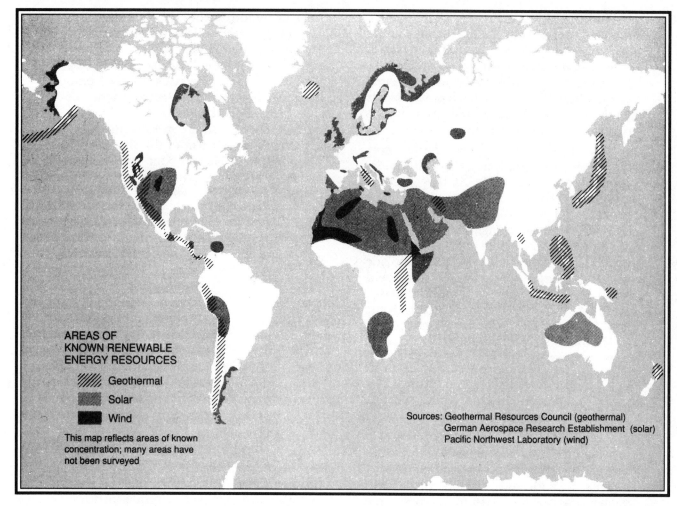

AREAS OF
KNOWN RENEWABLE
ENERGY RESOURCES

////// Geothermal

Solar

Wind

This map reflects areas of known
concentration; many areas have
not been surveyed

Sources: Geothermal Resources Council (geothermal)
German Aerospace Research Establishment (solar)
Pacific Northwest Laboratory (wind)

Wind power could provide many countries with one-fifth or more of their electricity. Some of the most promising areas for wind energy are in North Africa, the western plains of the United States, and the trade wind belt around the tropics—including the Caribbean, Central America, and southeast Asia. In Europe, the largest wind farms will likely be placed on offshore platforms in the turbulent North and Baltic seas.

U.S. government studies show that one-quarter of the country's power could be provided by wind farms installed on the windiest 1.5 percent of the continental United States. A windy ridge in Minnesota, located less than 400 miles from Chicago, could provide one-quarter of the power the city now uses.

Most of the best land for wind power in the United States is grazing land in the western high plains—costing no more than $40 an acre. If "planted" in wind turbines, an acre of this land could generate $12,000 worth of electricity annually while cattle still graze below. One reason it's not being developed is that the region already has more elec-

trical generating capacity than it can use.

Of course, any energy source has its drawbacks, and wind power development, with its rows of towering turbines, will need to be limited in scenic areas, particularly on coastlines. Further efforts are also needed to reduce the incidental bird kills that plague some wind farms.

The energy captured from burning crop residues, wood, and other forms of "biomass" also could play a role in certain locales. It is already the main energy source in scores of countries. In the future, 50,000 megawatts of generating capacity, 75 percent of Africa's current total, could come from burning sugarcane residues alone. However, biomass energy development is likely to be constrained by environmental issues, including the heavy demands already being placed on much of the world's forests and croplands.

Another potential source of power is geothermal energy—the heat from the earth's core. Already, El Salvador gets 40 percent of its electricity from the earth's natural heat, Nicaragua 28 percent, and Kenya 11 percent. Most Pacific Rim countries, as well as those along East Africa's Great Rift Valley and around the Mediterranean Sea, sit atop geothermal "hotspots." Virtually the entire country of Japan lies over an enormous heat source that one day could meet much of the country's energy needs.

Geothermal energy is not without its environmental drawbacks, including the underground sulfur it tends to release, and development will have to be limited in ecologically sensitive areas. However, this still leaves a vast resource potential, particularly as engineers develop techniques to drill deeper and deeper into the earth's crust.

The Hydrogen Solution

If renewable energy is abundant and economical, then why isn't it being harvested on a larger scale? The answer stems in part from the difficulty of storing and moving energy from ephemeral, intermittent sources such as the sun and the wind. While oil can be moved from remote areas by tanker, and coal by barge, sunshine is hard to carry to far-off cities.

Electric power produced by renewable energy could be stored and transported to the user, but at some cost. Electric batteries are expensive, heavy, and must be recharged frequently. Power lines also are costly, generate potentially dangerous electromagnetic fields, and lose energy over long distances due to resistance in the lines. Nonetheless, extended transmission of electricity is already common: California, for example, relies on hydropower produced nearly 1,000 miles away in British Columbia.

Given the limits on moving electricity, it makes more sense to convert renewable power to a gaseous form that is cheap to transport and easy to store. Hydrogen is an almost completely clean-burning gas that can be used in place of petroleum, coal, or natural gas. It releases none of the carbon that leads to global warming. And it can be produced easily by running an electric current through water—a process known as electrolysis.

Hydrogen can be transported almost any distance with virtually no energy loss. Over distances greater than 400 miles, it costs about one-quarter as much as sending electricity through a wire. Gases are also less risky to move than any other form of energy—particularly compared with oil, which is frequently spilled in tanker accidents and during routine handling.

Hydrogen is much more readily stored than electricity—in a pressurized tank or in metal hydrides, metal powders that naturally absorb gaseous hydrogen and release it when heated. Years' worth of hydrogen could be stocked in depleted oil or gas wells in regions such as the U.S. Gulf Coast.

Moreover, hydrogen can provide the concentrated energy needed by factories and homes. It can be burned in lieu of natural gas to run restaurants, heat warehouses, and fuel a wide range of industrial processes. Around the home, new hydrogen-powered furnaces, stove burners, and water heaters can be developed that will be much more efficient than today's appliances.

The gas can also be used to produce electricity in small, modular generators that turn out heat and power for individual buildings. Such co-generating plants would produce far less pollution than today's power plants do in getting a similar amount of electricity to individual users. Hydrogen can be used to run automobiles, using either an internal combustion engine or, more efficiently, a fuel cell [see "Green Machines"].

Eventually, hydrogen fuel could be even more prevalent than oil is today. The gas could become cost-competitive as a transportation fuel within the next two decades. Solar or wind-derived electricity at 5 cents per kilowatt-hour—achievable by the late 1990s—could produce hydrogen that would sell at the pump for about the equivalent of a $3 gallon of gasoline. While this is more than Americans are now charged to fuel their cars, it is less than the price most Europeans pay.

The transition to hydrogen as a major energy source likely will be eased by the ability to mix hydrogen with natural gas up to a one-to-ten ratio with minimal alteration of the existing infrastructure of gas pipelines, furnaces, and burners. Thus, as natural gas reserves are gradually depleted and prices rise in the early part of the next century, hydrogen can gradually be worked into the mix.

The Global Solar Network

Solar energy—whether transmitted through electrical lines or used to produce hydrogen—could become the cornerstone of a new global energy economy. All of the world's major population centers are located within reach of sun- and wind-rich areas [see map]. The U.S. Southwest, for example,

Green Machines

In both industrial and developing countries, automobiles are now a major consumer of petroleum and a chief cause of the stifling air pollution that plagues many cities. Finding a way to power transportation systems on non-polluting renewable energy sources is essential to a sustainable energy future.

Electric automobiles are one way to achieve this end. Electric cars produce virtually no air pollution in cities and, of course, they require no oil. While electric cars have been around since before the advent of the gasoline engine, the size of their batteries and the need for frequent recharging have limited their appeal.

But during the past year, spurred by tightening air-quality standards, a half-dozen auto manufacturers have announced plans to develop electric cars and vans during the 1990s. Fiat already has its Elettra on the market, the first mass-produced electric passenger car sold by a major manufacturer.

General Motors announced last year that it will introduce an electric sports car, known as the Impact, in 1993. Aerodynamic and sporty, the Impact accelerates rapidly and has a top speed of 100 miles per hour. But its batteries need to be recharged every 120 miles—making the car impractical for anything but city driving. More convenient electric cars are likely to be produced later in the decade as better batteries come along.

The U.S. Congressional Office of Technology Assessment estimates that the United States has sufficient existing power plant capacity to operate more than 10 million of these vehicles. Even if electric cars were used for one-quarter of U.S. auto travel, total electricity use would rise only 7 percent and overall air pollution emissions would drop significantly.

The battery-powered automobile is not the ultimate in "green" cars, however. Automakers are also considering vehicles fueled with hydrogen, the cleanest burning fuel, which produces only water vapor and small amounts of nitrogen oxides. The German companies BMW and Mercedes-Benz have built and test driven prototype hydrogen-powered cars; French and Japanese automakers are launching projects of their own.

Only minor modifications are needed to make a gasoline-powered engine run on hydrogen, although storing hydrogen remains a problem. Tanks that contain compressed gas are bulky, and hydrides—a chemical storage system—are expensive, more suitable for large trucks or buses than passenger cars.

Fuel cells offer a solution. They chemically combine hydrogen or natural gas with oxygen to produce electricity, which then can be used to run an electric motor. This process is, in effect, the reverse of electrolysis, which likely would be used to produce the hydrogen in the first place.

Because fuel cells are more than twice as efficient as today's internal combustion engines, cars won't need to carry as much hydrogen, greatly easing storage problems and reducing fuel costs. Fuel cells are currently being developed in both public and private R&D programs; several prototypes are in use. Researchers are still at work on reducing the cost of fuel cells and developing a compact version that could be used in cars.

A new generation of natural gas-powered vehicles will make a first step toward hydrogen-powered fleets, since the technologies behind both are very similar. A half-million natural gas vehicles are now on the road worldwide, and the United States alone could have 4 million by 2005 due to clean air laws. Researchers also have found that hydrogen mixed with natural gas produces dramatically less pollution than natural gas alone. In recent tests, a one-to-seven mixture lowered hydrocarbon emissions by more than 50 percent and nitrogen oxide emissions by 75 percent.

Still, "green" cars will not solve all the problems created by over-reliance on individual automobiles. To effectively deal with congestion, traffic fatalities, unlivable cities, and other impacts of car-dominated transportation planning, nations will need to move toward public transit systems. Subways, light-rail systems, and high-speed inter-city trains are efficient modes of transport that already are electrically driven. These transit modes—powered by renewable energy—could be greatly expanded in the years ahead.

could supply much of the country either with electricity or hydrogen.

Although renewable energy sources are regionally concentrated, they are far less so than oil, where two-thirds of proven world reserves are found in the politically unstable Persian Gulf area.

Wherever renewable resources are abundant, hydrogen can be produced without pollution and shipped to distant markets: from the windy high plains of North America to the eastern seaboard; from the deserts of western China to the populous coastal plain; and from Australia's outback to its southern cities.

For Europe, solar-power plants could be built in southern Spain or in North Africa. From the latter, hydrogen would be transported along today's natural gas pipeline routes into Spain via the Strait of Gibraltar, or into Italy via Sicily. Within Europe, today's expanding pipelines and electrical networks would make it relatively easy to distribute the energy.

To the east, Kazakhstan and the other semi-arid Asian republics might supply much of the Soviet Union's energy. In India, the sun-drenched Thar Desert in the northwest is within 1,000 miles of more than a half billion people. Electricity for China's expanding economy could be generated in the country's vast central and northwestern desert regions.

While pipelines must be sited to avoid ecological damage or accidents, their overall environmental impacts are minimal, especially where natural gas pipelines already exist, as between Wyoming or Oklahoma and the industrial Midwest and Northeast. The pipelines themselves will need to be modified or rebuilt to accommodate any shift to hydrogen, since the gas has properties that corrode some metals.

Germany leads the effort to develop solar-hydrogen systems. It has demonstration electrolysis projects powered by photovoltaic cells already operating in Germany and Saudi Arabia. Germany spends some $25 million annually on hydrogen research projects, according to Carl-Jochen Winter, a scientist with the German Aerospace Research Establishment (by contrast, the United States devotes less than $3 million). Experience in transporting hydrogen comes from a 120-mile pipeline in Germany that transports hydrogen produced from fossil fuels for use in industry.

Getting There from Here

In the end, major energy transitions tend to be driven by fundamental forces, either the evolution of new technology or problems facing society, such as population growth, resource depletion, or climate change. These forces, it can now be argued, are pushing the world toward a solar-hydrogen economy.

But the pace of change will inevitably be determined in part by government policies, most of which are now biased to favor today's energy systems. Will political leaders cling obstinately to the status quo or will they begin encouraging the development of new energy systems? In California, for example, it was a series of state policy changes made in the late 1970s and early 1980s that cleared the way for a new era.

Many state and local governments now encourage use of renewables through regulations and incentives. National governments are also moving toward new policies—under strong pressure from voters. Some European countries have raised gasoline taxes, other governments are taxing cars that pollute, and still more are forcing change via regulation. These are policies that work best in consort.

Higher taxes on fossil fuels is one way to accelerate the energy transition. The carbon taxes levied so far in countries such as the Netherlands and Sweden are an important step, but they haven't been set high enough to cause major shifts in the choice of energy technologies.

To make a real difference in energy habits, a carbon tax would have to be large enough to replace at least a quarter of today's taxes. One possibility is to lower income taxes as carbon taxes are raised. The voting public might accept higher taxes on gasoline, coal, and other fuels in exchange for more take-home pay.

Another approach that can level the playing field is including environmental costs within the electric-utility planning process. If environmental costs were added to construction costs in considering what kind of power plants to build, new coal-fired capacity would become economically unattractive compared with renewable energy sources.

The state of Nevada, for example, decided in early 1991 to tack on a hefty environmental charge when utilities license new coal-fired power plants. Part of the charge is for the potential costs of climate change, the rest is attributed to pollution costs, such as for the extra medical care required when air pollution damages people's lungs. As a result, the coal-based power that now dominates the state is likely to shift in the future to energy-efficiency programs, natural gas, geothermal heat, and sunshine.

National governments also can speed the transition to a sustainable energy future by providing modest incentives for the building of renewable energy systems. New energy technologies have in the past been subsidized by governments—hydropower and nuclear power are obvious examples. In this case, the subsidy can be justified due to the avoided environmental damage that results from renewable energy development. It is worth remembering that California state tax credits for renewable energy development helped spark the renewables boom of the early 1980s.

The U.S. Congress is now considering a subsidy for the generation of renewable electricity. An incentive of just 2 cents per kilowatt-hour—equivalent to about 25 percent of the average retail price of power in the United States—would be sufficient, according to market analysts, to spark a boom in renewable energy development. The cost to the taxpayer would be about $1 billion over five years—a fraction of 1 percent of the nation's annual power bill.

A re-orientation of research and development programs is also called for. In 1989, the leading industrial-country governments spent just 7 percent of their $7.3-billion in energy research funds on renewable technologies. Most of the rest went to nuclear energy and fossil fuels. Drastically trimming breeder-reactor and nuclear-fusion programs would free billions of dollars to accelerate the commercial development of new technologies.

Most countries still have a long way to go in reforming their outdated energy policies, but there is a new sense of urgency about future energy sources as the public reacts to the threats posed by greenhouse gases building in the atmosphere.

There are, for example, 23 countries committed to limiting carbon emissions. More are likely to follow as a United Nations-sponsored treaty to protect the world's climate is readied for signing in Rio de Janeiro in 1992.

Some of the biggest obstacles blocking change in many countries are caused by the politicians who are captives of today's energy industries. The halls of the U.S. Congress, for example, are filled with lobbyists for powerful energy industries—ranging from oil to coal to nuclear power—and their policy agenda predominates. Ironically, while their political power remains intact, these industries have been automated and no longer provide many jobs. As more such positions are eliminated in the 1990s, the political position of these industries is likely to weaken.

In the end, the key to overcoming political barriers is to demonstrate that a solar economy would have major advantages over today's fossil fuel-based systems. California again provides a good example. It already has greatly reduced its fuel bills and begun to clear its skies as a result of the energy policy changes begun more than a decade ago.

As California's political leaders seem to understand, a solar future is just too attractive to be ignored. Indeed, a solar economy would be healthier and less vulnerable to oil price gyrations of the sort that have shaken the world in recent decades. And a solar future is the only practical energy future that would be environmentally sustainable—eliminating the greenhouse gases now threatening the planet's health.

Energy Crops for Biofuels

Research now under way aims to make cultivated trees and grasses an important source of fuel for transportation and electricity generation.

Janet H. Cushman, Lynn L. Wright, and Kate Shaw

Janet H. Cushman, Lynn L. Wright, and Kate Shaw work in the Biofuels Feedstock Development Program, Environmental Sciences Division, Oak Ridge National Laboratory, Oak Ridge, TN.

The "amber waves of grain" in "America the Beautiful" may be replaced by "rows of poplar trees" or "silver waves of grass," if energy crop researchers meet their goals. Rows of trees may replace rows of corn, and fields of soybeans or cotton may be interspersed with fields of switchgrass. Growing crops for energy may become as important as growing crops for food.

In laboratories and fields across the United States, common trees and grasses are beginning to receive the same kind of intensive study as that directed toward solar collectors and fuel cells—and for much the same reasons. Scientists have recognized that trees and grasses can be biologically "engineered" to become more efficient collectors and storers of solar energy. They have also recognized a wonderful potential versatility of plant matter ("biomass"[*]) as an energy source—it can either be burned

directly to release heat or be converted to a variety of readily useable fuels, including methane, ethanol, and hydrogen. Biomass conversion technologies include both those of direct combustion to produce heat and those for producing liquid or gaseous fuels. [See "Biomass Conversion Technologies."]

Burning biomass to produce heat is an ancient and still widespread practice. At present, however, only small-scale applications that have access to very low-cost biomass burn it to produce electricity.

For conversion of biomass to a liquid fuel, especially ethanol, existing technologies use as feedstock only select components of a plant, such as the starch from corn kernels or the sugar from sugarcane. The stalks of both corn and sugarcane, made primarily of cellulose, have been waste materials.

However, a new approach to better using the full potential of biomass energy is rapidly developing. This approach is expected to offer a near-term, relatively low-cost, and renewable alternative to coal or oil on a significant

scale. The idea has been generated both by new discoveries about the growth potential of trees and grasses and by technological developments that promise enhanced efficiencies in converting biomass to useable energy.

Of special interest is the research to develop cost-effective processes for converting cellulosic material, including wood and grass as well as corn and sugarcane stalks, to ethanol. The mastery of this technology would make possible the production of liquid fuels for transportation from a wide range of fast-growing plants. Other research is developing combustion processes with enhanced efficiencies.

A new urgency for developing biofuels has resulted from the recent concerns about the possible role of increased levels of atmospheric carbon dioxide in causing global warming. In 1987 almost 30 percent of the carbon dioxide released in the atmosphere in the United States came from transportation fuels. Coal-fired electricity-generating facilities contributed another 36 percent. If biofuels were to replace a significant portion of the fossil fuels being used, the rate of carbon dioxide buildup would be substantially reduced. Indeed, carbon diox-

[*] Whereas *biomass* in its general definition includes both plant and animal matter, in the field of biomass energy the term refers only to plant matter.

■ **Experimental plots of cultivated hybrid poplars in the midst of Wisconsin farmland foreshadow a time when farmers will raise crops both for food and for fuel.**

ide would still be released by the burning biomass, but that carbon dioxide would have been removed from the atmosphere through photosynthesis during the short lifetime of the plant being used as feedstock. In contrast, the carbon dioxide released by burning fossil fuels was removed from the atmosphere millions of years ago. Burning biomass thus yields no net change in the level of atmospheric carbon dioxide, except possibly for carbon dioxide released by the machines that tend energy crops or carbon dioxide released in the production of pesticides and fertilizers used to raise the energy crops.

The DOE initiative

In 1978 the U.S. Department of Energy's (DOE) Biofuels Feedstock Development Program, managed by the Oak Ridge National Laboratory (ORNL) in Oak Ridge, Tennessee, started ex-

ploring the variety of plants that could be used as biomass feedstocks and the types of land most suitable to cultivate such crops. Today the program coordinates an extensive network of research on crop development, maintains a unique data base on energy crop production, and performs resource and economic analysis regarding energy crops ranging from trees to grasses. This work is also coordinated closely with the cellulosic conversion technologies being developed by DOE's Solar Energy Research Institute in Golden, Colorado.

The oil crises of the 1970s drove home the fact that wood might provide transportation fuel. In 1977 the DOE called for research proposals "to improve the productivity and increase the cost efficiency of growing and harvesting woody plants for fuels and petrochemical substitutes." This initiative was able to build on the pioneering experiments of foresters and university researchers in the late 1960s and early 1970s to grow trees as crops.

Twenty-seven proposals, most from universities, were selected for the new Fuels and Chemicals from Woody Biomass Program. Before long, investigators across the country were scrutinizing 140 different species, trying to learn how to establish energy crop plantations—which trees

■ **Four-year-old sycamore trees fall before the "scythe" of a tree harvester in research trials conducted by North Carolina State University and Scott Paper Company.**

grew best, where, and under what conditions. In the Pacific Northwest, Reinhard Stettler, a geneticist at the University of Washington, and Paul Heilman, a soil scientist at Washington State University, joined forces on poplar research. Their research has also gained the support of the U.S. paper industry as a few farsighted companies have already initiated programs to begin to meet their pulp and energy needs from sustained-

■ **The U.S. Biofuels Feedstock Development Program today includes research on the energy crop potential of grasses (at HECP sites) and trees (at SRWCP sites). After widespread screening trials, research at many less-promising tree-growth sites has been discontinued (inactive SRWCP sites).**

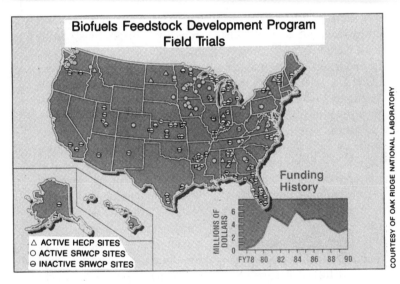

Biofuels Feedstock Development Program Field Trials

△ ACTIVE HECP SITES
○ ACTIVE SRWCP SITES
⊖ INACTIVE SRWCP SITES

Funding History

MILLIONS OF DOLLARS
FY78 80 82 84 86 88 90

COURTESY OF OAK RIDGE NATIONAL LABORATORY

yield tree farms using Stettler and Heilman's hybrid fast-growing tree varieties.

By 1984 the first phase of DOE biofuels research had pinpointed 25 tree species with high growth potential. Two years later, in response to the need to focus on fewer trees, investigators selected poplar, eucalyptus, black locust, sycamore, sweetgum, and silver maple as models for woody crops. Along the way researchers learned that successful "tree farming" requires genetically superior clones or seedlings, careful site selection, inten-

sive site preparation before planting, and a variety of weed control techniques. Densities of between 800 and 1,600 trees per acre were settled on as being most cost-effective for harvest cycles of 5 to 10 years. The concept name was purposely changed to "short-rotation woody crops" to signify that the biomass being produced should be thought of as a standard agricultural product.

Herbaceous energy crops

Also, in 1984 the DOE decided to expand its energy-crop research efforts to include nonwoody herbaceous plants. The department, in collaboration with ORNL, sought proposals for herbaceous energy crops that might complement and supplement wood energy crops, and by 1985 five herbaceous energy crop screening studies were set up in the Southeast, the Midwest, and the Lake States. Mindful of the debate over the use of cropland, DOE concentrated on species suitable for marginal cropland. Thereby an increase in energy crop production would not significantly reduce food production.

The researchers had originally plunged into the herbaceous energy crops work assuming that the effort would focus on the selection of a few species for

a standard production system. But they soon discovered that herbaceous plants, with their diverse growth forms, offer an array of interchangeable production options. Appropriate crops could be developed for any kind of land that becomes available for energy feedstock production.

Land quality is a variable from place to place and climatic conditions change from year to year. In order to optimize land use in any given region and year, from both an economic and environmental perspective, it is desirable to have a variety of energy crop choices. Cellulosic herbaceous energy crops greatly expand the choices. Like short-rotation woody crops, which are perennial and therefore require that specific areas be dedicated to their production, many herbaceous energy crops are also perennial. These plants, because they can effectively minimize soil erosion, are also good choices for land prone to erosion. Further, the equipment and farming practices for these crops are similar to those used for forage crops.

As a complement to perennial herbaceous energy crops, annual herbaceous energy crops, like most food crops, are desirable because they allow flexibility in changing crops annually to adapt to market demands or climatic variation. Again, the equipment and farming practices needed for annual energy crops are similar to those used for many food crops. Sorghum is the most promising of the annual cellulosic energy crops tested to date. However, like all row crops, sorghum should be grown only on relatively flat cropland to reduce soil erosion. Leguminous plants are also good cellulosic crops that help optimize land use, and, being nitrogen-fixing plants, they replenish the soil with nitrogen. In order to select a full complement of choices, several species are being grown on sites in four areas of the eastern United States with different soil qualities, fertilization rates, and weed-control measures. The results of the first phase of those trials reveal that herbaceous crops hold tremendous potential as biofuels.

Other characteristics of herbaceous crops that need to be considered when developing management systems is when and how they should be harvested and how they should be handled after harvesting. Crops like sorghum, cane varieties bred for bio-

mass rather than for sugar, and other tall tropical grasses grow best during summer and should be harvested in late summer or early fall. These crops also have thick stems which means they will not dry easily and thus must be stored green if storage is required. If storage costs are to be avoided, other varieties of cellulosic feedstocks should be considered. For instance, thin-stemmed grasses can be harvested and stored like hay in large round bales that may be left for several months in the field.

While switchgrass, the most promising of the thin-stemmed grasses tested, grows best in summer, many other thin-stemmed grasses grow best in spring and fall. By including several crops in the feedstock supply mix for a conversion facility, including short-rotation woody crops, the supply of fresh material can be extended.

Although farmers have experience with several herbaceous crops, their successful production for energy purposes will require a different mind-set. Farm-

Biomass Conversion Technologies

Biomass conversion technologies are both old and new, traditional and innovative. Biomass, the world's first fuel for heating and cooking, meets almost 90 percent of the energy requirements of many developing countries even today. Worldwide, biomass provides 14% of the total energy needs, and in the United States, according to the National Wood Energy Association, it provides 5 percent of all energy consumed.

The nation's approximately 200 biomass energy facilities predominantly use residues from forestry and agriculture to produce steam or electricity through conventional combustion processes. Although the existing technology may be cost-effective when biomass is cheap or free, it is often only 20 to 25 percent efficient in converting biomass to energy. New, electricity-generating combustion systems, which may have biomass to energy conversion efficiencies of 35 to 45 percent, are under de-

velopment. These systems are better suited for cellulosic biomass, including wood, grasses, and the stalks of corn and sugarcane.

In the whole-tree-burner concept, entire trees, dried with waste heat, would be batch-cut into four- to eight-feet-long segments and fed into a large furnace for electricity production. Steam-driven turbines, similar to ones that propel jet engines, could produce electricity efficiently and economically using gasified biomass feedstocks. These new biomass electric developments could be demonstrated and proved cost-effective within the next 5 to 10 years.

Liquid fuels from cellulosic biomass may not be cost-effective for another 10 to 20 years, depending on oil prices. Methanol production from wood is already possible using a high-temperature gasification process, but it hasn't been attempted on a large scale. A nearly commercial-size biomass gasifier demonstration project, test-

ing wood and herbaceous crops, will begin construction in 1992.

The technology for commercially fermenting cane sugar and corn starch to make ethanol has long been practiced. However, processes for converting cellulosic biomass, only a part of which is sugar, to ethanol are still under development. The sugar portions (cellulose and hemicellulose) must first be separated and broken down into simple sugars before traditional fermentation processes will work. The most promising methods for separating and breaking down the components appear to be enzyme-catalyzed processes in which the enzymes are produced by yeasts or bacteria. These enzyme-catalyzed processes present tremendous opportunity for technology improvement through genetic engineering, which could make ethanol's market price competitive with that of existing fuels.

—J.H.C., L.L.W., K.S.

COURTESY OF OAK RIDGE NATIONAL LABORATORY

■ "Energy cane," bred not for sugar but for biomass, is a prime energy crop candidate in regions of the southern United States where it can be grown. Harvest here is by students from Auburn University who will carefully collect all biomass as a step in evaluating biomass productivity.

ers will have to grow the crops in ways that will optimize cellulose production rather than the nutrient content or seed production. This means, for instance, that thin-stemmed grasses will need to remain longer in the field before being harvested as opposed to grasses for standard forage crops. Because herbaceous energy crops can be interwoven into standard farming operations, the farmers' interest could come swiftly if energy markets were developed.

There are also other noncellulosic energy crops that farmers may choose to grow. For instance, rapeseed, whose oil (canola) is normally sold for human consumption, can also be used to replace diesel fuel. It is particularly valuable as a crop in the Southeast since some varieties can be grown over the winter on the same fields that support a summer food crop.

Future research

Energy crop researchers contin-ue to look for genetic factors that can improve total biomass productivity. Scientists are trying to determine if the solar collection efficiency of a stand of trees can be enhanced by altering the size and orientation of the tree leaves, or by changing the overall shape of individual trees. Evaluations of the structure and growth rate of roots are answering questions about the ability of different trees to optimally utilize soil nutrients. Investigations into how some trees survive better under drought conditions involve looking at the chemistry of the molecules that the trees manufacture. This research is providing information about tree biology that has implications far beyond energy issues. Similar research is also being done on herbaceous plants.

A major emphasis of genetic research on poplars will be to develop a large number of clones

■ A thick-stemmed, annual grass (corn) on the right and a thin-stemmed, perennial grass (switchgrass) on the left illustrate the wide variety of grasses that can be grown as energy crops. While fuel can presently be produced from the starch in corn kernels, recent research indicates that corn stalks may also be a valuable energy source.

COURTESY OF OAK RIDGE NATIONAL LABORATORY

Strides in Technology

Between 1980 and 1990, short-rotation woody crop investigators made some remarkable strides in developing new research techniques. Several short-rotation woody crop species were among the first trees to be propagated from tissue cultures in order to replicate the exact genetic makeup of superior trees.

Tissue culture, called micropropagation or microcloning, is important to the development of energy crops because it is a precursor of new methods of genetic improvement. In taking a tissue culture, a researcher extracts a piece of plant tissue—a half-inch stem, perhaps—and stimulates it to grow a new complete plant. All model short-rotation woody crops have been successfully tissue-cultured. And in a first for tree research, genetic screening field trials were established with tissue-cultured silver maple plantlets.

Hybrid poplar, often referred to as "the white rat tree," has been in the forefront of the emerging field of biotechnology. Even researchers who prefer to work on conifers have used hybrid poplars as a "model" species for developing new techniques.

Hybrid poplars have been regenerated from single cells by using culture methods developed for regenerating plants from tissues. The ability to regenerate plants from single cells has allowed scientists to test individual cells for possible genetic differences. For example, clumps of cells of poplar clones are being subjected to such treatments as exposure to toxic compounds and those surviving are regenerated into complete plants. After further testing of the young plants to determine which of them has truly achieved improved tolerance to the compounds, tissue culture techniques can be used to clone several thousand copies of the tolerant plants.

Genetic transformation, which involves the transfer of DNA, or genes, from one species to another, was also first achieved in trees with a hybrid poplar clone. Currently, at least three genetically transformed poplar genotypes are being grown in the lab or field to test whether the transferred genes remain active throughout the plant's growth.

The first complete gene map for a tree is expected to be completed in 1992 and again, it will be a hybrid poplar. This work is being cofunded by private industry, the Department of Energy, and the state of Washington. One of the products of the work will be the development of a method for DNA fingerprinting. "We want to take a leaf's fingerprints and be able to tell what clone it's from," says Chuck Wierman of Boise Cascade, a paper and wood products company cofunding the research. This is a very practical goal since it is easy to mix up clones in nursery operations, and planting the right clone can make the difference between a profit and a loss.

—J.H.C., L.L.W., K.S.

that have resistance to pests and diseases. This will be done both through traditional breeding procedures and by experimenting with the insertion of specific genes for insect and pathogen resistance.

A major thrust of future energy crop research will be to ensure that crop management systems will be environmentally benign and sustainable. By testing species mixes, researchers hope to discover which planting patterns reduce erosion and which combinations create favorable wildlife habitats. Short-rotation woody crops or perennial herbaceous energy crops may offer greater biological diversity in both plant and animal species than in most cropping systems, particularly if appropriately landscaped.

It is clear that if biofuels are to meet a significant portion of the nation's energy demand, large amounts of land will be needed to cultivate biofuel feedstocks. Preliminary analyses of current and future projections of overall land use trends indicate that sufficient land may be available in the United States to produce enough feedstock to supply about 20 percent of the nation's energy needs. These analyses need to be expanded and improved by performing test case studies for a number of specific locations around the United States.

Of course, the United States is not alone in its biofuel research efforts. Several European countries are examining biofuel production and conversion systems. Sweden has already made a national commitment to increase the use of biofuels to produce electricity instead of building nuclear facilities. Programs aimed at working directly with farmers are already under way in Sweden.

Developments in biotechnology, though presently unrelated to energy crops, may have a major impact on the use of energy crops of the future. As we know, plants can produce a wide range of chemicals in small amounts. If genes responsible for producing specific chemicals can be isolated and inserted into the fast-growing energy crops, then trees and

COURTESY OF OAK RIDGE NATIONAL LABORATORY

COURTESY OF OAK RIDGE NATIONAL LABORATORY

The lush growth of grasses and trees converting solar energy through photosynthesis to a form of chemical energy that lends itself to conversion to fuels holds great promise for the world's energy future.
■ *Above:* Switchgrass, a perennial that is harvested annually like hay.
■ *Right:* Hybrid poplars in their second year of growth.

grasses may become not only sources of biofuels, but also chemical manufacturing plants of the future. The possibilities offered by biological engineering open whole new realms of research and practical application.

TILTING TOWARD WINDMILLS

Long seen as a quixotic energy source, wind power is making a comeback, shattering myths in the process.

Jon G. McGowan

JON G. McGOWAN is a professor of mechanical engineering and co-director of the Renewable Energy Research Laboratory at the University of Massachusetts at Amherst. He has studied, designed, and participated in the siting of wind-turbine installations for the past two decades.

EMERGING from the shadow of an energy crisis in the 1970s, a wind-power industry flourished briefly in the United States. Part of an ambitious U.S. government program to support research and development on renewable energy sources, the Department of Energy (DOE) and the National Aeronautic and Space Agency (NASA) sponsored the construction of a wide variety of large wind turbines—most accompanied by exaggerated claims from promoters. Many believed this program would provide the spark for a global enterprise featuring a pollution-free, renewable source of energy. But by the early 1980s, with the exception of the California wind farms, U.S. interest in wind energy as a large-scale source of electricity almost disappeared.

What happened?

World oil prices dropped and the Reagan administration dramatically curtailed funding for renewable-energy research. Also, the initial work on large wind turbines yielded disappointing results—their design was almost always based upon ill-suited engineering codes developed for helicopters and other aerospace applications. These problems were compounded by overly optimistic economic projections, siting snags, and difficulties connecting wind-generated electricity to utility power grids. Because progress was stymied so early, the results gathered from the early wind projects never reflected the energy source's full potential. Wind power's kinks had yet to be worked out.

Even a separate effort in California in the 1980s—to develop wind farms employing large groups of smaller,

mass-produced, easy-to-install wind turbines—brought bad reviews. Often hastily erected as tax shelters to take advantage of federal and state tax credits, the earliest California wind farms were plagued by poor engineering design, leading to several cases of structural failure. Thus a significant market for small, backyard wind turbines never developed.

Despite initial problems, California's largest wind-farm developers and operators actively tried to improve their wind-turbine systems, and their persistence paid off. Today the California wind farms have a capacity of 1,500 megawatts—comparable to a large nuclear power plant—and produce about 1.5 percent of the entire state's electricity. Their peak production approaches 8 percent of the electricity generated by the major California utility that buys their power. All told, the state's three wind farms generate enough power to meet the residential needs of a city as large as San Francisco.

Nevertheless, the U.S. public's view of wind power remains distorted by its early failures. But technological progress over the past decade is prompting resurgent interest in wind energy, laying to rest lingering misconceptions that have hampered the technology's acceptance as a reliable and important source of electricity.

MYTH # 1

WIND POWER IS NOT A SIGNIFICANT ENERGY RESOURCE.

Most people find it hard to imagine that a steady breeze could ever provide enough power to generate significant amounts of electricity, at least compared with

WIND RESOURCE IN THE UNITED STATES

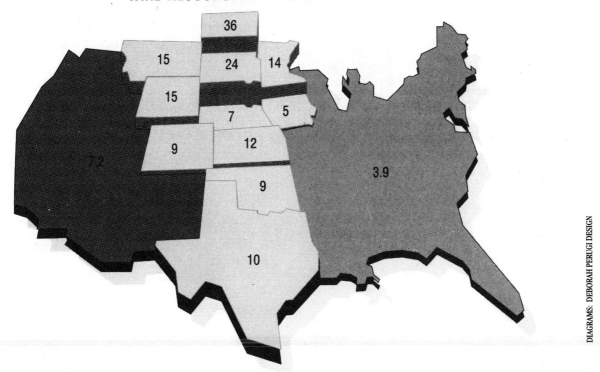

DIAGRAMS: DEBORAH PERUGI DESIGN

Like prototype race cars, large-scale, vertical-axis wind turbines never managed to economically capture the vast U.S. wind resource. Today's wind farms are poised to do better. The numbers on the map represent the percentage of 1990 electricity demand in the continental U.S. that each state or region could theoretically meet using today's wind technology. North Dakota alone, for example, could supply 36 percent of the electricity needs of the lower 48 states.

a large nuclear power plant or other massive generating station. To be sure, wind power is dispersed, but a huge wind resource is there to be captured.

According to recent studies of wind energy feasibility, some 14 states—especially those in the nation's midsection, from Texas to North Dakota—possess a wind resource at least as great as that found in California. In fact, if wind farms were built today on all the available sites in North Dakota, that state alone could supply more than a third of all the electricity consumed in the continental United States. (Such an assessment assumes that land used for urban and residential areas, parks, and wetlands as well as a substantial fraction of forest and agricultural lands are exempt from consideration.) A study by the Battelle Pacific Northwest Laboratory estimates that today's turbine technology could supply 20 percent of the country's electrical needs, even if the wind resource were exploited only in locations ranked class 5 or higher—where wind speed averages at least 16 mph at a height of 30 meters. If class 3 areas could be economically developed (where the average wind speed is 14 mph at a height of 50 meters), approximately 13 percent of the contiguous U.S. land area could be eligi-

ble for wind turbines—enough to supply roughly four times the nation's current electricity consumption.

Even more helpful, though, are examples of wind power's local potential. A recent study supported by Minnesota's energy commission determined that the wind resource at just one of the state's southwestern ridges could supply almost all of Minnesota's electricity.

It is little wonder, then, that the Electric Power Research Institute (EPRI) reports that numerous electric utilities have begun to launch wind-farm projects within the past year. For example, Niagara Mohawk, a New York utility strongly committed to nuclear power just a decade ago, has joined a consortium developing the next generation of wind machines and has already installed New York State's first utility-scale wind turbines near Lake Ontario, replete with advertisements dubbing them the "Niagara Mohawk Air Force."

Even populous regions—in the northeastern United States, for example—need not write off wind power's potential. In the early 1970s, researchers from the University of Massachusetts concluded that it would be possible to produce roughly 650 billion kilowatt-hours

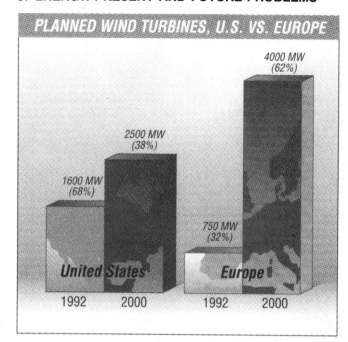

PLANNED WIND TURBINES, U.S. VS. EUROPE

4000 MW
(62%)

2500 MW
(38%)

1600 MW
(68%)

750 MW
(32%)

United States   Europe

1992   2000   1992   2000

As of 1992, more than two-thirds of the world's wind turbines could be found in the United States. But European countries are scheduled to install some 3,250 megawatts of new capacity—more than four times the amount on line in the European Community today.

WIND-GENERATED ELECTRICITY IS EXPENSIVE AND UNRELIABLE.

MYTH # 2

The persistent misperception that wind power is too costly results largely from early NASA/DOE research focusing on turbines that featured vast blade diameters, stood hundreds of feet tall, and were rated in megawatt sizes. For example, the NASA/DOE program produced four series of machines ranging from a 0.1-megawatt model with a blade diameter of 38 meters to a 3.2-megawatt model with a blade diameter of 98 meters. In many respects, these machines were technical successes, but they were akin to high-speed, prototype race cars: they were not designed for ease of production or maintenance and they were enormously expensive. Because the major factors influencing the delivered cost of wind power are the cost of the turbine and supporting systems (including land), as well as operating and maintenance costs, it is no great surprise that the program failed to yield a system that could supply electricity to utilities at a competitive price.

Yet the modern efforts epitomized by the California wind farms have dramatically changed the economic picture for wind energy. These systems, like other such installations in Hawaii and several European countries, have achieved economies of scale through standardized manufacturing and purchasing. The efforts have led to a dramatic drop in the installed cost of new wind turbines from roughly $4,000 per kilowatt in 1980 to approximately $1,000 per kilowatt today (in constant dollars).

each year—about 22 percent of today's total U.S. electrical energy use—from offshore wind systems. Like so many wind studies of that era, the results were largely ignored. But recent studies sponsored by the European Community are prompting researchers and utilities alike to think again about wind power's potential in crowded areas. The EC studies estimate that at least 10 percent of Europe's electrical power could be supplied by land-based wind turbines using current turbine technology. Spread over the densely populated continent, the total land requirements would be no greater than the size of the island of Crete. Furthermore, most of this land could still be used for agriculture, with crops growing beneath and between the elevated turbines.

Several European countries and developers have recently turned their attention as well to exploiting the continent's enormous offshore wind resource. While offshore wind farms are comparatively inaccessible and costly, they have the advantage of avoiding siting problems in populous areas and benefiting from average wind speeds that are higher than those found among their land-based counterparts. Two prototype offshore turbine installations now operating are designed to investigate the problems encountered in constructing and operating wind farms in a hostile ocean environment. One system, two kilometers off the coast of Denmark in water depths varying from 2.5 to 5 meters, features nearly a dozen 450-kilowatt turbines.

Improvements in machine design and efficient maintenance programs for large numbers of turbines have similarly reduced operating costs. The cost of electricity delivered by wind-farm turbines has thus decreased from about 30 cents per kilowatt-hour to roughly 7-9 cents per kilowatt-hour—generally less than the delivered cost of electricity supplied by conventional power plants. Even lower costs are reported at wind-farm sites with above-average wind resources. The California Energy Commission estimates that investor-owned wind-power plants now generate electricity at a rate ranging from 4.7 to 7.2 cents per kilowatt-hour.

According to recent studies by the National Renewable Energy Laboratory as well as by EPRI, numerous U.S. wind-farm sites can now be expected to deliver electricity at a cost of roughly 5 cents per kilowatt-hour, ranking wind as one of the least costly sources of new electric power in many locations. And the benefits of full-scale mass production have yet to be reached; capital costs are likely to fall further as the industry develops. Meanwhile, the reliability of wind turbines has shown similar dramatic improvement. Recently installed turbines are available to generate electricity over 95 percent of the time, compared with roughly 60 percent in the early 1980s.

NEW AND IMPROVED MACHINE DESIGNS ARE NEEDED TO MAKE WIND POWER FEASIBLE.

MYTH # 3

Despite the above-mentioned progress, many people believe that the future of wind energy will depend on the development of improved turbines that will look very different from today's designs. Until recently, that argument was used to delay the installation of available wind turbines. The fact is, however, that today's state-of-the-art wind turbines are already highly efficient. They will surely improve incrementally over time, benefiting from new materials and other innovations. But, like the bicycle, today's wind turbines represent a simplicity of design that is already well developed and likely to last for a long time.

Today's conventional utility-scale wind machines have propellor-like blades that rotate on a horizontal axis, perpendicular to the wind stream. The number of blades may vary (usually two or three), as can the position of the rotor (upwind or downwind), but despite numerous other designs proposed or built by wind turbine inventors or developers, the propellor-blade design has emerged as the predominant type in the more than 20,000 utility-scale wind turbines now in operation worldwide.

As emphasized in a 1991 study by the National Academy of Sciences, this configuration is based on detailed analytical and computational models as well as an extensive experimental database. Stresses and materials have been studied in great depth, as have the effects of wind shear and turbulence. And the technology has been tested in the field for more than a decade, providing billions of kilowatt-hours of electricity to consumers.

Today's utility-scale turbines, like the gas-driven turbines that power jet airplanes, represent a sophisticated piece of rotating machinery. There is no reason to believe that other configurations such as vertical-axis machines will produce major improvements because the efficiency of the horizontal machines' ability to convert wind power to electricity is already so high. Despite this advanced state of development, however, the design of rotor blades can still be streamlined and, with the advent of stronger and lighter materials such as carbon-based composites, taller and larger-diameter wind machines (which could generate more electricity per machine) can surely be developed.

Advances in power electronics, especially the widespread incorporation of a variable-speed design, could also boost efficiency. This innovation, currently under development in the United States and Europe,

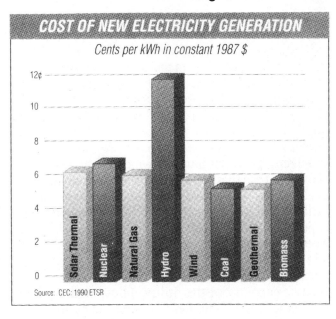

Wind power is one of the least expensive sources of new generating capacity, yielding electricity that costs half that produced by new hydropower plants. Oil-fired plants, not represented above, would be even more costly and are rarely sought for baseload power today.

uses state-of-the-art electronics to generate a constant-frequency AC power despite varying wind (and turbine rotor) speed. If the designs prove as commercially successful as expected, they will increase energy productivity (and hence lower energy-production costs) by allowing turbines to usefully capture electricity at lower and fast-changing wind speeds.

THE TECHNOLOGY IS IMPRACTICAL FOR USE BY UTILITIES BECAUSE OF PROBLEMS CONNECTING WIND MACHINES TO THE ELECTRICITY GRID, AND BECAUSE WIND ITSELF IS INTERMITTENT.

MYTH # 4

There is nothing about the hardware required to connect wind turbines with the utility grid that differs markedly from that used by conventional power plants. Early protoype wind farms did have some transmission and interface problems. The power generated by the individual turbines was sometimes difficult to synchronize, causing difficulty in tightly controlling the current delivered to the grid. But advances in power electronics and wind-turbine control systems have virtually eliminated these problems.

Variability of its power sources can be a problem for a utility, however, and wind power is a variable source of electricity. But this just means that utilities must incorporate this feature of wind energy into their planning. Utilities deal with short-term variations in power input (over seconds to minutes) the same way they deal with short-term variations in energy demand: by exploiting the ability of the electricity grid itself to absorb some fluctuation in the amount of current it carries at any given moment. Supply and demand can never be matched perfectly; the key for a utility is simply to make sure enough current is flowing so that the next application can be met.

Numerous industry studies have also shown that the short-term power output of a wind farm varies significantly less than that of a single turbine, so if anything the problem has diminished. Still, the effect on an electric utility of all variations in the output of a wind farm depends on the so-called penetration level of wind-generated electricity on the grid. Today, wind energy generally has a small effect on utility operations because it normally represents only a tiny fraction of the total electric capacity of the utility. Thus, medium-term fluctuations (over minutes to hours) can be addressed simply by adjusting the amount of electricity produced by other sources.

In an important assessment, England's Central Electricity Generating Board concluded recently that as much as 20 percent of the country's electricity could be produced from wind power without major modifications to the present grid. At penetration levels much above 20 percent, however, it is widely believed that wind-generated electricity would lose some of its economic benefits. Energy storage systems like batteries or pumped hydro would likely be required to buffer the variability of the energy source.

To provide electricity when the utility needs it, utility planners must determine how much of a generating plant's theoretical capacity it can reliably be expected to provide (whatever the source). Here again, wind power presents few problems that differ substantially from other sources of electricity. Wind farms' so-called capacity factor can be determined by the predictability of the winds at the site, then correlated with the utility's needs.

To Compete with Europe

The reality is that wind-produced electricity is now less expensive than electricity produced by conventional fossil- or nuclear-powered generating plants in many parts of the world. And unlike some of the proposed renewable electric-power sources like photovoltaics, wind power's future is not dependent on further breakthroughs in engineering or materials technology. What's more, many energy experts argue that energy sources like wind power should be given credit for avoiding the added costs to society—such as the price of controlling

PHOTO: MIT MUSEUM COLLECTION

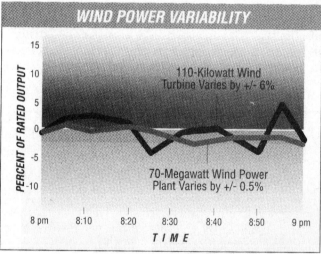

The output of individual wind turbines fluctuates significantly. Groups or "farms" tend to smooth out such variations.

air pollution, managing nuclear waste, or avoiding global warming—associated with conventional electrical-energy production.

On the other hand, the siting requirements of wind farms have caused numerous regulatory and public interest groups to take a hard look at their adverse environmental effects. U.S. and European studies have identified a host of potential problems, ranging from telecommunications interference to impacts on wildlife and natural habitats. However, when these effects are balanced against those associated with other forms of electricity generation, it is clear there are no major barriers to large-scale implementation of wind power.

While wind power has been comparatively ignored in the United States—in no small part because of the persistent myths discussed above—the center of gravity of the wind industry and expertise has shifted from the United States to Europe over the past few years. For proof, one has only to compare the attendance or technical content of a U.S. vs. European wind-energy conference, look at the national source of utility-scale wind turbines, or note the level of wind-farm development in many areas of Europe. The interest and enthusiasm for wind power in Europe is much higher—not surprising in light of the fact that the European Community is presently spending about 10 times more for wind-energy R&D than the U.S. government.

During the 1980s, progress in European wind energy proceeded at a steady pace, with increasing commitments from several countries for R&D and market incentives. As a result of this European Community program, member nations have over 25 wind-turbine manufacturers offering utility-scale wind machines. For the past two years, European wind installations have exceeded those in the United States. And according to the American Wind Energy Association, Denmark, England, Spain, and the Netherlands will each surpass the United States in new installations of wind-turbine capacity during the rest of the decade. The goal of the European Community is to install more than 3,000 megawatts of new wind energy by the year 2000, and there is every indication that this goal will be met or exceeded. Meanwhile, U.S.utilities expect to add a total of roughly 900 megawatts of new wind energy over the same time period.

In many ways, the European experience is instructive for U.S. energy planners. The market stimulation carried out by the European countries, acting cooperatively through subsidies, production incentives, and tax credits, has contributed to the growth of wind-energy development. Even though the huge wind resource in many parts of the United States is now well documented, and small, independent power producers have shown interest in using wind to contribute to the nation's electricity grid, it is clear that the widespread development of wind energy cannot succeed piecemeal. Farsighted energy policies are first needed to help guide the efforts of utilities, various zoning authorities, and state and federal agencies.

PHOTOS: NASA

Much of our knowledge—and many lingering myths—about generating electricity from wind derive from experience with large turbines like these, sponsored by the U.S. Energy Department and NASA in the 1970s and 1980s.

3. ENERGY: PRESENT AND FUTURE PROBLEMS

On the federal level, the U.S. government should institute a production incentive for wind-generated electricity to help develop the industry. Such an incentive could take the form of a tax credit to developers or utilities for every kilowatt of new, renewable-energy generating capacity they build. At the state level, regulatory reforms are needed to overcome lingering barriers to renewable energy: utilities must be encouraged to incorporate the external costs of power generation (such as pollution) into their decision making and pricing. Meanwhile, states can offer tax incentives and other support to small manufacturers of wind-energy components to help remove the biases in the energy industry that favor nonrenewable sources and large manufacturers.

Developers of wind turbines need to continue to improve the cost-effectiveness of new machines, and engineers will require tools such as computer-aided design to better understand machine fatigue and incorporate the use of new, lighter, and stronger carbon-based composite materials. Because some of this work is likely to be beyond the capability of small firms, government support will be essential. Such support could also provide a powerful stimulus to diverse industries. For example, it could open a gold mine of opportunity for a wide range of engineering disciplines, including mechanical, materials, aeronautical, electrical, and civil, with consequent payoffs not confined to wind power alone.

Although countries in Europe and elsewhere around the world have already begun to awaken to the potential of wind power, the United States is fortunate to have one of the world's largest resources of wind energy. And the world's major supplier of wind turbines and developer of wind farms continues to be a California-based firm called U.S. Windpower, currently operating more than 4,200 wind turbines and involved in manufacturing and developing wind farms at a number of European sites. But one firm cannot provide an industrial base for a growing field. With a coordinated effort, guided by an enlightened government energy policy, it is not too late for the United States to resume its role as the world's largest developer of this resource, and to regain a leading role in wind-energy technology and manufacturing.

ALL THE COAL IN CHINA

Unless this giant nation embraces a new strategy for producing and using energy, its fast-growing economy could overwhelm international efforts to control greenhouse warming.

NICHOLAS LENSSEN

Nicholas Lenssen is a senior researcher and the author of Worldwatch Paper 111, Empowering Development: The New Energy Equation.

The mushroom cloud, for four decades a haunting and omnipresent symbol of the greatest threat ever faced by humanity, has finally begun to fade from the global consciousness. Today's generation of children may be the first since World War II to grow up not instantly recognizing—and shivering at—its apocalyptic shape.

But even as it fades, it is being replaced by another kind of cloud—one that is all but invisible. And the new threat is not a symbol, but a physical reality: the steadily building cloud of greenhouse gases that scientists say is likely to lead to a massive disruption of the earth's climatic system, making the planet warmer than at any time in the last million years. Just how hot the planet ends up getting will be partly determined by the actions of one country—China.

In the popular literature of global warming, China has received only passing mention. After all, since 1950, it has been the industrial countries of the West and former Soviet bloc that were responsible for 79 percent of the fossil fuel-derived emissions of carbon dioxide, the leading greenhouse gas. But in the future those proportions are expected to shift dramatically. Future growth is expected to come more from China and

other developing countries than from all the industrial nations combined.

Carbon emissions in China have increased 65 percent in the past decade, largely due to a sharp rise in its burning of coal. This boosted the country's share of global carbon emissions to 11 percent—still less than its share of global population. And China still emits only half as much carbon as the United States, and only one-ninth as much per capita. But this is already more than the total amount generated by Russia, and in China the real boom may be just beginning.

China's fast-growing economy is what's driving its carbon dioxide emissions upward. Bolstered by economic reforms, the country's economy grew by nearly 10 percent annually over the last decade and, after a government-induced slowdown late in the decade, is back in high gear. Analysts forecast that high growth will continue far into the next century.

Unfortunately, all this growth could have dire consequences for the planet's atmosphere. Indeed, China is forecast to emit more carbon dioxide by 2025 than the current combined total of the United States, Japan, and Canada, according to a United Nations panel studying climate change. It is now apparent that redirecting China's energy economy may be as important to the global atmosphere as changing those of the United States and Europe.

China's leaders don't share this view. They readily dismiss the notion that concern over global warming should alter their energy strategy. Besides, they argue, since carbon dioxide remains in the atmosphere for well over 100 years, it is largely the industrial West that caused the accumulation now hanging over us—through decades of burning coal in Manchester, or oil in Los Angeles. The Chinese government says the country has little to do with the 27 percent increase in atmospheric concentrations of carbon dioxide that has taken place since the dawning of the industrial revolution. Environmentalists say this argument is specious, however, since any increase will take the world further away from the 80 percent *reduction* scientists believe necessary to stabilize the climate.

While they don't seem particularly worried about global warming, China's leaders are thinking hard about what it will take to provide adequate living standards for the 1.1 billion people who live within the country's borders. If increased carbon dioxide emissions are what it takes to meet human needs—so goes the argument—then increase they must.

The salient question, though, is not whether the Chinese have the right to follow a carbon-intensive energy path, but whether it's in their interest to do so. The promise of development based on heavy industry, often fueled by coal, is a mirage that was pursued with a notable lack of long-term success by the Soviet bloc countries in the past—and is still being pursued by China. Industrial facilities in China, like those in the former Eastern Bloc, are less productive than factories in Europe or Japan. They also use more energy and emit far more pollution.

Rather than attempt to resolve a false dilemma between economic growth and the environment, China would better serve its own interests—and thereby the world's—by simultaneously improving living standards and reducing its growth in carbon dioxide emissions. Indeed, the country may really have no choice but to do this. Fortunately, some younger planners are acutely aware of this imperative.

Economic Boom, Energy Brake

In environmental and energy circles, China is notorious for its heavy dependence on coal, which generates 76 percent of all its energy. Only Poland and South Africa rely so heavily on this highly polluting and inherently inefficient fuel—a fuel selected not through economic competition, but through centralized government planning. In fact, the energy production system in China is still largely based on the Stalinist model of production quotas, enormous government investments, and subsidized prices, which result in gross economic inefficiencies.

Low energy prices have long been the keystones to centrally planned economies, including China's. In 1987, according to the World Bank, the Chinese government directly subsidized energy prices by $17 billion, and many of the subsidies appear to be continuing. Most oil in China, for example, was sold for slightly more than $5 a barrel in 1992, far below the international price of roughly $18 a barrel. And coal marketed by state-owned mines, which produce nearly half of the country's total, was priced below the cost of extraction—costing the treasury some $2 billion in 1991.

Low prices encourage—or at least fail to discourage—wasteful use of limited supplies. In fact, just one-third of all fuel burned in China ends up as useful energy, according to Vaclav Smil, a geographer at the University of Manitoba. That's far below the 50 to 60 percent levels found in the United States and Japan. The inefficiencies further exacerbate shortfalls in energy supplies, since larger supplies are needed to accomplish the same amount of work. At last report in 1987, a shortage of coal had led to electrical power deficits that idled one-fourth of the country's factories.

The Chinese government has typically responded to energy shortages by pouring scarce capital into building more mines, power plants, and oil wells. More than half of the industrial capital expenditures in state-owned enterprises in 1989 went to energy production. Yet despite this Brobdingnagian investment, energy shortages are expected to ease little if at all over the next two decades, as efforts to boost energy production face daunting obstacles.

China already produces more energy than Saudi Arabia and is the world's largest producer of coal—accounting for 25 percent of global output. Yet efforts to further boost coal production face imposing hurdles. Coal already accounts for more than 40 percent of the country's railway shipments by weight. Hauling endless trainloads of coal from mines in the north to the eastern economic heartland has led to transport gridlock and supply shortfalls. On the other hand, the obvious alternative—burning coal at power plants near the mines, and transmitting energy by wire instead of by train—has been stymied by shortages of water (needed for coal-fueled power production) in the coal-rich but arid northern and northwestern provinces.

The government hopes to reduce coal's importance for electricity generation by promoting large hydroelectric dams and nuclear power, but those too face formidable obstacles. A decade ago, China planned to have 10 nuclear power plants running by the turn of the century, but nuclear power is now estimated to be four times as costly as coal-generated power—cutting the likely number of nuclear plants back to the three now nearing commercial operation. For the time being, it is unclear how many more plants China is likely to complete.

Hydropower holds more promise, especially since less than 10 percent of the country's potential has been tapped. Cracks have begun to appear, however, in the consensus for building dams. In April 1992, the Chinese People's Congress, usually a rubber-stamp body, finally approved the Three Gorges Dam project—first proposed for the Yangtze River in the 1920s. But nearly a third of the delegates abstained or voted against the project, largely due to its cost—as high as $100 billion—and its plan to resettle more than 1 million people by force if necessary.

Efforts to boost oil production, which accounts for 17 percent of the country's commercial energy use, continue to face difficulties as well. Despite large government expenditures, guided by the expertise of dozens of foreign oil companies, China's oil output has failed to keep pace with its consumption since the mid-1980s. One result has been a 70-percent decline in net proceeds from international oil sales since 1985. Within another three years, China could find itself a net oil importer.

By now, it is clear that simply pouring money into expanding conventional energy supplies—even if such money were available to pour—would be unlikely to solve China's energy dilemma. Even if its central planners somehow managed to patch together an adequate supply of coal, there is a whole host of reasons beyond the economic for not burning it in such immense quantities. China's people, cropland, forests, and waterways simply could not survive the air pollution that would result.

In Beijing and other major cities, sulfur dioxide concentrations—mainly from coal—regularly violate international guidelines. The level of suspended particles in Chinese cities—also from coal—is 14 times that in the United States. Acid rain falls on at least 14 percent of the country and is spreading, not just through China but into Japan and South Korea, damaging forests, crops, and water ecosystems.

Global warming portends other serious problems, according to a 1992 study by the Chinese State Meteorological Administration (SMA) and the World Wide Fund for Nature. A warmer atmosphere would raise sea levels, which in turn would flood China's coastal regions, especially in river deltas, where rice production is concentrated. Rice production would also fall if the predicted reduction in available water due to warmer temperatures holds true. Indeed, Luo Jibin, deputy director of SMA, asserted last year that global warming was already exacerbating a severe drought in the northern part of the country. The environmental and health costs of China's energy production, particularly with so much of it driven by coal, are placing a mounting burden on the Chinese people—and on the world.

The Efficiency Revolution

Raising energy efficiency is something China can't afford *not* to do, especially if it hopes to continue boosting living standards for its citizens. That may be true of other countries too, but it is particularly true in China. It takes far more energy—and pollution—to produce goods or services in China than elsewhere. A concerted move toward efficiency would not only reduce exorbitant costs but would lead to greater employment—a major benefit in an economy that is labor-rich and capital-poor. And of course, an efficient energy system would slow the increase in carbon emissions.

More than two-thirds of China's commercial energy consumption goes to industry. Yet for each ton of steel or cement produced, the typical factory in China uses from 7 to 75 percent more energy than its Western counterpart (see table). China's 300,000 small industrial boilers operate at only 55 to 60 percent efficiency, as compared to an 80-per-

Table 1. Additional Energy Requirements of Energy-Intensive Products in China

Product	Difference[1] (%)
Steel	70
Electrolytic aluminum	7
Synthetic ammonia	75
Caustic soda	39
Cement	63
Plate glass	46
Paper	21

[1]Additional amount of energy Chinese industry uses to produce the same amount of product as industrialized countries.

Source: Wu Zongxin and Wei Zhihong, "Policies to Promote Energy Conservation in China," Energy Policy, December 1991.

cent average for the 24 member countries of the Organization for Economic Cooperation and Development, according to the World Bank.

The lower efficiency of the Chinese plants, which results partly from poor maintenance and operating procedures, can be readily improved given sufficient information and incentives—including market prices for energy. Efficiency is further impaired by reliance on old or obsolete industrial processes—often purchased at bargain prices from industrial countries.

China has the opportunity to base future economic growth not only on more efficient industrial processes, but on more efficient products as well. Building a $7.5-million compact fluorescent light bulb factory, for example, would eliminate the need to build $4.9 *billion* worth of coal-fired power plants and transmission facilities, if the bulbs (which need 75 percent less power than incandescent ones) were used domestically. Each dollar invested in efficiency would save $650 in capital expenditures—before the savings on energy use even begin. China's factories already have the capacity to produce 10 million compact fluorescent bulbs a year, but this is still only a small portion of the 2.8 billion light bulbs it produced in 1991.

Likewise, Chinese buildings use three times as much energy for heating as U.S. buildings, even though their buildings are still colder. By making boiler improvements and using insulation and double-glazed windows, the Chinese could raise average winter building temperatures in the northern part of the country from 52 degrees to 64 degrees Fahrenheit—and consume 40 percent less coal. In Harbin, for example, such improvements could pay for themselves in six and a half years even with energy costs artifi-

cially low because of subsidized coal, according to Yu Joe Huang, a scientist at the U.S. government's Lawrence Berkeley Laboratory in Berkeley, California. With unsubsidized coal, the payback period would be quicker—around four years.

In combination, the efficiency potential now within reach for industry, buildings, agriculture, and transportation could provide an enormous boost to China's economy. By investing $3 billion a year, the country could cut future growth of its energy demand by nearly half, eliminating the need for $16 billion worth of new power plants, oil refineries, and other energy infrastructure each year, according to a study prepared by the U.S. Working Group on Energy Efficiency. And instead of pumping 1.4 billion tons of carbon into the atmosphere in 2025, it would emit less than 1 billion tons.

Wide-scale savings are not simply paper prophecies—nor are they entirely unknown to Chinese planners. In 1980, the government launched an ambitious efficiency program to improve energy use in major industries. By directing about 10 percent of its energy investment to efficiency, the nation cut its annual growth in overall energy use from 7 percent to 4 percent by 1985, without slowing growth in industrial production, according to energy analyst Mark Levine and his colleagues at Lawrence Berkeley Laboratory.

Levine discovered that efficiency improvements accounted for more than 70 percent of the energy savings during the 1980-1985 period, with shifts toward less energy-intensive industries yielding the remainder. And efficiency gains were found to be one-third less expensive than comparable investments in coal supplies. One result was that China's energy consumption expanded at less than half the rate of economic growth from 1980 through 1988. Had the nation failed to make such progress, either energy consumption in 1990 would have been 50 percent higher than it actually was or—more likely—economic output would have grown far more slowly, as China would have been unable to import the $80 billion of energy the difference represented (see figure).

Unfortunately, since the mid-1980s, China has poured money into expanding its energy supply, thereby *reducing* its spending on efficiency from 10 percent to just 6 percent of total investment in the energy sector. That could soon change, however, since

Figure 1. Energy Consumption in China,
Actual and with Constant Energy Intensity, 1950-1990[1]

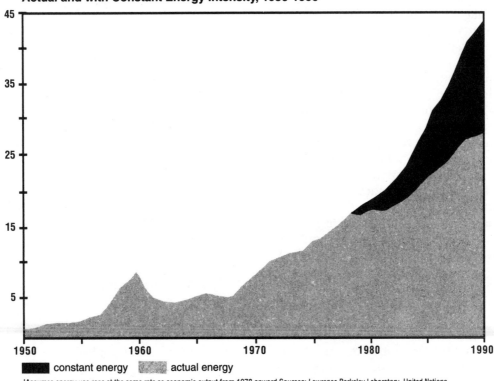

constant energy actual energy

[1]Assumes energy use rose at the same rate as economic output from 1978 onward. Sources: Lawrence Berkeley Laboratory, United Nations

China's energy and economic planners are beginning to recognize efficiency's past success in facilitating strong economic development. There may be, too, a growing recognition among these planners that exacerbating global warming in the interests of short-term economic gains would not be in the country's long-term interests.

Alternatives to Coal

If China proceeds to maximize its efficient use of energy, it will greatly reduce—but not eliminate—the need for increased energy supplies as its overall demand more than doubles over the next 35 years. Unfortunately, government planners and international lending institutions still assume that China has to follow the energy path the West blazed a century ago—a strategy that relies primarily on expanding supplies of coal and oil to meet people's needs.

Over the long haul, though, China will need to develop its own alternatives to polluting coal and costly oil, partly to alleviate supply problems, but also to reduce air pollution. The country evidently has extensive, unexploited reserves of natural gas, which could supplant oil and coal use in buildings, transport, industry, and power generation,

while emitting about 60 percent less carbon dioxide. And renewable resources, other than large hydroelectric dams, are increasingly viable today, even before factoring in what they save by not polluting.

Natural gas accounts for only 2 percent of China's current energy use, though the country has started to reconsider gas as part of its effort to slow the growth in oil and coal use. In 1986, the government formed a gas research institute, and in early 1992 it decided to build a pipeline from a large offshore gas field that had been discovered in the South China Sea during an unsuccessful search for oil years earlier.

Gas commonly accompanies not only oil but also coal, suggesting that China, with its almost limitless coal reserves, is well endowed with natural gas too. One multiagency Chinese group estimates the country's gas resource to be about half as big as its enormous proven coal reserves (more than 100 years' supply). In one region in north-central China, every well drilled in the first five months of 1991 struck natural gas.

China could also look to import natural gas, even swapping gas for pipeline rights-of-way being planned by its neighbors. Russia and South Korea signed an agreement last

113

November to study the possibility of building a gas pipeline from Siberia to the Korean peninsula—a project that could cut across northern China. And Japanese engineers have drawn up plans for pipelines criss-crossing China to move gas from Turkmenistan, Indonesia, and Australia to Japan.

Beyond the increased use of natural gas, China has an enormous potential to draw on solar, wind, biomass, and geothermal energy resources. Western countries are increasingly pursuing these options as technological advances, and cost reductions make them more attractive. The opportunities are even greater in China, because so much of its energy infrastructure has yet to be built. Decisions the country makes today will determine how readily it can tap these resources in the future.

The country has already tapped many renewable sources in rural parts of the country. For example, small hydroelectric generators supply roughly half of the electricity used in rural areas. Electrical output from these small dams nearly doubled between 1979 and 1988. Also, more than 110,000 small wind turbines churn away, mainly in Inner Mongolia.

China also appears poised to electrify at least a portion of its rural households with photovoltaic technologies. More than 60,000 households in other developing nations receive their electricity this way. The government of Gansu Province plans to electrify 1,500 homes with domestically manufactured systems in a project being organized by the Washington, D.C.-based Solar Electric Light Fund.

During the past decade, the cost of solar and wind electricity systems has fallen 66 and 90 percent respectively, and these renewables are emerging as the least expensive route to electricity in some developing countries. China recently announced a program to install 200 megawatts of wind turbines over the next three years, and 1,000 megawatts by the turn of the century.

The final step for China would be to use wind and solar power plants to produce hydrogen in the generally sunny and windy areas of the vast central and northwestern desert regions, then pipe it to the populated eastern coast. Larger investments in renewables in the next few years would stimulate development of the technology and business infrastructure—and the in-country expertise—needed to replace fossil fuels on a large scale.

Policies for Clean Energy

China has relied on market-based reforms for much of the boost in its economy since 1978. Unfortunately, the energy sector has been lagging. In getting its energy industry up to speed, it will need to adopt some of the innovative policies that other countries have used to encourage both more efficient energy use and more serious reliance on non-polluting renewables. The logical starting point for these policies is the adoption of real market prices.

Energy prices are critical because if they are too low, there is little incentive for the consumer to use energy efficiently or for industry to invest in alternative supplies. China has already made a promising move on this front, having planned to triple the price of oil sold domestically in January. The key now is to raise prices on coal and electricity as well.

Experience in industrial countries, however, shows that higher energy prices eliminate only one of the many obstacles to reducing wasteful energy use. To work, substantial price hikes may need to be accompanied by investments in end-use efficiency. That way the final cost of the energy service—whether manufacturing cement or lighting a home—remains unchanged or declines. For example, subsidizing the use of efficient lighting technologies would cost less than providing below-cost electricity to households and building new power plants.

As China learned in the 1980s, investing in energy efficiency can yield rewards that the country can hardly afford to miss. The first challenge is to once again boost efficiency's share of the total energy investment to levels above the 10 percent that brought the earlier gains. And such investments should no longer be limited just to major industries, but expanded to transportation, buildings, and consumer products as well.

Likewise, the country could profitably begin to shift the money it now spends obtaining coal and oil to the pursuit of natural gas and renewables. As part of a new environmental policy announced in 1992, the government said it plans to encourage renewable energy resources, including solar, wind, geothermal, tidal, and biomass—though it left its plans unclear.

The Twain Shall Meet

China is likely to find that building a sustainable energy system won't be possible without international support. China already receives a good sum of aid for energy development, only that money is probably doing as much harm as good, not just for the global climate but for China's long-term economic health. The World Bank, for example, loaned China nearly $1 billion for energy projects in 1992, yet nearly all of that money went to coal or large hydroelectric projects. That's not surprising, in view of the fact that 80 percent of the bank's energy loans around the world since 1948 have been aimed at supplying electric power. But now, for China to make the changes it needs, the lending agencies will need to shift their focus as well.

There are promising signs that this will happen. In Manila, the Asian Development Bank (ADB) has started to invest directly in efficiency. Last September, the ADB approved a $107 million loan for improving energy efficiency in Chinese fertilizer, cement, and steel industries. The ADB also has started to incorporate into its energy planning a novel technique—known as integrated resource management—that compares efficiency investments to expansion of energy supplies in determining the least expensive, and least environmentally destructive, energy option.

Investment in education is another neglected area that aid programs can bolster. One successful program, run by the Lawrence Berkeley Laboratory, has trained a cadre of energy specialists in critical efficiency and environmental issues. More than a dozen Chinese scientists, engineers, and other specialists have conducted research at Berkeley over the past four years, while scores more, including the director of China's Global Climate Change program, have participated in study tours and joint conferences.

Taken one step further, establishing an energy efficiency center in China would allow even more Chinese citizens to tap into the country's expanding expertise and knowledge. Based on past success in launching three such centers in the former Czechoslovakia, Poland, and Russia, William Chandler of Battelle Pacific Northwest Laboratories has now set his sights on China. "Such a center could help cut carbon emissions by half a billion tons a year over 20 years," says Chandler.

It is critical, too, that support from the wealthier industrial nations include not only the hundreds of millions of dollars annually provided as foreign assistance, but the power of example as well. That has already occurred to some degree in the development of more energy-efficient industrial processes and consumer products, but it will become far more persuasive when it includes more substantial shifts to renewable, non-polluting energy sources, compliance with stringent targets to reduce carbon emissions, and the pursuit of less energy-intensive lifestyles.

Pollution: The Hazards of Growth

Of all the massive technological changes that have combined to create our modern industrial society, perhaps none has been as significant for the environment as the chemical revolution. Although the chemical industry reminds us that life itself is not possible without chemistry, the fact remains that the largest single threat to environmental stability is the proliferation of chemical compounds for a nearly infinite variety of purposes, including the universal use of organic chemicals (fossil fuels) as the prime source of the world's energy systems. The problem is not just that literally thousands of new chemical compounds are being discovered or created each year, but that the long-term effects of these compounds in an environmental system are often not known until an environmental disaster involving humans or other living organisms occurs. The problem is exacerbated by the time lag that exists between the recognition of potentially harmful chemical contamination and the cleanup activities that are ultimately required.

A critical part of the process of dealing with chemical pollutants is the identification of toxic and hazardous materials, a problem that is intensified by the myriad ways in which a vast number of chemicals can enter environmental systems. Governmental legislation and controls are an important part of the correction of the damages produced by toxic and hazardous materials such as DDT, PCBs, or CFCs in the environment or in the limitations on fossil fuel burning. Unfortunately, as evidenced by most of the articles in this unit, we are losing the battle against these harmful chemicals regardless of legislation, and chemical pollution of the environment is probably getting worse rather than better.

The first three articles in this section deal with specific components of the pollution problem—nuclear waste, agricultural chemicals, and municipal solid wastes. Nicholas Lenssen, a research associate with the Worldwatch Institute, discusses the difficult question of nuclear waste disposal in "Facing Up to Nuclear Waste." Lenssen notes that recent events such as the breakup of the Soviet Union's military machine and the development of a new safety emphasis on nuclear power generation by the U.S. Nuclear Regulatory Committee have created a belief among some that the atomic-age fears of the past are now safely behind us. Nothing could be farther from the truth, according to Lenssen. In the former Soviet Union, the dismantling of the military has left the handling of nuclear materials in chaos. And although the prospects for continued

military buildups between superpowers have grown dimmer, the proliferation of nuclear materials among Third World countries continues. These problems pale, according to Lenssen, when compared with the remaining problems of what to do with the nuclear waste materials left over from the generation of electricity by nuclear generators.

Persistent chemicals are also the subject of the next article, "A Place for Pesticides?" by another Worldwatch researcher, Peter Weber. While the persistence of agricultural chemicals—such as pesticides—in the environment is nowhere near as dramatic as that of radioactive materials, many pesticides and their relatives do accumulate and remain at high levels in soil and water long after their application has been halted. In many parts of the world, farmers have become so dependent upon the use of agricultural chemicals that their very production systems are based on massive annual applications, a practice that is costly in both monetary and environmental terms.

Moving from farm to city, the third article in this section addresses the question of municipal wastes. In "Stewing the Town Dump in Its Own Juice," environmental engineer Wendall Cross notes that, through leachate recycling, municipal landfills can be built as solid-waste treatment centers for harvesting valuable gases, containing runoff, and stabilizing the site of the landfills themselves.

In the next three articles, the scene moves from component-specific discussion to regional-specific examples of pollution problems. In the fourth unit article, writer Pete Hamill describes the unique pollution problems of Mexico City. Once a region of remarkable air clarity and purity, partly because of its high elevation, Mexico City by the late 1980s had what was arguably the world's worst urban pollution problem. The blessings of industrialization and a rising standard of living, which made possible the proliferation of internal combustion vehicles, combined with the particular atmospheric conditions, brought all of the fruits of the energy and chemical revolutions to bear in a negative way on Mexico City. The problem has been exacerbated by population growth rates that led to the conversion of Mexico City from a relatively small metropolis to one of the world's largest.

In the next discussion of a regional pollution problem, R. Dennis Hayes describes what may well be—on a regional level—the world's most polluted region. When the two independent nations of the Czech Republic and the Slovak Republic emerged from the ashes of the former Czechoslovakia, they inherited serious problems of air and water pollution. These problems were caused by the availability of very cheap and highly polluting brown coal and by the failed policies of a government that held economic development and industrialization to be the primary goal without regard for environmental protection. The pollution problems in the two new republics of eastern Europe are so severe that the region's environmental future probably depends upon outside forces and funding.

The next article of the unit deals with a pollution problem that began as a local condition and quickly became a regional one. It was produced by many of the same governmental policies that held industrialization supreme and environmental costs as insignificant. In "Chernobyl's Lengthening Shadow," historian David Marples describes an environmental disaster that is becoming more manifest, rather than less, with the passage of time. The earliest reports from the 1986 Chernobyl nuclear disaster in the Ukraine suggested that the accident was contained, that only 31 people died, and that the viability of surrounding agricultural land was not affected. All of these reports, it now seems, were examples of doublespeak. Nearly nine years after the Chernobyl disaster, air, water, and soil are so polluted with radioactive isotopes that public health problems will persist for generations.

Finally, Michael Kowalok of the Carnegie Commission on Science, Technology, and Government attempts to tie together the various threads of the pollution issue. In "Common Threads: Research Lessons From Acid Rain, Ozone Depletion, and Global Warming," Kowalok takes a historical perspective by looking at the way in which science has increasingly begun to view pollution problems holistically. Pollution is not just a local or regional problem but a global one.

The pollution problem would appear to be one that is nearly impossible to solve. Yet solutions do exist. We now possess the knowledge and the tools to ensure that environmental cleanup is carried through. It will not be an easy task, and it will be terribly expensive. It will also demand a new way of thinking about humankind's role in the environmental systems upon which all life forms depend.

Looking Ahead: Challenge Questions

What specific characteristics of nuclear waste make its disposal such a difficult problem? How do these charac-

teristics conflict with the economic and political institutions responsible for establishing environmental policies related to the safe disposal of radioactive materials?

What are the economic and environmental reasons for the tendency of farmers to become dependent upon agricultural chemicals such as pesticides, and why has chemical agriculture been referred to as a "treadmill economy"?

Why are the leachates of municipal landfills such a serious environmental problem, and how can leachates and other materials in the landfill be managed to produce both environmental and economic benefits?

What factors, both environmental and socioeconomic, have combined to make the air pollution problem of Mexico City so severe? What steps are being taken to resolve the problem?

What are some of the most serious pollution problems in the former Czechoslovakia, and what steps have been taken to alleviate these problems?

What are some of the more serious long-term effects of the nuclear disaster at Chernobyl?

What common threads link three of the world's most severe pollution problems: acid rain, ozone depletion, and global warming?

FACING UP TO NUCLEAR WASTE

The question of what to do with nuclear waste tears at nation after nation—and may never be resolved until a consensus is reached on the issue of nuclear power.

NICHOLAS LENSSEN

Nicholas Lenssen, a research associate at the Worldwatch Institute, is author of Worldwatch Paper 106, Nuclear Waste: The Problem That Won't Go Away, *on which this article was based.*

A series of rapid-fire events has recently swept the nuclear field: amidst an almost celebratory atmosphere, *The Bulletin of the Atomic Scientists* rolled back the minute hand of its famed Doomsday Clock; the Soviet Union peacefully closed shop; the U.S. Nuclear Regulatory Commission's new chief pronounced an emphasis on "safety, safety, safety"; and the nuclear power industry made plans for a happy-days-are-here-again ad campaign proclaiming the virtues of the "new" nuclear power. The cumulative effect suggests that the atomic-age fears of the past are now safely behind us.

In fact, nothing could be farther from the truth. While the fall of the Soviet Union may have ended the superpower nuclear arms race, it has left the management of thousands of bombs, bomb-making facili-

ties, nuclear materials, and radioactive waste sites in a state of near-chaos. And while the prospect of nuclear confrontation between superpowers has faded, the likelihood of more Saddam Husseins getting their hands on nuclear bomb-making materials is a growing concern.

But the most underestimated dangers of all may be those of the civilian nuclear power industry, which—after 50 years of costly research—has yet to find a safe and permanent way to dispose of its radioactive waste. In the United States, and possibly in the world as a whole, 95 percent of all radioactivity emitted by nuclear waste comes from the civilian sector—primarily from nuclear electric power plants. The cumulative discharge of irradiated fuel from these plants is fast growing; it is now three times what it was in 1980 and twenty times what it was in 1970.

Despite this increase, not a single one of the more than 25 countries producing nuclear power has found a solution that stands up to close scrutiny. The central problem is that nuclear waste remains dangerous for hundreds of thousands of years—meaning that in producing it, today's governments assume responsibility for the fate of thousands of future generations. Yet, neither technically nor politically has any way been found to assure that those generations—not to mention the present one—will be protected. The most prudent policy under such circumstances—to store waste in long-term temporary storage while searching for more responsible and permanent solutions—is being proposed by environmentalists and independent scientists in numerous countries; but even as that happens, the governments of the major nuclear nations continue their pursuit of more grandiose strategies for deep geologic burial, which also may entail greater long-term risks.

The nuclear waste issue has been marked by a series of illusions and unfulfilled promises. Like mirages, safe and permanent methods of isolating radioactive materials seem to recede from reach as they are examined closely. No government has been able to come up with a course of action acceptable either to advocates or opponents of nuclear power. Proponents insist that adequate permanent burial options have been developed and proved—and that it's time to jump-start the industry. Anti-nuclear advocates have identified flaws in every burial option proposed so far. They feel that in lieu of a commitment to abandon nuclear power, any waste site will become an excuse to start up the nuclear engine again. A political stalemate of this nature has formed in nearly every country.

Ironically enough, out of concern for the threat of global warming has come a political juggernaut that is being used to try to revive the nuclear power industry. Government officials and nuclear industry executives around the world believe that to achieve this "jump-start" will require a fast resolution of the nuclear waste problem. But just as earlier nuclear power plants were built without a full understanding of the technological and societal requirements, a rushed job to bury wastes may turn out to be an irreversible mistake.

Bury the Problem?

Since the beginning of the nuclear age, there has been no shortage of ideas on how to isolate radioactive waste from the biosphere. Scientists have proposed burying it under Antarctic ice, injecting it into the seabed, or hurling it into outer space. But with each proposal has come an array of objections. As these have mounted, authorities have fallen back on the idea of burying radioactive waste hundreds of feet below the earth's crust. They argue, as does the U.S. National Research Council (of the National Academy of Sciences), that geologic burial is the "best, safest long-term option."

The concept of geologic burial is fairly straightforward. Engineers would begin by hollowing out a repository roughly a quarter of a mile or more below the earth's surface. The repository would consist of a broadly dispersed series of rooms from which thermally hot waste would be placed in holes drilled in the host rock. When the chamber is ready, waste would be transported to the burial site, where technicians would package it in specially constructed containers made of stainless steel or other metal.

Once placed in the rock, the containers would be surrounded by an impermeable material such as clay to retard groundwater penetration, then sealed with cement. When the repository is full, it would be sealed off from the surface. Finally, workers would erect some everlasting sign post to the future—in one U.S. Department of Energy (DOE) proposal, a colossal nuclear Stonehenge—warning generations millennia hence of the deadly radioactivity entombed below.

Geologic disposal, though, as with any human contrivance meant to last thousands of years, is little more than a calculated risk. Future changes in geology, land use, settlement patterns, and climate all affect the ability to isolate nuclear waste safely. As Stanford University geologist Konrad Krauskopf wrote in *Science* in 1990, "No scientist or engineer can give an absolute guarantee that radioactive waste will not someday leak in dangerous quantities from even the best of repositories."

According to a 1990 National Research Council report on radioactive waste disposal, predicting future conditions that could affect a burial site stretch the limits of human understanding in several areas of geology, groundwater movement, and chemistry. "Studies done over the past two decades have led to the realization that the phenomena are more complicated than had been thought," notes the report. "Rather than decreasing our uncertainty, this line of research has increased the number of ways in which we know that we are uncertain."

They Call it Disposal

In Germany and the United States, where specific burial sites have been selected for assessment and preparation, the work to date has raised more questions than answers about the nature of geologic repositories. German planners have targeted the Gorleben salt dome, 85 miles from Hanover in north-

In most countries, even the study of a location for a potential nuclear waste burial site brings people to the streets in protest.

ern Germany, to house the country's high-level waste from irradiated fuel. But groundwater from neighboring sand and gravel layers is eroding the salt that makes up the Gorleben dome, making it a potentially dangerous location.

Groundwater conditions at the U.S. site at Yucca Mountain, a barren, flat-topped ridge about 100 miles north of Las Vegas, Nevada, are also raising concerns. According to the current plan, the waste deposited in Yucca Mountain would stay dry because the storerooms would be located 1,000 feet above the present water table, and because percolation from the surface under current climatic conditions is minimal.

But critics, led by DOE geologist Jerry Szymanski, believe that an earthquake at Yucca Mountain, which is crisscrossed with more than 30 seismic faults, could dramatically raise the water table. Others disagree. But if water came in contact with hot waste containers, the resulting steam explosions could burst them open and rapidly spread their radioactive contents. "You flood that thing and you could blow the top off the mountain. At the very least, the radioactive material would go into the groundwater and spread to Death Valley, where there are hot

springs all over the place," University of Colorado geophysicist Charles Archambeau told the *New York Times.*

Other geologic forces could threaten the inviolability of underground burial chambers. For instance, in 1990 scientists discovered that a volcano 12 miles from Yucca Mountain probably erupted within the last 20,000 years—not 270,000 years ago, as they had earlier surmised. Volcanic activity could easily resume in the area before Yucca Mountain's intended lethal stockpile is inert. It is worth remembering that less than 10,000 years ago, volcanoes were erupting in what is now central France, the English Channel did not exist 7,000 years ago, and much of the Sahara was fertile just 5,000 years ago.

Political Hot Potato

Since the early days of nuclear power, scientists have issued warnings about the long-lived danger of radioactive waste. In 1957, a U.S. National Academy of Sciences (NAS) panel cautioned that "unlike the disposal of any other type of waste, the hazard related to radioactive wastes is so great that no element of doubt should be allowed to exist regarding safety." In 1960, another Academy committee urged that the waste issue be resolved *before* licensing new nuclear facilities.

Yet such recommendations fell on deaf ears, and one country after another plunged ahead with building nuclear power plants. As government bureaucrats and industry spokespeople went about promoting their new industry, they attempted to quiet any public uneasiness about waste storage with assurances that it could be dealt with. However, early failures of waste storage and burial practices engendered growing mistrust of the secretive government nuclear agencies that were responsible. For example, three of the six shallow burial sites for commercial low-level radioactive waste in the United States—in Kentucky, Illinois, and New York—have leaked waste and been closed.

Trust also faded around the world as the public came to view government agencies as more interested in encouraging the growth of nuclear power than in resolving the waste problem. Grass-roots opposition has sprung up against nearly any attempt to develop a radioactive waste facility.

The United States has perhaps the most dismal history of mismanaging waste issues; from the 1950s onward, the U.S. Atomic

Energy Commission (AEC) and its successors have swept nuclear waste problems under the rug. Only following a stinging 1966 NAS critique of the AEC's waste policy (suppressed by the AEC until Congress demanded its release in 1970), and a 1969 fire at the U.S. government's bomb-making facility at Rocky Flats, Colorado (which created vast amounts of long-lived waste in need of storage), did the AEC concoct a rushed attempt to solve the problem by planning to bury nuclear waste in a salt formation in Lyons, Kansas.

By 1973, the AEC was forced to cancel the Lyons project because serious technical problems had been overlooked. For example, the ground around the site was a "Swiss cheese" of old oil and gas wells through which groundwater might seep. The Lyons failure set off a decade of wandering from potential site to potential site, and of growing opposition from apprehensive states.

A number of states, led by California in 1976, responded by approving legislation that tied future nuclear power development to a solution of the waste problem. Suddenly, the future of nuclear power seemed threatened. The nuclear industry pushed the AEC's successor, the Department of Energy (DOE), to bury waste quickly. But DOE had no better success in finding a state amenable to housing the nation's waste.

The department's repeated failures prompted Congress to pass the Nuclear Waste Policy Act of 1982. A product of byzantine political bargaining, the law required DOE to develop two high-level repositories, one in the western part of the country and the other in the east.

From the outset, the department was hampered in its response by an unreasonable timetable and its own insistence on considering sites that were technically and politically unacceptable. As DOE failed to gain public confidence, the process became embroiled in political conflicts at the state level. Finally, when the uncooperative eastern states forced the cancellation of the unsited eastern repository in 1986, the legislation fell apart. With the whole waste program in jeopardy, and over the strong objections of the Nevada delegation, Congress ordered DOE in 1987 to study just one site—Yucca Mountain, adjacent to the federal government's nuclear weapons test area.

While the federal government is determined to saddle Nevada with the country's waste, the state is vigorously seeking to disqualify the site, claiming in part that DOE—given that Yucca Mountain is the only site being investigated—cannot conduct research objectively. So vehement are the objections of Nevadans that the state legislature in 1989 passed a law prohibiting anyone from storing high-level waste in the state. Former U.S. Nuclear Regulatory Commissioner Victor Gilinsky describes Yucca Mountain as a "political dead-end."

Going in Reverse

Although most of the countries using nuclear power are now preparing for geologic burial of their waste, almost every disposal program is well behind its own schedule. In 1975, the United States planned on having a high-level waste burial site operating by 1985. The date was moved to 1989, then to 1998, 2003, and now 2010—a goal that still appears unrealistic. Likewise, Germany expected in the mid-1980s to open its deep burial facility by 1998, but the government waste agency now cites 2008 as the target year. Most other countries currently plan deep geologic burial no sooner than 2020, with a few aiming for even later [see table next page].

So charged is the atmosphere surrounding the waste disposal issue that it's questionable whether any government has the political capacity to build and operate nuclear waste repositories. In most countries, even the study of a location as a potential nuclear waste burial site brings people to the streets in protest, as in South Korea and the former Soviet Union. So far, most governments have made short-term decisions on waste while leaving their long-term plans vague, hoping to muddle through.

Even in France, the acknowledged leader in European nuclear power generation, the waste issue defies ready solution. In 1987, the French waste agency, ANDRA, selected four potential sites for burying high-level radioactive wastes. Officials in those locales, disturbed that they had not been consulted, joined with farmers and environmentalists to paralyze the research program. Blockades obstructed government technicians at three of the four sites, and work proceeded only with police protection. In January 1990, in one of the country's largest anti-nuclear demonstrations since the late 1970s, 15,000 people marched in Angers against one site. By February, then-Prime Minister Michel Rocard had imposed a nationwide morato-

Table : Selected Programs for High-Level Waste Burial

Country	Earliest Planned Year	Status of Program
Argentina	2040	Granite site at Gastre selected.
Canada	2020	Independent commission conducting four-year study of government plan to bury irradiated fuel in granite at yet-to-be-identified site.
China	none announced	Irradiated fuel to be reprocessed; Gobi desert sites under investigation.
Finland	2020	Field studies being conducted; final site selection due in 2000.
France	2010	Two sites to be selected and studied; final site not to be selected until 2006.
Germany	2008	Gorleben salt dome sole site to be studied.
India	2010	Irradiated fuel to be reprocessed, waste stored for twenty years, then buried in yet-to-be-identified granite site.
Italy	2040	Irradiated fuel to be reprocessed, and waste stored for 50-60 years before burial in clay or granite.
Japan	2020	Limited site studies. Cooperative program with China to build underground research facility.
Netherlands	2040	Interim storage of reprocessing waste for 50-100 years before eventual burial, possibly in another country.
Soviet Union	none announced	Eight sites being studied for deep geologic disposal.
Sweden	2020	Granite site to be selected in 1997; evaluation studies under way at Äspö site near Oskarshamn nuclear complex.
United States	2010	Yucca Mountain, Nevada, site to be studied, and if approved, receive 70,000 metric tons of waste.
United Kingdom	2030	Fifty-year storage approved in 1982; exploring options including sub-seabed burial.

Source: Worldwatch Institute, based on various sources.

rium on further work, providing the government a cooling-off period to try again to win public support.

The French Parliament approved a new plan in June 1991. The number of sites to be investigated was reduced from four to two, and the government claims the selection process will be more open this time around. Also, ANDRA officials have a new approach for winning support. They will pay the two communities contending for the site up to $9 million a year for "the psychological inconvenience" of being studied, according to then-Industry Minister Roger Fauroux. However, parliament has delayed any decision on a final burial site for 15 years. In that time, the country's high-level waste inventory will more than triple.

In Germany, the controversy over radioac-tive waste mirrors that surrounding nuclear reactor construction, which has come to a standstill. Local opposition to any nuclear project appears deeply entrenched, and there is a general inability of the major political parties to agree on nuclear policy.

The German waste controversy erupted in 1976, when the federal government's investigation of three sites in Lower Saxony created such an uproar among local farmers and students that the state government rejected every one. The following year, the federal government selected another site in Lower Saxony—the salt dome at Gorleben.

Large protests erupted even before the official announcement; 2,500 people took over the drilling site for three months before police hauled them off and set up a secure camp from which scientific work could be conducted. Although the federal government has put all its bets on Gorleben, continuing technical problems and strong opposition from the Lower Saxony government make plans to bury waste by 2008 highly improbable. Critics have warned that the site's geology is unstable. Public confidence in the project sank even lower when a worker was killed by collapsing rock during a 1987 drilling accident.

In Sweden, nuclear issues have been erupting since the 1970s, when two governments were thrown out of office following attempts to promote nuclear energy. Only after a national referendum in 1980 to limit the number of reactors in the country to 12 and to phase even these out by 2010, was the country able to focus on the waste issue. One immediate dividend from the agreed phase-out was a clarification of exactly how much waste would eventually need to be dealt with: 7,750 tons of irradiated fuel, and 7.2 million cubic feet of other radioactive waste.

Sweden's high-level waste program has won international praise for relying not simply on deep burial but on a system of redundant engineering barriers, starting with corrosion-resistant copper waste canisters that have four-inch thick walls with an estimated lifetime of 100,000 years or longer if undisturbed by humans or geologic forces. Even with the announced phase-out and international scientific praise, Swedish public support has not been forthcoming for burial. Protests halted attempts to site a permanent high-level burial facility 12 years ago, and efforts to explore other sites have met determined local opposition.

The Japanese government also has run into public opposition to its burial plans. In 1984, planners selected an amenable village, Horonobe, near the northern tip of Hokkaido island. But opposition from the Hokkaido Prefecture governor and diet and from nearby villages and farmers has blocked the government from constructing a waste storage and underground research facility there.

There are signs that Japan is now looking beyond its borders for a high-level waste disposal site. Since 1984, China has shown interest in importing irradiated fuel or waste for either a fee or in return for assistance with its own fledgling nuclear program. In November 1990, China and Japan agreed to build an underground facility in China's Shanxi province, where research is to be undertaken on high-level waste burial.

Walk, Do Not Run

Because of the scientific and political difficulties with geologic burial, above-ground "temporary" storage is likely to remain the only viable option well into the 21st century. Fortunately, there need be no rush to bury nuclear waste, other than for public relations reasons. As a result, rather than continue focusing on developing controversial and potentially dangerous burial sites, governments can still choose a course of action that will buy time—and gain public support—to continue the search for a dependable long-term solution.

To choose this course requires, first, that nations employ safer methods of temporary storage for radioactive waste, particularly irradiated fuel. For instance, most spent radioactive material is stored in cooling ponds at nuclear power plants—an inherently risky proposition, since electric pumps are needed to circulate cooling water to prevent the fuel from overheating. Yet both **governments and independent analysts believe that storage technologies such as dry casks, that rely on passive cooling and are capable of containing materials for at least a century, are safer than water-based systems. Such storage would allow radiation levels to fall 90 percent or more.**

Even with improved temporary storage systems, however, an institution for the careful monitoring and safeguarding of the waste will be needed to prevent catastrophic accidents or even terrorism. But no government can guarantee the durability of an institution whose responsibilities must continue many times longer than any human institution has ever lasted.

Temporary storage does not solve the problem of nuclear waste, but it could allow time for more careful consideration than is now witnessed in many countries of longer-term options, including geologic burial, seabed burial, and indefinite storage. It also could permit long-term, in-situ experiments with promising technologies such as the Swedish copper canisters.

But addressing the waste problem demands much more than a reduction of technical uncertainties. It also requires a fundamental change in current operating programs as well as new measures to regain public confidence. A lack of credibility plagues government nuclear agencies in most countries. Public distrust is rooted in the fact that the institutions in charge of nuclear waste cleanup also promote nuclear power and weapons production—and have acquired reputations for equivocation, misinformation, and secrecy.

In the United States, the U.S. Office of Technology Assessment, the National Research Council's Board of Radioactive Waste Management, and private research groups have called for an independent government body to take over the task of managing the country's nuclear wastes from DOE. So far, Congress has responded merely by requiring more oversight of DOE. Forming autonomous and publicly accountable organizations to manage nuclear waste would go a long way toward regaining public support, and getting waste programs on track.

Beyond Illusion

In the end, some observers believe that the nuclear waste issue is a hostage of the overall debate on nuclear power, which increasingly tears at nations. "If industry insists on generating more waste, there will always be confrontation. People just won't accept it," believes David Lowry, a British environmental consultant and coauthor of *The International Politics of Nuclear Waste*. Because the political controversies are so intense, true progress on the waste issue may only come about when nations come to decisive resolutions, once and for all, of the nuclear power issue.

Sweden, which has perhaps the broadest (though not universal) public support for its nuclear waste program, made a national decision to phase out its twelve nuclear power reactors by 2010. Without such a decision, public skepticism toward nuclear technologies and institutions could grow only stronger.

While most countries do not have formal policies requiring phase-out of nuclear power, there is a de facto phase out of new plants imposed by rising costs and concern over safety. Worldwide, roughly 50 nuclear power plants are under construction today—fewer than at any other time in the last 20 years.

Despite this trend, nuclear advocates continue to call for a rapid expansion of atomic power. They've seized upon the threat of global warming and public anxieties about dependence on Middle Eastern oil, aroused by the Gulf War, to push their point.

Yet a world with six times the current number of reactors, as called for by some

---

N₀ government can guarantee the durability of an institution whose responsibilities must continue many times longer than any human institution has ever lasted.

---

nuclear supporters, would require opening a new burial site every two years or so to handle the long-lived wastes generated—a gargantuan financial, environmental, and public health problem that nuclear power proponents conveniently continue to ignore. President Bush's 1991 National Energy Strategy, for example, proposed a doubling in the number of U.S. nuclear power plants over the next 40 years, but did not discuss the need for future waste sites.

As experience with nuclear power plants has demonstrated, it will not necessarily become any easier to site and construct future geologic burial facilities once the first is opened. A single accident could set back government and industry efforts for decades. While waste is but one of the problems still facing nuclear power, it is clearly the longest-lasting one.

A PLACE FOR PESTICIDES?

With an eye on ecology, the world's farmers can control pests and reduce the mounting hazards of pesticide dependence. They'll also liberate themselves from the treadmill economics of chemical agriculture.

PETER WEBER

Peter Weber researches agriculture, technology, and resource issues at the Worldwatch Institute.

Farmer Ron Rosmann is no longer content to go along with conventional wisdom. In his home state of Iowa, most corn and soybean farmers act on the notion that without pesticides, they'd be out of business. They know that pesticides get into their drinking water and threaten their health, but in a competitive market they don't think they have any alternative.

Rosmann, who lives near Harlan, subscribed to that staunch belief for 10 years after taking over the farm from his father. Then, in 1982, he visited a couple of operations where the farmers didn't use pesticides. "Their fields were cleaner than mine," Rosmann recalls. The two farmers didn't have weed problems even though they didn't use herbicides, the pesticides that kill weeds. "I said wait a minute, something is wrong here."

That next year Rosmann started experimenting with alternatives to pesticides. Weeds had been his biggest problem, and herbicides had been his biggest pesticide expense. He found that he could virtually eliminate sprayings with the new planting and cultivation techniques he had seen the summer before. In the nine years since, he has used herbicides only once, last year, when he made some mistakes and let the weeds get out of hand in one field. Even then, he said, "I used them in small amounts, only if absolutely necessary." Rosmann calculates that he saves between $4,000 and $5,000 a year on pesticides, and his yields have actually gone up.

Rosmann isn't alone in successfully bucking conventional wisdom. He's a member of Practical Farmers of Iowa (PFI), a group of some 400 ordinary farmers turned trendsetters. Although members span the spectrum from organic to conventional, all are interested in cutting costs, raising income, preventing soil erosion, and using fewer chemicals—in short, farming more sustainably. They blend innovation with hard-nosed practicality in a way that any farmer could respect. Groups similar to PFI are starting up in other farm states across the U.S.

Insights from PFI and other farm groups could make big differences in farming practices, both in the United States and abroad. For instance, Rosmann says that corn and soybean farmers could cut their herbicide use by half or more if they sprayed only where weeds are hard to control—in the soil directly under the crop. In the areas between the rows, weeds can be cleared out with a cultivator for the same or less than it costs to spray.

This cultivation practice could topple corn and soybeans from their standings as the top pesticide-using crops in the world. On U.S. farmland alone, pesticide use would fall by 30 percent. Other techniques used by PFI—such as innovative tilling practices—would make even greater reductions possible. The environmental benefits of these cuts would be enormous. Atrazine, a herbicide commonly used on these two crops, shows up frequently in drinking water supplies in the United States and in Europe. Germany, Italy, the Netherlands, and Sweden have banned Atrazine, which they suspect is a carcinogen. Other corn and soybean pesticides kill birds and other wildlife.

Even on the energy side of the environmental ledger, sustainable agriculture appears to be a net gain. PFI coordinator Rick Exner at Iowa State University calculates that the extra diesel fuel farmers use to drag around their one-ton cultivators represents less energy than would be used to make the herbicides for the same acreage.

Applying PFI's basic approach to other crops would lower pesticide's environmental costs without losing the battle against hunger. Small but growing numbers of farmers are discovering that sustainable practices can actually boost harvests and income. That discovery goes against conventional wisdom, but it is a revelation that more farmers—and policymakers—can use.

Good Idea at the Time

For a long time, no one seemed to question the safety of pesticides. Weeds, diseases, insects, and other animals had always waged a relentless battle on farmers' fields, and through the centuries, it seemed that any compound, no matter its hazards, was a welcome weapon against pests.

The first records of pesticides come from the ancient Greeks. Pliny the Elder (A.D. 23–79) compiled a list of common compounds like arsenic, sulfur, caustic soda, and olive oil used to protect crops. The Chinese later recorded using similar substances to combat insects and fungi. In the 19th century, European farmers began using heavy metal salts like copper sulfate and iron sulfate to fight weeds.

With the invention of DDT in 1939, the war against pests escalated sharply. This new synthetic made earlier pesticides appear crude and dangerous. DDT was effec-

tive, relatively cheap, and apparently safe for people—a miracle chemical. Paul Müller, its Swiss discoverer, received a Nobel Prize for his find.

DDT launched what Robert Metcalf, an entomologist at the University of Illinois, calls the Era of Optimism. Popular periodicals of the day envisioned a world without flies, and entomologists pondered the future of their discipline once insect pests were eradicated. Under the stream of new insecticides, herbicides, and fungicides, scientists reported unprecedented crop yields. Farmers began using pesticides intensively in industrial countries and, when money allowed, everywhere else. The notion that pesticides are essential to modern, high-production agriculture was born. Farmers now apply about one pound of pesticides per year for every person on the planet, 75 percent of it in industrial countries.

Pesticide use continues to grow despite a number of problems that have surfaced since the Era of Optimism. Severe poisonings in developing countries—perhaps only the tip of the iceberg of pesticide-caused health problems—affect around 1 million people per year, according to the United Nations World Health Organization (WHO). Of these, about three-quarters suffer chronic health problems such as dermatitis, nervous disorders, and cancer. Between 4,000 and 19,000 people die. WHO expects poisonings to increase as pesticide use in developing countries, now concentrated on cash crops like rice, cotton, and coffee, intensifies and expands to other crops.

In the former Soviet Union, *glasnost* brought to light a pesticide-induced health crisis of sickening severity in the cotton-growing region adjacent to the Aral Sea. Pesticide exposure up to 25 times the national norm had brought on a rash of chronic medical problems, including liver and kidney disease, chronic gastritis, rising cancer rates, birth defects, and infant mortality four times the Soviet average.

In industrial countries where people are exposed to a mix of chemical hazards, the lines of cause and effect aren't so clear. Isolated illnesses can be impossible to link to a cause. According to Susan Cooper, an ecologist with the National Coalition Against the Misuse of Pesticides, based in Washington, D.C., scientists still hardly understand the effects of exposure to one chemical, and they have little idea what happens when people are exposed to many. The risks could be worse than we think.

At a time when some forms of cancer are on the rise in industrial countries, pesticides add to the risks of chronic illness, especially among farmers and farm workers. One study from the U.S. National Cancer Institute found that farmers who use 2,4-D, a common herbicide, were more likely to contract non-Hodgkin's lymphatic cancer. The U.S. Environmental Protection Agency lists 55 possible carcinogens legal for use on domestic food crops.

In recent years, improved safety testing has given rise to tougher standards for the latest generation of pesticides. The best of these are applied in ounces instead of pounds per acre and show few indications of causing health problems. The new generation of pesticides also costs more, which has led chemical companies to develop relatively few of them. This leaves many farmers using pesticides from previous generations, especially in developing countries.

Even new-generation pesticides aren't completely safe, especially not for wildlife. Ultra-low volume, low-toxicity sulfonylurea herbicides can drift from a field and kill rare plants (and susceptible crops), and ultra-low volume synthetic pyrethrum insecticides, which mimic natural plant toxins, are deadly to honey bees, other beneficial insects, and fish.

Wildlife also continues to suffer the effects of older pesticides. While DDT is now banned in most industrialized nations, its continued use in the developing world threatens eagles and other birds of prey. In the United States, carbofuran, one of the toxic granular insecticides used around the world on corn and other crops, annually kills an estimated 1 to 2 million birds that mistake the granules for food. Last summer, insecticide runoff from Louisiana sugar cane fields contributed to the death of an estimated 750,000 fish in nearby waterways. The U.S. Fish and Wildlife Service finds that pesticides menace about 20 percent of the country's 681 endangered and threatened plant and animal species.

A Necessary Evil?

Despite the health and environmental risks, farmers are still hooked on many of the worst offenders. Without these pesticides, they say, their costs would skyrocket, harvests would plummet, and more people would go hungry. It is widely assumed that for big harvests, pesticides are essential. After all, it is noted, farmers in industrial countries apply more pesticides, lose less to pests and reap higher yields than farmers in developing countries.

A closer look at the data, however, indicates that pesticides are not as essential as many people think, say researchers from the International Rice Research Institute (IRRI), based in the Philippines. As one of the original "Green Revolution" research centers with a mandate to help Third World farmers feed growing populations, IRRI through the years supported increased pesticide use as a way to raise

N*ot only do insects take a larger percentage of the U.S. harvest now than they did before the DDT revolution, but damage from fungi and other plant diseases is also higher.*

yields. Over the past two decades, however, the respected research center has shifted its perspective 180 degrees. Its recent studies have found no correlation between yields and pesticide use in several Southeast Asian countries. K.L. Heong, an entomologist at IRRI, concludes that "Governments should stop treating pesticides as a critical input."

The IRRI results, along with those of the Practical Farmers of Iowa, raise an obvious question: If pesticides created a revolution by increasing yields, how can it be that cutting back (or cutting out) their use can also increase yields? Part of the answer is that pesticides are most effective when first applied.

A classic example was that of Indonesia,

where the government promoted pesticides as part of its push in the 1970s for rice self-sufficiency. Indonesia offered pesticides to farmers for 15 percent of the market price, and farmers responded by rapidly increasing their pesticide use to one of the highest levels in the developing world. Partly as a result of Indonesia's subsidy, rice soon became the third biggest pesticide market in the world,

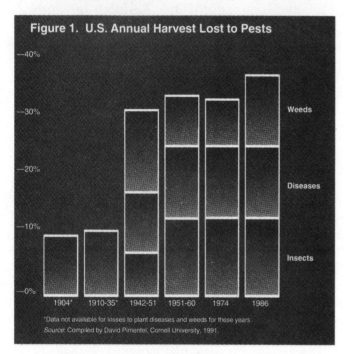

Figure 1. U.S. Annual Harvest Lost to Pests

—40%

—30% Weeds

—20% Diseases

—10%

—0% Insects

1904* 1910-35* 1942-51 1951-60 1974 1986

*Data not available for losses to plant diseases and weeds for these years.
Source: Compiled by David Pimentel, Cornell University, 1991.

behind corn and soybeans.

Following IRRI's Green Revolution recommendations, Indonesia's farmers transformed the country from the largest rice importer in the world to a net exporter by 1984. But the country also became lashed to an accelerating pesticide treadmill. Indiscriminate spraying killed more predatory insects than crop-eaters, unleashing so-called secondary pests that had previously been held in check naturally. With wolf spiders, water striders, and other natural enemies out of the way, a juice-sucking insect called the brown planthopper emerged as the most destructive rice pest in Indonesia.

Over time, the brown planthopper adapted to the insecticides used against it, and some of the chemicals actually spurred it to reproduce faster. Heavier spraying only led to heavier hopper infestation, until by 1986 the country risked losing its newfound self-sufficiency. IRRI scientists studied the situation and came to the conclusion that overuse of insecticides was the problem.

Faced with a potential disaster, Indonesia banned 57 problem insecticides for use on rice. In time for the next season, it deployed crop advisers to teach farmers to spray more selectively and protect natural predators, essential elements of a system known as integrated pest management (IPM). The government also phased out pesticide subsidies over the next two years.

"The results were dramatic," according to Roland Roberts, a policy analyst in Indonesia for the U.S. Agency for International Development, as quoted in a recent issue of the agency's magazine *Front Lines*. "Pesticide use has dropped by 65 percent since 1987, and rice production is up 15 percent," says Roberts. The country saves around $120 million per year on pesticides, most of which were manufactured overseas. This money can now be used to cover the cost of the IPM program.

No matter where or how pesticides are used, the treadmill will start spinning. Resistance to pesticides is as natural as evolution. In fact, resistance is natural selection in fast forward, provoked by the very chemicals meant to eliminate the pest. Kill 99.9 percent of one pest, and the survivors are super-arthropods. Fred Gould, an ecologist and evolutionary biologist from North Carolina State University in Raleigh, says that resistance to DDT was found shortly after its introduction: "Since the DDT case the insects have, as a group, never met a chemical they couldn't take to the mat."

In the United States, overall crop losses have actually gone up since DDT's introduction. David Pimentel, an entomologist at Cornell University in Ithaca, New York, found that insecticides first caused crop losses to fall, but over time, pests began to rebound. Not only do insects take a larger percentage of the U.S. harvest now than they did before the DDT revolution, but damage from fungi and other plant diseases is also **higher, and weed losses are on the rise [see Figure 1]**.

Pimentel cites a number of contributing factors, from planting crops that are more susceptible to pests to neglecting traditional pest-control practices such as crop rotations. Prominent on his list are the same treadmill problems Indonesia suffered. Of the United States' 300 most destructive insect pests, about 100 are secondary pests. Worldwide, the number of resistant pests continues to climb. Some

504 arthropods (insects and arachnids), 150 plant pathogens (fungal and bacterial diseases), and 113 weeds have developed resistance to at least one pesticide [see Figure 2].

Pest Prevention, Pesticide Reduction

Tainted groundwater and resistant pests are not news to farmers. Yet, most are skeptical of the alternatives. They dismiss the no-pesticide route of organic agriculture as going back to the "bad old days" before pesticides. They associate this path with heavy toil and the threat of being wiped out by infestation.

But there is a middle road. An emerging movement of farmers like Iowa's Rosmann and Indonesia's rice farmers are blending time-honored practices with new insights into the ecology of farm fields. Rather than attempting what is now recognized as an ecological impossibility—eliminating pests—they try to strike a profitable balance.

Integrated pest management is the foundation of this new movement. Devised in the 1960s by entomologists wary of the pesticide treadmill, IPM combines pesticides with protection of predator insects and other pest-control strategies. At the heart of IPM is the practice of scouting. Farmers spend time in their fields counting insects, weeds, or diseased plants in a test area. They spray only if pests really threaten the crop. Because natural enemies often keep pests in check, farmers can usually spray less frequently than recommended.

Rosmann takes IPM's basic concepts and pushes them to the limit. Instead of simply focusing on particular pests, he plans his entire operation around minimizing all purchased inputs—from pesticides to fertilizers—for both economic and environmental reasons.

Rosmann describes what he does as bringing conventional and organic agriculture together to spawn sustainable agriculture. "Organic farming is high cost to consumers. Conventional farming is expensive because of its environmental and health problems," he says. "The third option is what I'm talking about. It would be the cheapest."

While the recent trend is for farms to get bigger and more specialized, Rosmann's is basically the same moderate-sized and diverse 480-acre farm his father ran: He plants 250 acres in corn and soybeans; the rest is devoted to hay and pasture fields, cattle, hogs and chickens, plus a new tree nursery to finance

his three young sons' college educations. With this range of projects, Rosmann, his wife Maria, and one employee can spread out the work load over the entire year.

Rosmann needs to avoid time crunches because his cultivation takes longer than conventional techniques. He uses ridge-till, a soil-conserving practice that makes the field look like a giant washboard with the crop planted on top of the ridges. Because he doesn't stir up the soil, fewer weed seeds germinate, and the ones that do, he wipes out with a pass of his tractor. Yet, if he didn't have the money from his livestock, his corn

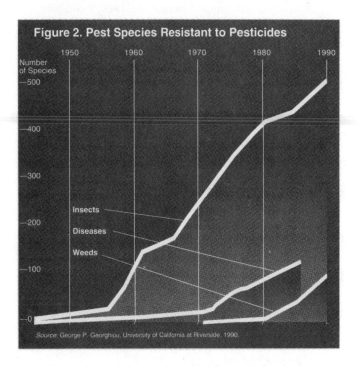

Figure 2. Pest Species Resistant to Pesticides

Source: George P. Georghiou, University of California at Riverside, 1990.

and soybean acreage would have to be more extensive, and in peak seasons, he might not have time to run the tractor for weed control. In fact, with more corn and soybean acreage he might have to drop ridge-till altogether, even though it tends to raise yields and lower costs.

Farmers like Rosmann use a variety of other tactics to cut back on pesticides, from old-fashioned crop rotations to new crop varieties that resist pest attacks and biological pesticides made possible by biotechnology. Orchestrated together, these sustainable practices can bring dramatic results.

The Necessity of Alternatives

The farmers most interested in alternatives are the ones that have no choice. Growers of corn in Egypt, rice in China, cocoa in Malaysia, coconuts in the South Pacific Islands, bananas in Costa Rica, soybeans in Brazil, vegetables in Holland, and cotton in Texas have slashed pesticide use to save money, to replace banned pesticides, and to sustain productivity.

Cotton farmers have been forced in the greatest numbers to find alternatives. That's because cotton is vulnerable to boll weevils, bud weevils, and other insects, which farmers have tried to eliminate with heavy doses of insecticides. Cotton farmers use more pesticides per acre than almost any others, making their crop the world's fourth largest pesticide consumer. The insecticides, however, have unleashed a host of secondary cotton-eaters—some of which are resistant.

Cotton farmers in the Cañete Valley of Peru jumped on the pesticide treadmill in the 1940s with arsenic and nicotine-based insecticides. By 1949, resistant insect booms had caused crop failures. Then in the 1950s, DDT and its generation of synthetic insecticides came to the valley, taking the farmers through the cycle once more. Resistant insects and record losses returned in 1956.

Under a government-led IPM program, the Peruvian farmers cut sprayings from an average of 16 per season to 2.35 per season and cotton yields recovered to record highs in two seasons. The program worked because the government banned the problem pesticides, taught farmers to protect beneficial insects, and prescribed pest-preventive farming practices such as fixed planting dates and cleanup campaigns to throw off pests' reproductive cycles.

While successful for cash-croppers, these policies—and IPM—would seem to have nothing to offer poor farmers in developing countries who can't afford to use pesticides in the first place. However, the same methods farmers use to cut back on pesticides can work to raise yields for small-scale farmers and keep them off the pesticide treadmill.

The most dramatic example comes from sub-Saharan Africa where a new pest, the cassava mealybug, was attacking cassava, a root crop that is the primary staple for some 200 million Africans. In the early 1980s, the mealybug was causing $2 billion in damages annually in the cassava belt that stretches across the continent's midsection.

A sister organization of IRRI, the International Institute for Tropical Agriculture (IITA), based in Nigeria, started a type of sustainable agriculture program in 1977 to assist the stricken farmers. Researchers scoured the mealybug's South American homeland for its natural enemies, hoping to bring them to Africa as a kind of arthropod liberating army marching under the banner of "biological control."

They found a tiny parasitic wasp, *Epidinocarsis lopezi*, that killed mealybugs by injecting its eggs into them. After checking that *E. lopezi* would survive in Africa's tropics and wouldn't cause any ecological problems, IITA's biological control program started distributing the wasp across Africa's 28-country cassava belt in 1986. *E. lopezi* now provides cassava protection in 24 of the countries, at no cost to farmers, with a return of $149 in saved crops for every dollar spent by IITA.

Slow Going

Despite success stories around the globe, and ample reason to cut back, farmers have been slow to adopt even IPM unless they have to. Where farmers can get them, pesticides have become as much a part of farming as seeds and fertilizer. Farmers seem to see in pesticides at least the illusion of a guarantee, which they can never get from the weather, markets, or politicians. In fact, some banks require farmers to use pesticides to qualify for crop loans.

Understandably, farmers are reluctant to change their ways, especially since switching to alternatives to pesticides can amount to a high-wire walk without a net. In the United States, for instance, the Department of Agriculture dedicates just 1 percent of its $1.6-billion research and extension budget to IPM and sustainable agriculture programs. In the private sector, there are no deep pockets backing research like that of the Practical Farmers of Iowa. Pesticide companies, on the other hand, give chemical agriculture an additional push of $1.7 billion annually in research and development worldwide.

Other barriers to sustainable agriculture are:

• Heavy spending by pesticide companies on advertising and lobbying;
• Lax pesticide regulations in developed and developing nations;
• Absence of funds and expertise in the Third World;
• Government subsidies and pricing policies that encourage pesticide use; and

• Development agency loans and grants for pesticides.

Despite these formidable hurdles, sustainable agriculture is gaining ground wherever farmers—and policy makers—are able to free themselves from the belief that pesticides are always essential. Sweden is a case in point. Out of concern over health and environmental risks, this Scandinavian country decided in 1985 to try to cut pesticide use by 50 percent in five years. By 1990, pesticide use had fallen by nearly half.

Swedish leaders helped farmers achieve their target with three basic policy tools. First, they tightened pesticide regulations and banned hazardous compounds, then they provided farmers with proven and practical ways to produce more by using less pesticides, and they taxed pesticides to discourage their use. On the heels of its first success, Sweden hopes to cut pesticide use by another 50 percent by 1997.

Denmark and the Netherlands have also set 50 percent reduction goals, and the European Parliament has proposed a less ambitious program for the European Community as a whole. In the United States, the University of Illinois' Metcalf and Cornell University's Pimentel, among other researchers, estimate that U.S. farmers could cut their pesticide use by up to 50 percent and still not lower harvests or significantly raise costs.

The situation differs somewhat in developing countries, where pesticide use is now low but growing. These nations need dual programs: one to reduce pesticide use among cash-crop farmers, and the other to improve harvests for small-scale farmers without making them pesticide-dependent. But the basic policy tools are the same. Cuba began using biological control on sugarcane and other crops in the 1970s. Thailand is taking the first steps by funding research into sustainable agriculture. Vietnam has a nascent rice IPM program modeled on Indonesia's and supported by IRRI. The program's future, however, hangs on a pending development loan from the World Bank that might fund either pesticides or IPM.

Pesticide resistance is natural selection in fast forward, provoked by the very chemicals meant to eliminate the pest. Kill 99.9 percent of one pest, and the survivors are super bugs.

Countries with sustainable farm policies, however, are still rare—despite the facts that pesticide problems show no signs of going away and that more farmers will be forced to take up alternatives as older pesticides are banned or lose effectiveness and fewer new ones become available. Official policies may not acknowledge the new era in farming that is dawning, but, according to Michael Fitzner, national program leader for IPM at the U.S. Department of Agriculture's Extension Service, "Farmers can see the handwriting on the wall."

Out in front, breaking new ground, are farm groups like Practical Farmers of Iowa, whose members are standing conventional wisdom on its head. Ron Rosmann may not talk or act like a revolutionary, but what he's doing in his fields could change the way the world farms.

Stewing the Town Dump in Its Own Juice

Wendall Cross

Wendall Cross is senior research scientist of environmental engineering at Georgia Institute of Technology in Atlanta.

Through leachate recycling, municipal landfills can be built as solid-waste treatment centers for harvesting valuable gases, containing runoff, and stabilizing the site.

Since the first well-gnawed bone along with a worn-out animal-skin garment was thrown on the ground outside the mouth of the cave, land disposal has been the most convenient, popular, and cost effective method of "treating" and disposing of solid waste. For this reason and others the town dump has always been one of the archaeologists' most fertile sources of information.

From individual trash piles outside the cave we graduated to community dumps where any and all refuse was simply dumped. These facilities were an eyesore as well as a hazard for both people and the environment.

In the early twentieth century larger metropolitan areas began to design and operate trash dumping sites for disposal of large quantities of solid waste in a more efficient and regulated manner. This marked the creation of the municipal or sanitary landfill. However, these landfills for many years accepted all types of waste materials and it wasn't until environmental concerns began to emerge in the late 1960s and early 1970s that much thought was given to more stringent regulation of these disposal sites.

Current and pending regulations allow us to divide landfills into three broad categories: the municipal or solid waste landfill, the secure or hazardous waste landfill, and repositories for radioactive materials. Landfills are as vital a part of an industrialized country's infrastructure as are the drinking water systems, the sewers and sewage treatment systems, the highways, and the gas and electrical distribution systems.

This article introduces research devoted to improving the design and operation of municipal landfills, which are the disposal sites for the wastes generated by individuals or households in the ordinary conduct of life. A landfill, inevitably is the site of microbial reactions that convert available organic matter into basic constituents. One major research and development challenge is to develop means of managing these reactions so the products can either be harvested or contained, providing some financial return while minimizing potential environmental contamination.

Out of sight, out of mind

In common landfill practice, once the garbage leaves our house its only treatment prior to disposal is shredding in order to reduce as much as possible the volume of waste. At the disposal site a "cell," the active working area, has been dug in the earth. Until recently the practice usually has been to form the floor and the walls of the current active cell out of well-compacted fill (preferably clay) in order to form water-impermeable layers and prevent the seeping of liquids (leachate) out of the cell.

Solid waste is deposited in the cell and mechanically compacted to reduce its volume. In some instances, instead of being shredded and compacted, the garbage is baled before being placed in the cell. At the end of the working day the surface of the refuse is covered with a layer of soil approximately six inches thick. This soil, referred to as intermediate cover is to prevent the proliferation of vermin (flies, rats, mice, birds, etc.), minimize the escape of materials (paper, plastic) because of winds, and minimize the infiltration of water (rain).

Once a cell has been filled to

From *The World & I*, February 1992, pp. 284-289. Reprinted with permission from *The World & I*, a publication of The Washington Times Corporation. Copyright © 1992.

capacity it is capped with several feet of compacted soil, overlaid with topsoil, and planted with appropriate vegetation, usually grass. Usually several cells are operated somewhat in succession in a landfill. Years ago, it was generally assumed that once a cell was completed, nothing much occurred within the landfill. To a great extent that is what happens in many old landfills, which makes them prime hunting grounds for archaeologists.

On the other hand, completed landfill cells do contain considerable quantities of organic molecules that serve as a rich food supply for many microorganisms. Despite efforts to prevent microbial reactions within the landfill they do go on. Not only do microbes bring about various chemical reactions, but the products of these reactions initiate further reactions.

The by-products and end products of these many and diverse reactions give rise to the problems associated with landfills, such as their potential to contaminate a drinking water aquifer, adversely affecting the taste and odor of the water. Gaseous products such as methane and carbon dioxide are generated in approximately equal volumes during active landfill stabilization. These can present a hazard, especially from closed landfills if the gases accumulate and migrate below the surface to escape through an opening outside the landfill boundaries. The greatest threat is from methane, a flammable gas, which at low concentrations in air is explosive.

Recently, a new subdivision of about 40 homes was declared uninhabitable in the Savannah, Georgia, area. These homes had been built on a previously unrecorded landfill, and the presence of methane was considered a sufficient hazard to require abandoning most of the houses in the subdivision.

Dump dynamics

Microorganisms are ubiquitous on the planet. They occur in soils, garbage, air, and water, and it is virtually impossible to eliminate them from a landfill operation. Also, since landfills, because of their size and operating characteristics, are operated without covers, there is a continuous though variable input of moisture. The microorganisms along with the moisture and a substrate (the biodegradable component of the garbage) then naturally interact.

A very simple description of the biological conversion of garbage (organic substrate) can be as follows: Initially, the air in void spaces in the garbage furnishes the oxygen needed for a very rapid though short-lived aerobic microbial decomposition, which produces carbon dioxide and some heat. However, because of the depth of the landfill (anywhere from 10 to 200 feet), and the presence of the intermediate and compacted final cover, the oxygen is soon depleted. At this stage a series of reactions mediated by anaerobic microbes (capable of living without oxygen) begin to occur. Extra-cellular enzymes secreted by the microbes first convert the solid substrate to water soluble constituents. These constituents are then absorbed by other microorganisms and converted to an array of low molecular weight organic acids, which are eventually converted to acetic acid. Finally, the acetic acid is fermented by another group of microorganisms to result in the end products, methane and carbon dioxide. An almost infinite set of other microbe-mediated reactions occur simultaneously.

Although the sanitary landfill was designed to minimize the infiltration of water and thereby minimize both the volume of leachate and the production of gas, these goals are virtually impossible to attain because of the nature of the agents responsible for the process—water, microorganisms, and food source—all three of which are always present in a landfill. Conventional landfill designs greatly prolong the natural stabilization process since moisture is excluded as much as possible.

Various processes have been proposed to solve the leachate problem. They include collecting the leachate and treating it like other municipal wastewater. Problems associated with this approach occur with both the volume of leachate produced and the concentration of soluble organic components in the leachate. Both parameters fluctuate widely, and cause problems in the wastewater treatment process. Also, there often are no conveniently located wastewater treatment plants that can process the leachate.

The landfill as a treatment system

A more ingenious approach, initially proposed by F.G. Pohland, then at Georgia Institute of Technology and now at the University of Pittsburgh, makes the landfill its own treatment system. All of the additional design and operational features required by this approach satisfactorily meet the proposed EPA regulations on landfill design and restrictions.

The old design attempted to exclude moisture from the landfill, hampered the migration of leachate and gases through it, and attempted to minimize microbial activity.

In contrast, Pohland advocates optimization of the microbial process in order to stabilize the landfill as rapidly as possible. This approach also maximizes the generation of methane and carbon dioxide gases, and hence the amount that can be recovered if a gas recovery system is installed. Depending on a landfill's location, income from the sale of these gaseous by-products may help recoup some of the costs incurred in installing and operating the leachate recycling system. Recovering the gases also helps minimize potential environmental pollution.

Although many new landfill systems incorporate a leachate collection and control system in order to minimize the potential for groundwater contamination, the collected leachate is usually pumped to a sewage treatment plant. In less than 30 percent of the cases the collected leachate is reintroduced into the landfill. However, this is only as a means of waste disposal, not as a means of more rapidly stabilizing the landfill system.

The leachate recycling approach to landfill stabilization requires that the landfill be designed, operated, and maintained like any other waste treatment system. Both the leachate recycling designs and the EPA 1990 rules for municipal landfills require that a landfill cell start with a liner impervious to landfill gases and leachate. At present these liners are sheets of high density polyethylene. Above the liner a leachate collection system will be installed and the solid waste material will be placed over the leachate collection system. Waste compaction requirement will be less for these systems since complete permeation of the waste by moisture in order to promote biological activity is one goal in the successful stabilization of the landfill. As the cell is filled with solid waste a series of gas recovery wells are also installed.

In place of the daily soil cover, which impedes both liquid and gas flows, a layer of a urea-formaldehyde foam, one to two inches thick, is sprayed over the exposed working surface at the end of the working day. Though

the foam material hardens, forming a water-impervious layer, it is readily broken up during the next day's operations and does not impede the flow of gases or moisture through the landfill. When the cell has been filled to capacity, a leachate distribution system is placed over the solid refuse and then the final cover material(s) are placed over the leachate distribution system. The completed system is then operated as a waste treatment system where the ultimate goal is the rapid stabilization of the biodegradable waste materials along with maximum production of methane and carbon dioxide. In addition, the design offers complete containment of all residual materials.

Some of the most advanced full-scale work investigating the advantages of leachate recycling is being conducted by the Delaware Solid Waste Authority along with the EPA and the University of Pittsburgh. Two one-acre landfill sites in Delaware are assessing the advantages of leachate recycling as well as evaluating a variety of liner materials for their suitability in landfill operations.

Leachate recycling benefits

Advantages of the leachate recycle/gas recovery system are manifold. Foremost is the fact that we have a system in which all components are contained and we have greater operational control of the process. The waste stabilization process can be optimized by the controlled addition of supplementary nutrients to the leachate if necessary.

The time required for biological stabilization is markedly reduced. In typical landfill systems, the stabilization process can take anywhere from 30 to 40 years or more; the leachate recycle/gas recovery landfill can be stabilized within a period of five

to six years. The process can be controlled through both the volume of moisture allowed to enter the cell and the rate of recirculation of the accumulated leachate.

The leachate recycle system can also be used in arid areas by addition of water to the recycle system in order to produce sufficient leachate for operational purposes. With gas recovery systems, maximum amounts of methane and carbon dioxide can be recovered, compressed, cleaned, and sold, a current practice with many landfills, which originally were not even designed for this purpose.

Other benefits include the complete containment of all materials that are soluble in the leachate, including many metals. Under the same conditions that some microorganisms convert acids to methane and carbon dioxide, other microorganisms mediate reactions that convert sulfate to sulfide. The resulting sulfide reacts with many metals, such as lead, iron, mercury, and manganese, to form highly insoluble metal sulfides. These metal sulfides are then precipitated as solids and remain within the landfill mass.

After the complete conversion of the biodegradable organic materials, the residual components remaining in the landfill are primarily nonbiodegradable materials and leachate. The leachate can then be removed from the system, analyzed for contaminants, treated as necessary, and appropriately discharged. The leachate recirculation and gas recovery systems can also be recovered, leaving the stabilized landfill site safe for other limited uses.

Current landfill gas production for the United States varies widely in those systems collecting gas, from 40,000 to 32 million cubic feet per day, with a mean of approximately 2.5 million cubic feet per day. Total production is approximately 380 million cubic feet per day, enough to supply nearly all the natural gas needs of the greater metropolitan area of Washington, D.C. In fact, the majority of the gas (about 70 percent) is used to generate electricity while the remainder (about 30 percent) is processed and sold as methane.

The above scenario will not solve all of the current landfill problems. Since 70 to 80 percent of the contents of a landfill such as plastics, metals, glass, and newspaper to name a few, are not anaerobically biodegradable. However, these materials can either be recycled or perhaps incinerated (waste to energy conversion). If these materials were recycled, the degradable organic matter then remaining would constitute a much higher proportion of the solid waste. The result would be smaller landfills that would produce more methane and carbon dioxide per unit weight of waste, thus resulting in a reduction of required landfill space with a concomitant increase in saleable by-products.

Solving the municipal waste disposal dilemma requires a concerted multifaceted approach of source reduction recycling, reuse, incineration, and landfilling improved by the use of leachate recycling.

WHERE THE AIR WAS CLEAR

In the '50s Mexico City was still pristine. In just one lifetime it became an environmental hell. Now La Capital is cleaning up.

Pete Hamill

Pete Hamill is a columnist for the New York Post. *His articles have appeared in* Condé Nast Traveler, Esquire, *and* Vanity Fair. *His most recent book is* Loving Women.

"And some of our soldiers even asked whether the things that we saw were not a dream. . . . When we had looked well at all of this, we went to the orchard and garden, which was such a wonderful thing to see and walk in, that I was never tired of looking at the diversity of the trees, and noting the scent which each one had, and the paths full of roses and flowers, and the many fruit trees and native roses, and the pond of fresh water. . . . Of all these wonders that I then beheld, today all is overthrown and lost, nothing left standing. . . ."
—Bernal Diaz del Castillo, a soldier of Cortés, describing Mexico City in 1519

"The weather is lovely, the air fresh and clear, the sky one vast expanse of bright blue, without a single cloud. . . ."
—Madame Frances Calderon de la Barca, LIFE IN MEXICO, 1839

"The first impact of Mexico City is physical, immensely physical. . . . A dazzling live sun beats in through a window; geranium scented white-washed cool comes from the patio; ear-drums are fluttering, dizziness fills the head as one is bending over a suitcase, one is 8,000 feet above the sea and the air one breathes is charged with lightness."
—Sybille Bedford, A VISIT TO DON OTAVIO, 1953

"Here is Makesicko City. . . ."
—Carlos Fuentes, CHRISTOPHER UNBORN, 1987

When I first came to Mexico City, I wanted to stay forever. The year was 1956. I was 21, a student on the GI Bill, a half-formed product of the tenements of Brooklyn. To be sure, I was not completely raw. I'd grown up in the shadows of the towers of Manhattan; as a sailor, I'd seen other American cities. But Mexico City was magical.

The magic of that time lives within me still. Say "Mexico City" to me, and in memory I am walking into the cool morning in the clear, thin air of the high mesa. The shopkeepers are hosing the sidewalks of the Avenida Melchor Ocampo. The sky is a scrubbed blue. Trees rise high above the street, green and lush as they take the first rays of the sun. I turn a corner, and away off in the distance to the southeast I can see the two great volcanoes, Popocatepetl and Iztaccíhuatl. The light is brilliant. My blood tingles. I whisper, Nothing will ever be this good again.

I was as wrong as all 21-year-olds are wrong, of course; the world is full of good places and astonishing people. But this *was* true: Mexico City was never as good again as it was that blessed year in the 1950s.

That year there were about 3.5 million people in the city the Aztecs called Tenochtitlan ("Place of the Cactus"). The traffic was so light there were no stop signs on the Paseo de la Reforma, the broad and splendid boulevard that serves as the city's major northeast-southwest axis; drivers merely hit the horn and prayed for clearance. On most other major streets, trolley cars still shaped the traffic pattern.

The air was so exhilarating that we walked everywhere then. The balance of people, work, place, and climate was matchless. We knew it, but so did the Mexicans. A year later Carlos Fuentes published his first novel, a panoramic, multi-layered equivalent of a Mexican mural. The subject was Mexico City. The title was *Where the Air Is Clear.*

 From *Audubon*, Vol. 95, No. 2, January/February 1993, pp. 38–49. Reprinted by permission of International Creative Management, Inc. Copyright © 1993 by Pete Hamill.

At the end of that magical year, I had to go home. But across the decades, I've kept returning. I have seen the changes and the growing catastrophe. In 1964 a few of the old places were gone—a small, noisy cantina, a bookstore, a restaurant where I used to have squid in its own ink. Many of the streetcars had vanished too, replaced with buses that coughed black exhaust fumes along the 14 miles of the Avenida de los Insurgentes. I noticed more people on the noontime streets and more beggars in the Zona Rosa, but it was still basically the city of the mid-1950s.

I was back in 1968, on the eve of the Olympic Games. For the first time, my eyes teared when I walked out of the airport. The city was jammed that year, the streets denser, the traffic slower, louder, more bad tempered. These were people drawn by the Olympics, I thought; when the games ended, they would go away. They didn't. I still loved the city, but in the mornings, I could no longer see the volcanoes.

The capital had grown into a megalopolis of 9 million human beings, sprawling south out of the Federal District into the state of Mexico. Arriving by airplane, I could no longer see the town; it was hidden under a urine-colored smog. By the late '80s it had the worst air pollution in the world.

I didn't see the volcanoes again until the fall of 1985. On the morning of September 19, an earthquake smashed into the city, measuring 8.1 on the Richter scale. As many as 10,000 people were killed; 56,000 buildings were destroyed or damaged; and half a million people were left homeless. For a few days, everything stopped: work, factories, all traffic except emergency vehicles. At night the city was eerily silent. But after the huge aftershock, after the smoke of the fires had blown away and the dust of the broken buildings had settled, something extraordinary happened. People on rescue teams paused in their work and looked up, and there, blue and clean, was the sky. And away off, serene and majestic, their cones white with snow, were the volcanoes.

A week later, when traffic resumed and the factories opened again, they were gone. A year later, birds were falling dead out of the poisoned air.

I spent the winter of 1987 in Mexico City, editing an English-language daily newspaper, and began to understand the problem. Statistics weren't very reliable; some reporters suspected that the government was holding back details to avoid frightening its own people or scaring off tourists and investors. But the known facts were appalling. About 4.3 million tons of contaminants were emitted into the air every year—a pound and a half a day for every human being in the megalopolis. On an average day the air was polluted with 600 tons of particulate matter, much of it fecal dust, which led to widespread salmonella and hepatitis infections, particularly among infants. Respiratory infections also were common. On 80 percent of the days of the year, the air contained unacceptable levels of ozone.

The city I loved had become a megalopolis of 15 million human beings. By the late '80s it had the world's worst air pollution.

In addition, about 40 percent of the people were living without a sewage system. Wandering through the so-called *ciudades perdidos*—"lost cities"—on the outer perimeter of the city, I saw garbage rotting in the streets and kids hunting rats for sport. There was no running water in most houses and only scattered electrical service. I visited several huge mountains of garbage, including the notorious Santa Fe dump on the southwestern edge of the city, where platoons of the city's 15,000 organized garbage pickers, called *pepenadores*—most of them women and children—scavenged through rotting meat and fruit, dead dogs, broken furniture, plastic bottles, old dolls, and plump cats. The stench was sickening.

But even the corrupt odor of the city's enormous dumps was overwhelmed by the general contamination. Everybody knew where that contamination was coming from: On each weekday, burnt gasoline rose from 2.5 million automobiles, 200,000 buses, 35,000 taxis, and many additional thousands of trucks. Together they accounted for 70 percent of the pollution. The rest came from the city's 30,000 factories. On a regular basis, international levels for sulfur dioxide, carbon monoxide, lead, and suspended particles were surpassed. Just breathing was said to be the equivalent of smoking two packs of cigarettes a day.

There were no simple solutions. To begin with, nobody could change the facts of geography. At 7,500 feet above sea level, Mexico City is located in a vast natural bowl surrounded by a wall of mountains, some of which rise another 5,000 feet into the thin air. The air contains 23 percent less oxygen than at sea level; auto engines must work harder, producing twice as much hydrocarbon pollution and carbon monoxide.

Everywhere, man has wreaked havoc. When the Aztecs founded the city in 1325, most of the valley was covered by a shallow saline lake and its surrounding lagoons, marshes, and forests; today the lake is completely gone, the forests felled, the marshes drained, the valley covered with houses and factories, asphalt and concrete. During dry periods, great clouds of dust, heavy with bacteria and excrement, rise to blanket many square miles of shantytowns. The built-up city spread from 45 square miles in 1940 to 483 square miles in 1990, with a present population density twice that of the New York metropolitan area, four times that of London. Seventy percent of the trees have been destroyed in the past 30 years, thus removing a desperately needed source of oxygen. One new city, Nezahualcóyotl, didn't exist when I was a student; it now houses 3 million human beings. On its bald streets there are almost no trees.

New York and Tokyo have shown that large urban populations can exist under certain conditions. But Mexico City is one of the world's few major cities that is not on a mighty river or an ocean; no fresh wind travels down river valleys or comes whipping in from the sea. The city is locked into what geographers call a closed hydrologic system: Wind velocity is less than 3.4 miles per hour, too weak to propel a flow of clean air across the choking valley; the slowness of the wind actually traps some contaminants in the air.

Although Mexico City, at 19°25' N latitude, lies within the Tropics, its altitude puts the average mean temperature at 65 degrees Fahrenheit. In winter, the upper air is colder, and the warm, seething soup of polluted city air can't even escape into the atmosphere. These thermal inversions are scary. With direct sunlight blocked, the days are dark, all color muted. You feel nauseated, indoors and out. You have steady headaches, feelings of drowsiness, and occasional dizziness. As the inversion continues, day after day, the psychological pressures can be worse. From the first morning glimpse of the unmovable roof of filthy smog, you feel smothered, convinced that you will never again feel healthy, never again see the sky.

Air isn't the only problem. Water is almost as bad. The city requires 177 cubic feet of water every second. This can't be completely supplied from local wells and pumps, so an elaborate system of pumps, canals, and tunnels carries water from rivers, springs, and wells on the far side of the surrounding mountains. Some water travels almost 100 miles and must climb 2,000 feet or more. This is not a one-way problem. The city's sewage also must be carried away. None of this is easy. Because of the old lakes, the subsoil of the city is a spongy compost, five parts water to one part solid. Even the "solid" parts are often volcanic ash and lava, which turn into a grainy porridge during extended rains. That is why earthquakes can be so devastating; the 1985 shock caused a great rolling movement—forward, then backward, then forward again—like water sloshing in a bathtub. Tall buildings were heaved first in one direction, then in another, until they snapped.

As pumps have sucked groundwater from below, some parts of the city have begun to sink as much as 10 inches annually. Sidewalks crack; underground pipes break apart, wasting millions of gallons of fresh water or allowing raw sewage to seep into the ground. In the city's famous Chapultepec Park, the water table has dropped drastically. Trees assaulted by pollution above ground are also being starved for water from below. No wonder many of them are dying.

There are two obvious causes of the city's environmental calamity: overpopulation and the internal-combustion engine. Mexico City now contains almost one-fifth of the country's 84.5 million citizens, and more arrive every day by bus and on foot. There is some evidence that internal migration to La Capital has slowed from the late-1980s peak of 1,000 arrivals daily. Many of the nation's poor are now head-

The city's altitude requires high-octane fuel. Until 1986 gasoline made by Mexico's Pemex refinery was the world's most highly leaded.

ing farther north, to the rim of *maquiladora* assembly plants stretched along the border with the United States. When (and if) the North American Free Trade Agreement goes into effect, that demographic shift is likely to intensify. But even if the city's borders were sealed tomorrow, the pressure of too many people would be extraordinary. Mexico's birthrate is down to 1.9 percent, but half the city's population is under 18, and every year another 300,000 children are born. In addition, advances in education, medicine, and health have extended the lives of millions; in 1940 the average lifespan was about 40 years; today it is more than 60. Population growth might slow, particularly among the middle class, but it will not end, and it is unlikely to be reversed.

The writer Jonathan Kandell explored some of the overlapping causes of the immense demographic explosion in his masterful history of the city, *La Capital.* The trend started during the Mexican Revolution, from 1910 to 1920, when 1.2 million peasants were forced by the violence to leave the countryside. One-third came to Mexico City, and when the shooting stopped, they did not go home.

In the city, many believed, their children would have a chance to lead a better life. There were hospitals, schools, and, because of government subsidies, food that was cheaper to buy than it was to produce. Dozens of Mexican movies and novels, as well as such sociological studies as *The Children of Sanchez,* by Oscar Lewis, showed the price many of these internal migrants paid: Most were unprepared for life in the city; too many turned to crime or were forced to live on the margins of modernity. But virtually nobody went back to the countryside.

In addition, during the years after World War II, government policy made Mexico City the locomotive of national industrialization. In retrospect this decision seems bizarre: The city is 200 miles from any port and remote from sources of energy. But it was the financial capital of the nation, the hub of the railroad system, and, as its size exploded, it became the largest consumer market in Mexico. A ferocious push-pull rhythm was established: The countryside pushed people out and La Capital pulled them in.

As more factories were built, more workers were trained for industrial production, and their existence attracted still more factories that needed trained cadres. By the early

1960s there was a basic assumption about Mexico City and its adjoining suburbs: You built factories there because there *were* factories there. Industrial production was also booming in Monterrey and beginning to boom in Guadalajara, but Mexico City was the place to be. Every month, new smokestacks seemed to rise into the sky. No environmental atrocity was turned away. In one northern suburb, an oil refinery flooded the air with wastes; on the southern rim of the city, a paper mill sent up foul odors day and night.

There were other forces at work in the making of the catastrophe. For the growing middle class, Mexico City was virtually the only destination. It was the peak of the symbolic Mexican pyramid, the psychological imprint that so deeply shapes the Mexican character. As it had been since colonial days, the city was the nation's political and bureaucratic capital. It was the communications capital, too, the headquarters for the country's publishing, music, film, and television industries. If you wanted to be a singer or an editor, a movie star or the president, you had to be in Mexico City.

By the mid-1970s almost everyone realized that something terrible was happening. Anarchic growth had to be tamed, and with the discovery of vast petroleum reserves during the post-OPEC boom in oil prices, it seemed that Mexico could at last choose both jobs and the environment. Javier Caraveo Aquero, a young Stanford-educated urban planner, was asked to create a master plan to clean up the city. With a small, passionate staff, he went to work.

In 1979 the 182-page plan was ready. The megalopolis would be divided into nine "service concentration centers," each, in effect, a self-sufficient city. Each would have its special employment base: a light, nonpolluting industry or business or bureaucracy. Each would contain a commercial district, leisure centers, and residential areas. Within each minicity there would be self-contained transportation systems, with shuttles that went to other areas of the megalopolis. The government would plant 119 million trees to increase the oxygen supply.

The plan was brilliant. Everybody said so. Yes, it would cost billions. Yes, it would need time—at least a generation—to have an effect. But in the end, Mexico City could be saved. The plan was accepted by the city government in 1980, and a series of hearings to deal with specifics took place over the following year.

One problem was immediately obvious: the unwillingness of various bureaucracies to surrender their own narrow powers. Because there were overlapping responsibilities in competing agencies, the infighting was brutal. The power of bureaucracies to obstruct movement is immense in any country, and even more powerful in a "controlled democracy" like Mexico; hacks will do anything to hold on to their own jobs.

Sometimes the most difficult problem of all in Mexico is getting reliable information; the hacks don't want to give their bosses accurate measurements of their own failures. On environmental matters, they don't want to provide hard evidence of the calamity that most citizens know has already engulfed them. In analysis of the master plan, delay, obfuscation, and deliberate vagueness became the order of the day.

In Mexico City such bureaucratic tensions are even stronger because the mayor is appointed. He is called *el regente*—the regent—and he serves a constituency of one: the president. Too often, city developments such as highways or extensions of the metro are approved behind closed doors, without public hearings. Trying to bring about genuine change, supported by the people who are most affected by it, is further complicated because the adjacent state of Mexico—an inseparable part of the megalopolis and the location of much of the polluting industry—*does* have an elected governor and must raise its own revenues. The result is that two separate local governments, with two separate tax bases, are trying to solve what is essentially one problem.

In the usually docile Mexican media, the master plan was discussed in great detail and with much passion. Some newspapers even went out and did actual reporting, instead of dutifully printing government handouts, and more columnists and intellectuals began calling for the election of the mayor as an important first step.

And then all the sound and fury became irrelevant.

In 1982 world oil prices collapsed, and so did the Mexican economy. When Miguel de la Madrid became president at the end of that year, all government energies were directed toward preventing default on the enormous debts owed to foreign banks. With 80 percent of its budget picked up by the federal government, Mexico City itself was bankrupt, unable to pay almost $3 billion in debts. Without vast amounts of money, no master plan could be executed. The 1985 earthquake both underlined and deepened the crisis. Who could worry about the 21st century when families were living in tents?

And so Mexico City muddled along, imprisoned in a shroud of pollution and fatalism. The rich began wintering near their sequestered money, in Miami or Aspen. Others established second homes in Cuernavaca or Tepoztlan, saying they needed to clean their lungs on weekends. During the day, people held handkerchiefs to their faces as they walked the streets. A few lonely souls continued to protest.

One of them was a soft-spoken poet, novelist, and diplomat named Homero Aridjis. The son of a Greek father and a Mexican mother, he was born in 1940 in the small town of Contepec, in the state of Michoacán. The area was lush with natural beauty, including the sanctuaries of the monarch butterflies that arrived each year from as far away as Canada. Aridjis was to carry the amazements of nature with him on all his future journeys. The first such pilgrimage was to Mexico City in 1957, where he went to study journalism.

"Mexico City was *the* city," he said of those years. "It was full of opportunities. It meant cultural, educational, economic, and social opportunities. The countryside was where beauty was; Mexico City was where culture was."

4. POLLUTION: THE HAZARDS OF GROWTH

Aridjis was soon a fellow of the Centro Mexicano de Escritores, a remarkable writing center that helped nurture such talents as Carlos Fuentes, Juan Rulfo, and many visiting Americans. "In my poems, nature was always present," he said. "Especially the quality of light, the transparency of light, a feeling for trees and animals—everything that later developed into ecological activism. Later, a contemplative relationship with nature developed into an activist relationship with nature."

In 1966 Aridjis was awarded a Guggenheim Fellowship, and with his American wife—the translator Betty Ferber—he left Mexico City for Paris. He was to live abroad for 14 years, eventually spending almost a decade in the Mexican foreign service as cultural attaché and then ambassador to Switzerland and the Netherlands. He came back to Mexico City with his family in 1980. By then, the evidence of catastrophe was inescapable. Aridjis was on his way to becoming an activist. "It was as if you were writing poems for a beautiful woman, for a loved one, and then you saw a gangster coming to kill her," he said. "I said to myself, If I love nature I have to defend nature."

By the spring of 1985 he had founded the Group of 100, made up of Mexican artists and intellectuals. The group included Nobel Prize winner Gabriel Garcia Marquez, Octavio Paz, Fuentes, and other stars of the Mexican intellectual world. Some were political enemies; others had personal or artistic rivalries. They united in the name of cleaning up Mexico City. They signed the first manifesto, and later press releases were issued in their names; in practice, though, the Group of 100 was the work of Aridjis and his wife. "I opened a door," he said, "and a river came out called ecology."

Since the city was the product of many forces beyond its borders, Aridjis was concerned with a wide variety of ecological issues: the poisoning of rivers, the logging of forests, the devastation of the countryside. All of them fed the streams that ended in the capital. The Aridjis group was not the only Mexican environmental group to be born in the 1980s. "This movement is not about one man," Aridjis said. But he had great skill as a communicator along with diplomatic experience; he was often more effective than the others in calling attention to the vast problems and dealing with the politicians who must correct them. Along the way, he learned many lessons.

"In Mexico, in ecological terms, it's always one step forward and *three* steps backward," he noted. "I always say that the Mexican political system is the most ecological in the world because they recycle all the time. They recycle politicians. You see people who are very corrupt. They get in trouble. They are fired. One year, six months later, they are back. Usually in another position. But back. The government is like a huge employment agency. That's why it's very difficult to change anything in Mexico. You have good ideas, good projects. But they are approved and then put in the hands of stupid, corrupt, or incompetent people, and they are doomed. And you often feel, Why get involved in the solution if this person is to be part of the solution?"

Mexico, he said, is not alone in its subjection to the combination of politics and greed; this is true of most Third World countries, particularly those where political democracy is more promise than fact. "There is an absolute relationship between ecological destruction and the political system," he said. "If you have no democracy, no respect for civil rights, it's very difficult to have respect for nature."

He now believes that overpopulation is at the core of the problem and that the lack of a population policy will cause immense problems by the end of the century. Water will be the worst. "Our water supply is being polluted everywhere in the country," he said. "Where will we get clean water?" Water pollution will cause a crisis in the food supply and rationing in the cities, he says, perhaps even famine. "The worst poverty in the world is caused by the degradation of natural resources. You become Ethiopia. Your people become ecological refugees. It could happen to us. We are already seeing some of this—people whose countryside is ravaged, so they have no choice but to come to Mexico City or go to the United States.

"The city is the epitome of the whole country's ecological problems," he said. "Every problem of the country can be seen here. The lack of planning. The exploitation of natural resources. Corruption, incompetence, the refusal to work by rules, pollution of the air and soul. We have no choice but to struggle, even if you are like Sisyphus, forever rolling the rock up the mountain. There is no choice. This is a question of survival."

In 1988 Carlos Salinas de Gortari became president in the closest election in Mexican history. His Institutional Revolutionary Party (PRI) had been in power since the 1920s but was shocked by the returns. Among other setbacks, the PRI received only 28 percent of the vote in Mexico City and lost every election district there. One reason was the government's failure to deal firmly with pollution. The problem was, in fact, getting worse: A 1990 World Bank study showed that the city was three times as noxious as in 1986, with lead levels in the blood of the average citizen four times higher than for a resident of Tokyo.

In his first year, Salinas moved to repair the political and ecological damage. He appointed a young technocrat named Manuel Camacho Solis as the new mayor of the city. Under De la Madrid, Camacho had served as minister of urban development and ecology; if the PRI was not yet prepared to allow the election of the mayor, it at least chose a first-rate man to deal with the city's enormous problems. He began to make immediate changes. The huge March 18 refinery, sprawling over 430 acres north of the city, was closed. It would be replaced by parks. Said Salinas, "Let's plant trees where today there is nothing but pipelines."

In November 1989 the mayor also started the "Day Without a Car" program, using color-coded stickers to force motorists to leave their cars home one day a week. The colors were determined by license plate numbers. "*Hoy no circula*," was the order: "Today don't drive." Violators were fined about $115—a month's pay at minimum wage, the heaviest traffic fine in Mexican history—and had their cars impounded. The program applied to commercial vehicles too (but not taxis) and took 500,000 vehicles off the street every weekday. The air became noticeably cleaner.

On many days residents could see the volcanoes. Some government studies insisted that pollution had been reduced 2 percent from the year before, although ozone levels remained significantly higher than Mexico's own standards. But the program did give many people hope. "It's remarkable for such an anarchic city, especially one with a history of apathy and individualism," said Louis Manuel Guerra, director of the Institute for Ecological Research. "It marks a turnaround in the consciousness of the people." Aridjis was more skeptical but said it was a positive step: "Without it, the situation would be terrible."

His skepticism was justified. As Mexico's economy improved, cars became more affordable. This made it easier for some to circumvent the law, particularly those who didn't want to use the city's excellent but overcrowded metro, which was handling 5.3 million rides every day. The rich already had more than one car; now the middle class started buying second cars, an estimated 600,000 of them. These were often oil-burning heaps (the average age of a Mexican car is between 8 and 12 years) that began thwarting the brief period of progress. Between 1988 and 1991 gasoline consumption increased 30 percent, according to the Los Alamos National Laboratory. Some ecologists fear that if the free trade agreement with the United States and Canada is an economic success, an expanding Mexican middle class could add several million more automobiles to the city. That would make gas masks a new growth industry.

Camacho also overhauled the city's bus system, replacing or refurbishing the fleet of 3,500 city-owned buses by March 1991. Taxis that had been built before 1985 were taken off the streets. Trucks were ordered to convert from gasoline to LPG (liquified petroleum gas). Diesel fuel was desulfurized. From November 1992 on, all buses and trucks entering Mexico City from the provinces were forced to undergo emissions testing.

As part of a plan to order all new cars equipped with catalytic converters, unleaded Magna Sin gasoline was made available all over the city. Until 1986, gasoline produced by the state monopoly, Pemex, was the most highly leaded in the world. This was because of the physical difficulties of combustion in the city's high altitude; lead boosts octane. To replace or reduce the lead content, Pemex engineers came up with a secret mixture of hydrocarbons. Alas, the vast majority of Mexican cars had no catalytic converters, and the result was a new problem: a drastic increase in ozone levels.

At ground level, ozone is produced by a photochemical interaction of the sun with gases such as nitrogen oxides and unburnt hydrocarbons. High above the earth, it is no danger to human beings. When it is inhaled at ground level, though, it attacks the lining of the lungs. At least one study claims that ozone itself may be a "mutagenic agent," and according to the World Health Organization (WHO),

high concentrations increase the risk of lung cancer.

In Mexico City, the maximum allowable ozone level (which complies with WHO standards) is exceeded on 320 days every year—the worst ozone pollution in the world. By May 1991, a group called the Mexican Ecological Movement planned to install 25 oxygen booths in the downtown area, where for $1.60 pedestrians could get a one-minute belt. Cartoonists loved the idea. Nothing changed.

Last spring, in spite of all the reforms, Mexico City was subjected to the worst ozone pollution ever recorded. The maximum level acceptable by international standards is 100 points. On February 6, 1992, the Mexico City ozone level reached 342 points. It was 332 points on February 10 and an incredible 398 points on March 16. Factories were ordered to cut production 50 to 75 percent. Schools were closed. Citizens hid behind closed doors or wore masks.

Then the ozone levels dropped by 120 points. The factories reopened. Citizens went back to forcing leaded fuel into unleaded tanks or removing the catalytic converters. Life, such as it is, went on.

The office of the mayor is at the top of a flight of worn stone stairs, past a heroic mural inspired by the triumph of the 1910 revolution. The hallways and anterooms are full of the players in the city's drama: thickset, blocky women in peasant clothing; lawyers and engineers in tailored suits; tall men from the northern provinces with elegantly tooled boots, high-crowned cowboy hats, and fierce mustaches; cops, reporters, secretaries. They exchange petitions, requests, and demands, approvals and denials in a ceaseless ritual of need and reward.

I am ushered into a room where the only empty chair is still warm from my predecessor. Outside, on the great wide plaza called the Zócalo, all traffic has been banned. It doesn't matter. A thin blue haze still engulfs the National Palace along one side of the square, the Metropolitan Cathedral along another. It snakes around the pillars and columns of the Supreme Court and the National Pawn Shop and enters through the doors and windows of this building, the offices of the Departmento del Distrito Federal.

In sentimental 19th-century prints there are bucolic views of the Zócalo, then called the Plaza de Armas, where these same buildings are etched with a hard clarity. There are thick trees everywhere, and a splashing fountain and men in elegant *charro* costumes riding equally elegant horses in the company of gowned women in high lace mantillas.

Those men and women are long gone, of course, and the horses long dead. But the trees are gone too—cut down early in this century to provide a field of fire for defenders of the palace. Now the Zócalo is an arid stone plain. And the clear lines of the etcher's burin are blurred by smog.

From his office, Manuel Camacho Solis presides as mayor of Mexico City, *el regente* of the Federal District. Princeton-educated, touted as a future president, Camacho is part of the Salinas generation of technocrats who are trying to bring Mexico into a prosperous 21st century.

They honor the country's revolutionary past and its bruised idealism, but they resist ideology and sentimentality. They are, above all, pragmatists. They are for whatever works.

"Dealing with the problems of Mexico City is quite complex," Camacho says, seated at a long, polished conference table. "Everything happens here. . . . Here you see, to a certain extent, the result of what's happening all over the country. The problems of other places come to Mexico City. But it's nice to know that those problems can be solved. Although there are many conflicts and tensions, finally it is a country that can talk. And can bring about agreements that can become policies."

Camacho compares the problems of Mexico City with those of other large cities in the Third World and is encouraged. "In so many cities of the world, there are so many cleavages—so many things that don't seem to have any solutions," he says. "Here there are many problems, but we don't have so many deep cleavages. And people think there are solutions to the problems. That makes a big difference."

Like all Third World countries (and certain areas of countries of the so-called First World), Mexico must deal with the conflict between jobs and the environment. "The only argument that really works," Camacho says, "is to reduce the cost in jobs while improving the environment. If one follows a policy of losing jobs then one will not have the possibility of financing the control of pollution. And then other problems might grow so fast they might become more important—problems of social unrest, for example. What's the best mix? The best mix is the one that protects most of the jobs while cleaning up the environment. And if something doesn't have any possibility of solution, then one must pay the price in jobs—but those are the exceptions.

"It's extremely difficult to create new jobs. It can take decades to create one hundred thousand jobs, five hundred thousand jobs. So if you destroy jobs, then you will not have the jobs and you will have all of the pollution."

The ideal solution for all cities would be replacing the internal-combustion engine with some nonpolluting alternative, but that seems impossible for now. Still, Camacho has worked hard to control the automobile. "We have to reduce its attractiveness," he says. The costs of parking and fines have been increased. Public transportation has received a high priority: New subway lines are being built, comfortable and secure buses are coming on line. And he is proud of the progress of the past three years: the reduction of lead, sulphur dioxide, and carbon monoxide in the air. "We're now concentrating on ridding the air of particles," he says. "And we must deal with ozone. We are making progress."

He admits that population growth is a problem but has no faith in master plans. "I don't think it's possible to have master plans for population control in any city of any underdeveloped country or less-developed country," he says. "Remember all those plans for Cambodia? Well, even with a dictatorship like the Khmer Rouge it didn't work. If you have freedoms in a country, it's impossible to think that the size of a city can be controlled. Of course, we are doing some things. You can make decisions about where to grow, where not to grow. You can stimulate regional development. But a master plan for controlling a city like ours cannot be done anyplace in the world."

Camacho was encouraged by the 1990 Mexican census, which, although it was disputed by some critics, showed that the population was actually smaller than most people thought: about 15.5 million. The growth rate was 1.85 percent, adding about 300,000 people a year. "That's still too high," Camacho says, "but the trend is going down." He also said that as municipal services—water, electricity, schools, hospitals—have been extended to the city's poorest sections, the birthrate has decreased. "There is a direct relationship between poverty and population growth," he says. "In attacking the worst poverty we have done in three years what usually takes fifteen."

There are many projects ahead. One is to convert to the use of natural gas as quickly as possible (there are two basic problems: finding spaces for distribution and establishing absolute safeguards against calamitous accidents). The universities are being encouraged to expand degree-granting programs in environmental studies (Camacho recently gave jobs to the top 100 engineering graduates from the Politecnico). He has reached out to the city's environmental groups, too, understanding that theirs is a basically adversarial relationship but welcoming ideas and criticism. The Salinas government has also arranged for almost $1 billion in credits from Japan to help transform the city's environment. Camacho remains optimistic. But he must also live in the city he governs. I asked him how he deals with the pollution himself.

"Someone gave me an air filter as a present, for my house," he said, smiling dryly. "But I had it put away. I didn't think it would be just for the mayor to have better conditions than the rest of the people . . . . And it would be a very bad lesson for my children. At the same time"—he laughed out loud—"those things don't *work*."

The year 2010 will be the 100th anniversary of the outbreak of the Mexican Revolution, the crucible from which modern Mexico was born. I asked Camacho what kind of city the capital would be then. He paused, glancing around his high-ceilinged office, and answered: "I think we will have a city that will show the world that under the most adverse conditions, the inhabitants were able to agree upon long-term policies necessary to make the city of 2010 viable. A city that would not have been viable if decisions had not been taken in the 1990s. And also, with all the problems, the city will not be as big as was predicted." He paused. "I'm doing everything I can so that it will not be a paradigm of urban disaster; it will be a paradigm of how under adverse circumstances people can find a way to live together. And be part of government. To do things right. Not to lose what has been acquired in the past."

He smiled and shrugged. "I plan to be living here in the year 2010," he said. "Probably my children will be living here, too. I've never thought about leaving."

RAVAGED REPUBLICS

TWO MONTHS AGO TWO COUNTRIES EMERGED FROM THE ASHES OF ONCE-COMMUNIST CZECHOSLOVAKIA. BUT LEFT INTACT IS SOME OF THE WORLD'S WORST POLLUTION.

R. Dennis Hayes

Winter brings special discomforts to the people of Prague. High above the cobblestones of famed Wenceslas Square, a seasonal blanket of warmer air gathers. The inverted atmosphere traps some of Europe's worst pollution below, where a pall of caustic micro-grime accumulates. The air is thick with smoldering lignite, the low-calorie, high-trash brown coal that heats homes and powers factories everywhere.

Barely three years after the Velvet Revolution overthrew Communism, the 74-year-old federation of Czech and Slovak peoples known as Czechoslovakia has changed once again. On January 1 the federation dissolved into two republics: Slovakia, made up of the smaller, eastern third of the country, with Bratislava as its capital; and the Czech Republic, incorporating the regions of Bohemia and Moravia, with Prague as its capital.

Remaining intact, however, is what many say is the worst environmental degradation in the world. On good days air pollution exceeds safety thresholds for sulfur dioxide and airborne particulates; noses, throats, and lungs are fouled by sulfuric acid and traces of arsenic, mercury, and lead. On bad days emergency alerts cancel school recess; the increased pollution sends people flooding to the doctor and to "clean air" retreats in Slovakia's Carpathian Mountains.

The culprit is sulfur dioxide, with an assist from nitrogen oxide and heavy-metal fly ash. All are airborne by-products of coal-fired plants. When the inversions occur in the spring, pollen joins the fray; many people pack a pump-action medicated inhalant.

The two republics' record-setting pollution levels have been fueled by brown coal. Plentiful and cheap, coal drove old Czechoslovakia's relentless industrialization since the 1940s. Coal fired the weapons foundries, oil refineries, steel smelters, and chemical plants. Disavowing the environmental devastation, Czechoslovakian Communism indulged a cult of big production at any cost. The release of poisonous metals, pesticides, fertilizers, and chemicals—in concert with brown coal's direct emissions—foretold disaster.

Pollution has destroyed over 50 percent of the forests and damaged nearly 70 percent of the rivers, including the Labe, the Jizera, and the Ohře, all clean half a century ago. In Slovakia, petrochemicals and agrochemicals concentrate at unwholesome levels in food and livestock and percolate into underground drinking water. There are not enough sewage treatment facilities and still no approved hazardous waste sites. The situation at the Petrzalka complex in Bratislava is typical of conditions throughout the country. The highrise, thin-walled, prefabricated concrete *panelaks*—government apartment buildings and shops—are home to 130,000 people whose untreated sewage flows directly into the Danube River below Vienna.

Both republics' power grids help explain why the Communist-era energy policy is now widely regretted. Coal combustion supplies 60 percent of the total energy consumed. Supplementing the coal are aging and leaking Soviet-issue nuclear plants and big dams that threaten drinking water and croplands in Slovakia and in downriver Hungary. Besides drying up wells, a giant dam at Gabcikovo has damaged a natural water purification system by draining sensitive wetlands, drowning forests, and trapping industrial pollutants in the dam's newly created reservoir.

From air, soil, and water, the food chain gathers the pollution. A 1990 study found lead in children at levels more than three times the amount certified as neurotoxic in the United States. In the Czech region of Bohemia, where brown coal is king, life expectancy is three to four years shorter than that for the rest of the country, and respiratory disease rates for children are twice as high as in non-coal-burning regions.

The Bohemian Basin is the sooty heart of the Black Triangle, the world's largest brown coal mining region, which extends to eastern Germany and southern Poland. The basin mines three-quarters of old Czechoslovakia's lignite, burns over half of it, and traps enough pollution to bury the needle on any air quality gauge.

How grim is it? Consistently worse than the worst air pollution in the United States, which, according to the Environmental Protection Agency, occurs in Weirton, West Virginia. In Weirton, sulfur dioxide accumulation averages 102 micrograms per cubic meter of air. That makes Weirton the only site exceeding the U.S. safety threshold of 80. (By contrast, average SO_2 readings are 13 for Los Angeles and 42 for the Bronx.) In the Bohemian Basin, average SO_2 concentrations soar above the 100 mark, reaching a maximum of 2,440 in the town of Osek. The record for a single day is 3,193, set in Prague during a 19-day temperature inversion in 1982. (The maximum recorded in Weirton is 361.)

At an elementary school in the town of Mezibori, located in the foothills above the coal mines, the principal and children sport the latest in Bohemian Basin street wear: pollution masks. At the school entrance, a board displays the day's airborne SO_2 reading: 128 micrograms per cubic meter, actually low for Mezibori. Principal Milan St'ovicek is looking for a pollution filter so children can exercise *inside* on days when the concentrations exceed 500.

Nearby, in the town of Most, Jiri Kunes, a pediatrician at the local clinic, says, "I know of no other place like this in the world." In 40 years Czechoslovakia destroyed dozens of historic towns and villages to dig and burn more coal. Most is now a smoking crater (when excavated, coal beds often ignite spontaneously). Most's citizens were dispatched to the *panelaks* that dot the hills above the basin's moonscape.

Between school and *panelak*, all too many children visit Kunes and his colleagues at the clinic. It is a telling rendezvous with the area's pollution. In the children's ward, premature babies occupy incubators at nearly twice the national rate. In one bed a pallid, clammy two-month-old stares vaguely through a glass bubble, inhaling vapors to ease her distressed lungs.

"Fetuses and children are especially vulnerable to the SO_2 and toxic metals from coal," says Kunes. Toddlers absorb up to 50 percent more lead than do adults. The figure is similar for pregnant women, and once in the bloodstream, lead can cross the placenta. Lead poisoning in children causes anemia, mental retardation, and irreversible neurological disorders. Too much cadmium damages the kidneys. Mercury attacks the brain and nervous system. Kunes shows us a young boy recovering from bacterial meningitis. He notes that in northern Bohemia the incidence of bacterial meningitis is about 12 cases per 27,000 children each year. In less polluted southern Bohemia, almost no cases are reported. "It's immunodeficiency," Kunes says, which is caused by the constant struggle against toxic exposures. He regards a premature baby and shrugs his shoulders. "Of course, the main cause is pollution."

Most scientists agree that heavy metals can enter the body through food, drinking water, and air. In Bohemia, Kunes points out, industrial chemicals lay siege to the senses. When airborne sulfur dioxide comes in contact with respiratory mucus, sulfuric acid forms. The acid sears tissue, rupturing mucosa-forming cells and capillaries in the nose, throat, and lungs. The acid then leaches poisonous metals from lodged ash particles into the bloodstream. Kunes explains how the related chemistry of acid rain has decimated Bohemia's pine forests, the most denuded in Europe.

Yet what is to be done? Before the breakup of the two republics, the emerging national energy policy reflected the painful trade-off's confronting the country. For example, former federal environment minister Josef Vavrousek resisted demands to shut down Soviet-designed nuclear reactors because that would have meant increasing coal-fired plants and air pollution. In other decisions, nationalist politics muddied the debates. After Vavrousek had negotiated an agreement on a federal waste management law, Slovakia rejected it as a political encroachment into Slovakia's "national" affairs.

Now everything, of course, is up for grabs. The old policy endorsed energy savings and coal-cleaning technologies, projecting an unspecified increase in alternative energy sources that included heat from solar energy. It also envisioned the replacement of two coal-fired plants with two new nuclear reactors.

The potential policy tilt toward nuclear energy raises questions about the wisdom of trading the problems of coal for those of nuclear power. According to a poll in 1991, the populace is split, with no clear pro- or anti-nuclear majority but with a high percentage of people still afraid of nuclear power. They have good reason to be afraid: the people of Czechoslovakia did not learn about the severity of the world's worst nuclear accident at Chernobyl in 1986 until three years later; Communist authorities had suppressed the information.

Energy savings could go a long way, greatly reducing the need, if not the hue and cry, for expansions in energy output. By common reckoning, the republics now waste over half the energy they so imprudently extract, ranking close to last among industrialized countries in household and workplace energy efficiency. Factories and *panelaks* lack meters, pipe insulation, and variable thermostatic controls.

But the biggest problem could be a holdover from the old Czechoslovakia. "A lot of old ways and apparatchik technocrats still prevail in energy policy-making," as one government official put it. "There are a lot of words about environmental friendliness and energy savings. But the conclusions are for building new power stations with an accent on nuclear and large scale."

Much of the region's environmental future depends on the kindness of strangers—outside forces and funding. The question is whether the new republics can reverse a condition that is being widely experienced as a health and ecological catastrophe. This much is certain: in sheer future-shock impact, the environmental consequences will rival any unloosed by the Velvet Revolution.

CHERNOBYL'S LENGTHENING SHADOW

DAVID MARPLES

David Marples is professor of history at the University of Alberta in Edmonton, Canada.

The nuclear disaster at Chernobyl has been the subject of books, articles, dramas, and films. The April 26, 1986, accident spurred protest movements and jump-started political parties, new government agencies, and charitable institutions. It was also regarded as a key factor in the development of glasnost in Mikhail Gorbachev's Soviet Union. But after millions of words, there is no agreement on the ultimate outcome of the event.

The collapse of the Soviet Union and the formation of independent states has further complicated the issue—the new governments are unable to meet the myriad costs of the accident and have only recently begun to coordinate their actions. Meanwhile, tensions between Ukraine and Russia over a number of political and economic issues have not helped.

More than seven years after the accident, what do we know about the effects of the disaster at Chernobyl? Can one make a definitive statement about its consequences? What has happened to the nuclear plant itself and the irradiated zone around it? Have there been radiation-related illnesses among the population? And what are the current casualty figures? With these questions in mind, I made several trips to Belarus and Ukraine in 1992 and 1993 to talk to scientists and concerned citizens, to visit hospitals, and to gather information from local libraries. I discussed related

issues with politicians and the heads of some of the more prominent charitable organizations. I had last visited the Chernobyl nuclear power plant and the Center for Radiation Medicine in Kiev in 1989.

With the benefit of hindsight, the April 28–June 30, 1986, evacuations of the 30-kilometer zone around the reactor are now known to have been grossly inadequate (although the complete emptying of the reactor city of Pripyat was efficient and necessary). The radioactive plume from the burning reactor at Chernobyl moved north and then west, spreading radioactive iodine across two-thirds of Belarus and on into Poland and Sweden. Longer-lived radioisotopes such as cesium 137 and strontium 90 were dispersed over a very wide area—some 100,000 square kilometers. Russia was most widely affected by the fallout path, but the most dangerous fallout was in Belarus and immediately around the Chernobyl reactor in Ukraine.

For three years, information about the extent of Chernobyl's effects was withheld by Soviet authorities. Early stories in regional newspapers such as *Pravda Ukrainy* in Kiev or *Sovetskaya Belorussia* in Minsk were exercises in Soviet doublespeak.

The accident had been "contained," went the party line. The only long-term effect would be an estimated 200 excess cancer deaths over the next several decades—and they would be impossible to differentiate from other cancer-related mortalities. Only 31 people had died. The farmland in the affected areas in Kiev and Gomel provinces would soon be returned to cultivation (a statement supported by Hans Blix, the director-general of the International Atomic Energy Agency [IAEA] in Vienna). The government had acted promptly and efficiently in dealing with the evacuation, decontamination, and the sealing of the fourth reactor.

Incredibly, by October 1, 1986, the Chernobyl station was back in partial operation. Today, even the most avid of Ukraine's proponents of nuclear energy condemn the decision to go back on line as a propaganda ploy to convince the world that there would be no long-term after-effects of the accident.

In May 1986, military reservists replaced volunteer cleanup crews—a decision that would make it next to impossible for future researchers to learn more about the effects of radiation on cleanup workers. None of the original military reservists were officially registered. The Ministry of Defense, when confronted with names compiled from unofficial sources, denied that particular individuals had been in the zone. Workers' illnesses were never attributed to radiation. President Gorbachev said little about Chernobyl, other than to refer to the "heaps of lies" that were appearing in the Western press. He did not visit Chernobyl until almost a year after the event. Yet documents published in 1992 and 1993 by Alla Yaroshinskaya, a journalist and former Soviet people's deputy, show that Gorbachev was intimately involved in discussions about the effects of Chernobyl and about the plant's problems. He also authorized the classification of all medical information; access was restricted to the Soviet leadership and to health authorities.

Several months after the accident, the destroyed fourth reactor at Chernobyl was covered with a concrete shell or tomb called the sarcophagus which, authorities declared, would last "for eternity." It began showing signs of disintegration within five years.

By August 1986, the Soviet nuclear authorities had also provided an account of the disaster to the IAEA in Vienna. The plant's operators—who had dismantled seven protective mechanisms at the reactor during a safety test that ended in a partial meltdown—were blamed. In July 1987, the plant's director and chief engineer, along with four others, were put on trial and given stiff sentences. The hapless manager, Viktor Bryukhanov, had not even been there at the time; but when he arrived at the scene on the morning of April 26, he failed to realize that the reactor itself had been destroyed.

Chernobyl was a Soviet-designed graphite-moderated (RBMK) power reactor; the design was never exported. The RBMK design was pioneered at the Sosnovyi Bor station near St. Petersburg, and Chernobyl was the second such plant to be constructed. (Its Russian twin is at Kursk.) KGB archives reveal that throughout the 1970s, scientists were concerned about station flaws, from the combustible bitumen on its roof to serious faults with the control rods that had been manufactured with shortened tips. There had been several minor accidents before 1986, and the RBMK was known to be unstable when operated at low power.

The report to the IAEA referred only glancingly to these flaws. In fact, the Soviet authorities were proud of the reactor. But Valery Legasov, first deputy director of the Kurchatov Institute of Atomic Energy and the head of the Soviet delegation to the IAEA, had become bitterly disillusioned by the Soviet failure to confront the flaws. He committed suicide on April 27, 1988, the second anniversary of Chernobyl. The St. Petersburg plant suffered a serious accident in 1991. The Ukrainian authorities, after imposing a five-year moratorium on new reactors in 1990, announced that the Chernobyl station itself would be permanently shut down by 1995. The G-7 countries also condemned the RBMK as inherently dangerous.

Evacuation

A week after the Chernobyl accident, approximately 135,000 people had been evacuated, about 90,000 from northern Ukraine and 45,000 from southern Belarus. They had lived in the zone created by drawing an imaginary circle with a 30-kilometer radius, with Chernobyl as the center. They were moved by order of Politburo members Nikolai Ryzhkov and Yegor Ligachev. The evacuation was lengthy and disorderly and took until June to complete, mainly because there were few locations or housing available for resettlement.

For the next three years, there was a bizarre belief in the Soviet Union that the fence around the evacuated area somehow marked the outer limits of radioactive fallout. No restrictions were imposed, for example, on fishing in the Kiev Reservoir, or on collecting berries and mushrooms in the forests of Gomel province to the north. Somewhat arbitrarily, the official tolerance level for radiation was set at 35 rem over a natural 70-year lifespan, or 0.5 additional rem per year over the natural background. Environmentalists argued that the figure ignored the amount of radiation the residents had received during the first days after the disaster.

By 1989, *Pravda* issued the first maps of the fallout area, and in the republics, the newspapers followed suit. The maps showed that the contaminated region was much wider than previously thought, accounting for more than

40 percent of the total area of Belarus and about 14 percent of that in Ukraine. (Russia's affected area was large, but it was still a small percentage of the area of the vast Russian republic.) Each republic's government was obliged to establish committees to deal with the effects of Chernobyl. In Ukraine, the parliament created a commission for Chernobyl headed by Volodymyr Yavorivsky, a Rukh (Popular Front) party activist. The ministry for Chernobyl was led by Heorhi Hotovchits, formerly a Communist Party leader in Zhytomry province, one of the most contaminated zones.

About 28,000 square kilometers of territory had been contaminated with more than five curies of cesium per square kilometer. Of the total, 16,520 square kilometers (59 percent) were in Belarus, 8,120 square kilometers (29 percent) in Russia, and 3,420 (12.2 percent) in Ukraine. Belarus had the most serious cesium contamination, with 3,100 square kilometers of land with more than 40 curies of cesium per square kilometer. Ukraine had only 640 square kilometers and Russia 310 square kilometers with the same degree of contamination.

Today, Ukraine has completely evacuated its population from its most dangerous region. But in Belarus, 22 settlements remain in highly contaminated areas, and in Russia, eight. In the spring of 1993, about 70,000 people in Belarus were still awaiting evacuation, according to Evgeni Kanoplya, director of the Institute of Radiobiology of the Belarussian Academy of Sciences, and the chief scientific investigator into the effects of the disaster.

The problems created by the Chernobyl disaster are overwhelming. In Belarus, aside from the illnesses directly attributable to radiation, a glance at the official picture shows the scale of the predicament. By 1990, 27 towns and 2,697 villages with a total population of about two million had suffered radioactive fallout.

Between 1986 and 1989, 257,000 hectares of land in Gomel and Mogilev provinces were taken out of agricultural production because of contamination. More than 1.2 million hectares of forest were contaminated—20 percent of the total forest reserve. Immediate measures were taken to evacuate 24,700 people from the Bragin, Narovlya and Khoiniki districts of Gomel province, and almost 10,000 apartments were hastily constructed for these first evacuees. More than 900 million rubles (in 1989 prices) were spent on road construction and repairs, and another 500 million on social services for the evacuees.

Farms taken out of production or evacuated had produced about 40,600 tons of grain, 49,000 tons of potatoes, 8,600 tons of meat, and almost 40,000 tons of milk annually in the early 1980s. From 1986 to 1989, over 600 settlements were decontaminated, 214 cattle breeding farms had to replace their stock, more than 4,600 buildings were destroyed, and more than 250,000 hectares of land were treated with high acidity liming to help prevent topsoil wind erosion, thus lessening the potential spread of radiation. In addition to this program, the population had to be supplied with uncontaminated food and water. In the first two months after Chernobyl, more than 3,000 shallow wells had to be cleansed, and various deep artesian pools were linked to city water networks because the sources of existing water supplies, rivers and reservoirs, were contaminated. The scale of the problem is reminiscent of the rebuilding after the German-Soviet war. In Belarus—the chief homeland of the legendary Soviet partisans—the two events are often compared.

Repercussions

The Chernobyl disaster had far-reaching political and economic consequences. In Ukraine, the most politically organized of the affected regions, Chernobyl played a key role in the formation of the Popular Movement and the Green World environmental group and its political branch, the Green Party. In 1990, through Green Party efforts, a widespread protest against the development of nuclear energy in Ukraine led the government to declare a five-year moratorium on commissioning new nuclear reactors. In Belarus, a Popular Front party emerged, in part as a protest against official secrecy over Chernobyl. The entire Soviet nuclear power program, one of the most ambitious in the world, was brought to a halt. Plans for the construction of 60 new reactors were postponed or canceled as a result of public protests.

But the collapse of the Soviet Union created a paradox. It slowed progress in dealing with the effects of Chernobyl and created energy shortages that have strengthened the nuclear power lobby today. Russia has announced an ambitious nuclear program to preserve its valuable oil and gas reserves as sources of hard currency earnings. Its neighbors, Ukraine and Belarus, which are heavily dependent on Russian energy sources, have wavered on the nuclear power question. But there is powerful support to go the nuclear route.

The Chernobyl plant itself remains in operation. A second reactor has been decommissioned, but plans to shut down the station entirely by the end of this year have been shelved. Belarus has introduced a scheme to build its first two nuclear power stations, to be based on foreign technology. Neither Belarus nor Ukraine has adequate funds to deal with the effects of Chernobyl.

In 1991, the sarcophagus encasing the damaged reactor began to collapse. Studies were hastily conducted by the Ukrainian Nuclear Safety Inspectorate by the spring of 1992. Since it had no domestic resources, Ukraine decided to hold an international competition for a new and safe design. The winning entry, selected from thousands, came from Campegnon Bernard, a French firm. However, it is still unclear how a new covering is to be financed. Some Ukrainian scientists have criticized Chernobyl's continued operation while radiation levels around the plant remain high and while the sarcophagus issue remains unresolved.

Health

Questions about the long-term health consequences of Chernobyl's widespread radiation have been controversial from the outset. Could one predict the number of fatal cancers by measuring the levels of radiation in the air or soil? Could one predict the ultimate numbers of likely casualties (estimates of which have varied from 200 to 500,000)? There was also disagreement about the numbers of early casualties caused by direct exposure. The official Soviet figure was 31 (including two deaths unrelated to Chernobyl at the time of the explosion), and the number was never increased despite documented evidence of subsequent deaths from direct contact with Chernobyl radiation. Some 5,000 decontamination workers had died by the fall of 1990, according to estimates by local Ukrainian officials, while the Chernobyl Union, whose membership is made up of cleanup crew personnel, estimated from their records that 7,000 of their number had died. There is no way to determine which figure is closer to the truth.

Collaboration among the respective governments and the former Soviet authorities might yield more precise information about early casualties. But neither collaboration nor figures have been forthcoming, and the newly independent states are in no position to provide accurate statistics. Former cleanup workers and many evacuees are difficult to trace. Some were in the zone for only 30 days. Moreover, many cleanup workers have died from heart attacks, more likely caused by intense stress rather than by the direct effects of radiation. One medical investigation in Ukraine also revealed that many workers, mainly men under 40, were suffering from sexual problems, specifically a loss of libido. Others have suffered unknown and evidently incurable skin diseases.

Some medical authorities have been reluctant to attribute any illnesses to radiation. For example, in late 1992, Dr. Evgeni P. Ivanov, director of the Institute of Hematology at the Belarus Academy of Sciences, after completing a study of the subject, declared confidently that there had been no rise in the number of leukemias in the republic. As the assessment appeared to be somewhat premature, I tried to interview Dr. Ivanov during my visit to Minsk, but none of his staff could determine his whereabouts and he declined to respond to telephone calls. A similar kind of assessment was made in May 1991 by IAEA researchers who also concluded (without examining cleanup crews or evacuees) that no significant medical effects were thus far discernible.

A possible increase in the rate of thyroid cancer was also dismissed. A 1991 IAEA report concluded: "No abnormalities in either thyroid stimulating hormone (TSH) or thyroid hormone (free T4) were found in children examined. No statistically significant difference was found between surveyed contaminated and surveyed control settlements for any age group."[1] It would be hard to imagine a more misinformed conclusion. By 1990–91, the appearance of thyroid tumors among children in the contaminated zones had made a sudden and dramatic appearance, increasing in Belarus by more than five times between 1989 and 1990, according to the Institute of Radiation Medicine's authorities. If this was unknown to the IAEA team, it suggests a lack of communication with the proper authorities.

In April 1993, I interviewed Dr. Evgeni Demidchik, director of the Thyroid Tumor Clinic in Minsk. He has conducted the most detailed study of the development of thyroid tumors in Belarus, a study that began in 1966 when thyroid cancer was a rare disease. After the Chernobyl disaster, he began to see a marked rise in childhood thyroid cancers. The rise began in 1990 and has continued.

By 1993, Demidchik had found more than 200 cases. In the two decades before Chernobyl, Belarus averaged less than one case per year. Moreover, virtually all of the cases came from the most contaminated regions. Over 60 percent, for example, are from the affected parts of Gomel, and about 30 percent are from the eastern part of Brest province (around the town of Pinsk), which was seriously contaminated by radioactive iodine in the first days after the disaster. All the children were born shortly before or during the time of the Chernobyl disaster. There is a clear correlation between these cancers and Chernobyl-produced radiation.[2]

(One curiosity, however, was the low numbers of such cancers in the Mogilev region in the northeast. Demidchik explained that in Mogilev, the local health officials had promptly issued potassium iodide tablets to the children, an initiative that may have prevented the spread of cancers.)

Demidchik says thyroid tumors are highly aggressive and the cancer spreads rapidly to other parts of the body; one child has already died. The tumor must be discovered and removed at an early stage. The procedure practiced in Belarus involves the removal of only the affected area, not the entire thyroid gland, which is the common practice in the West.

Some children have been sent to Germany for medical treatment, but Demidchik favors home-based treatment because the children must be monitored for life.

In addition to thyroid cancer, there also has been a significant rise in general morbidity among children. Radiation specialists have cited persistent neuropsychic disturbances, various types of dystonias, anemia, chronic respiratory problems, and chronic diseases of the digestive system. The vast majority of these ailments have occurred in the heaviest-hit Gomel and Mogilev regions.

The effects of low-level radiation are not well known. For instance, it is not known if the unexpectedly high rate of juvenile diabetes (generally thought to be hereditary) that has appeared in Belarus is radiation-related. In one area of Gomel, researchers have estimated that more than 50 percent of the population is suffering from psychological disturbances, and indeed, stress levels are particularly high. The IAEA study noted that evacuation is a prime cause of increased stress. This stress is exacerbated first by the lack of suitable new housing and also by the scarcity of jobs in new settlement areas.

The provincial and district authorities in the Brest region—one of the areas where serious contamination was discovered late—have complained that Gomel, the "center of contamination," has received all the foreign aid and improvements to health facilities, while Brest lacks doctors and modern medical care.

Charities

Given the uncertainty and the scope of medical problems, it is not surprising that the republics have been heavily dependent on charitable aid from foreign sources for medical treatment and equipment. Before the collapse of the former Soviet Union, most aid was channeled though Moscow, and much of it failed to reach its destination. Many aid groups operated duplicate programs. In 1990, for instance, there were 13 officially registered charitable organizations in Belarus. Occasionally the communist-dominated government created organizations with the same name or a name similar to that of an unofficial group,

possibly to divert funds into its own coffers. Today, most of the aid is directed to children, and "Children of Chernobyl" is the most popular name for charitable institutions in the United States, Belarus, and Ukraine.

How effectively is the aid being used? First-hand evidence suggests that it has become a vital component of each of the republic's health systems, even for illnesses that may not be related to Chernobyl.

Many medical facilities have Western equipment. For instance, the Center for Radiation Medicine in Kiev has equipment from the United States, and the Hospital for Sick Children in Minsk has German equipment. One enterprising charity in Minsk has devoted itself to alleviating the problems of psychological stress and various ailments (including the effects of radiation) by organizing one- and two-month summer trips abroad for children from the contaminated zones.

Under the auspices of the Belarussian Charitable Fund for the Children of Chernobyl, more than 30,000 children have taken these trips, many of them over the objections of the authorities, who have tried in vain to control them. The mainly communist-dominated authorities originally had their own program to send children abroad, mostly to places like Cuba. These efforts were unsuccessful mainly because the host countries were not well prepared. The authorities have hindered the efforts of the unofficial associations, who tend to solicit help from the advanced capitalist nations of the West.

Nearly seven and a half years later, Chernobyl's effects on health are beginning to be manifest. The virtual epidemic of childhood thyroid cancer seen in Belarus has also begun to appear in Ukraine, leading medical authorities to predict an increase in leukemias and other cancers in the future. Current health problems are probably the result of exposure to radioactive iodine (iodine 131, with a half-life of eight days); meanwhile, the problems caused by longer-lived cesium and strontium radioisotopes in the soil, which have begun to penetrate the food chain, have yet to appear—although they almost certainly will. Major problems may arise in unprotected areas or among the elderly, who have generally refused to take precautions and have continued to grow their own food since the disaster. The 600,000 cleanup workers represent a source of concern still to be thoroughly investigated. And the majority of evacuees do not appear to be listed on official registers. In 1990, a computer disk containing an evacuee list was stolen.

The surprising surge of thyroid tumors suggests that the medical world is entering an unknown and uncharted realm in dealing with

Chernobyl. The consequences are more serious than anticipated, and they come at a time when the newly independent states are least equipped to deal with them economically. One economist noted that, given the current estimated costs of Chernobyl, it would take the government of Belarus 180 years to pay for them. Yet Chernobyl has become a secondary concern to governments that lack energy resources for the coming winter and are desperately introducing deep economic reforms in the hope of obtaining international credits.

Finally, researchers worldwide have disagreed over different aspects of the results of the Chernobyl disaster, sometimes deriding the public response as emotional or "radiophobic." In fact, the question of the psychological impact remains to be decided. And since Chernobyl appears to have been a tragedy without a precedent, its results are unpredictable. If Western scientists are to continue to provide assistance to their counterparts in the former Soviet Union, they should discard their misconceptions on the likely effects of radiation. Chernobyl is a disaster that is becoming more

manifest, not less, with time. It has been a tragedy for Russia, Belarus, and Ukraine, and it remains the dominant nuclear event of the past four decades.

The end of the Soviet regime has made the situation more complex. Many technical specialists have left the republics and returned to Russia, and there is a continuing fear of living in a contaminated zone. But above all, the focus has been on the affected children—the future generations for Belarus and Ukraine. Whatever the reality, the perception is that Chernobyl casts a shadow over an entire generation.

Notes

1. International Atomic Energy Agency, *The International Chernobyl Project: An Overview: Assessment of Radiological Consequences and Evaluation of Protective Measures*. Vienna: International Atomic Energy Agency, 1991, p. 10.

2. Vasilly S. Kazakov, et.al., "Thyroid Cancer After Chernobyl," *Nature*, vol. 359, (Sept. 3, 1992), pp. 21–22; and David R. Marples, "A Correlation Between Radiation and Health Problems in Belarus?" *Post-Soviet Geography*, vol. 35, no. 5 (May 1993), pp. 281–292.

Common Threads

Research Lessons from Acid Rain, Ozone Depletion, and Global Warming

Michael E. Kowalok

MICHAEL E. KOWALOK wrote this article as a consultant to the Carnegie Commission on Science, Technology & Government. He is currently employed as a policy analyst for the Office of Science and Technology Policy of the Executive Office of the President in Washington, D.C. The views expressed here are those of the author and not necessarily those of the OSTP.

The process by which environmental problems are identified and evolve as scientific issues cannot be taken for granted. Environmental threats do not just "pop up" overnight; rather, they are identified by individuals who are constantly interpreting the results of numerous research efforts in many different disciplines.[1] As governments and scientific communities organize their research efforts and determine their priorities, they must be sensitive to the process of scientific investigation, research interpretation, and problem identification. A discussion that highlights this process can provide insight into how critical environmental problems have been identified and how they grew into important issues for the scientific community. Reviewing the various steps in the transformation of the issues of acid rain, stratospheric ozone depletion, and global warming from elements of scientific curiosity to matters of public debate should illuminate the main events in the complex process of discovery, thought, and action that led to current programs.

Acid Precipitation

Much of the present understanding of what causes acid rain and how it affects the environment rests on a long history of environmental observations and research. As early as 1661, investigators in England noted that industrial emissions affected the health of plants and people and that England and France frequently exchanged windborne pollutants. These investigators suggested remedial measures that included placing industry outside of towns and using taller chimneys to dilute air pollutants and to "spread the smoke into distant parts."[2]

In the mid 1800s, many features of acid rain were discovered and detailed by Robert Angus Smith, who was a chemist and Britain's first Alkali Inspector, or public official who monitored pollution. In 1852, Smith published a detailed report of the chemistry of rain in the city of Manchester and noted that the city air became increasingly acidic the closer one came to town. He also noted that sulfuric acid in the air caused textiles to fade and metals to corrode.[3]

Twenty years later, Smith was the first to use the term *acid rain*. In his 600-page book entitled *Air and Rain: The Beginnings of Chemical Climatology* he set forth many of the fundamental principles of acid rain. On the basis of detailed studies in England, Scotland, and Germany, he demonstrated that the chemistry of precipitation is linked to such factors as wind trajectories, the amount of coal combustion, proximity to the sea, and the amount and frequency of rain or snow. Smith noted that acid rain damaged plants and materials and proposed detailed procedures for the proper collection and chemical analysis of precipitation.

From *Environment*, Vol. 35, No. 6, July/August 1993, pp. 12-20, 35-38. Reprinted with permission of the Helen Dwight Reid Educational Foundation. Published by Heldref Publications, 1319 Eighteenth St., NW, Washington, DC 20036-1802.

Unfortunately, Smith's early findings and ideas were ignored or overlooked by scientists for nearly 100 years. Not until the late 1950s did Smith's work inspire more detailed research by Eville Gorham, a Canadian ecologist.[4] Gorham earned a doctorate in botany in England, moved to Sweden to study the effects of rainwater on the health of ecosystems, and then continued this research in the Lake District of northern England. One of the first things that he found was that the quality of rainwater depended on whether it had been blown from the industry-dominated areas to the south or from the Irish Sea to the east. Rains blown from the east were dominated by sea salt, and rains blown from the industrial south were dominated by sulfuric acid.[5]

This work led Gorham to study the ecological effects of acid rain, but he found it difficult to proceed because little was known about how different organisms respond to different levels of acidity. To try to see some effect on biota, he started to compare the distribution of sulfuric acid-laden precipitation to the distribution of human mortality by respiratory diseases in British cities. In the late 1950s, Gorham found that increasing amounts of acidic pollutants corresponded to a greater incidence of bronchitis. He also found positive relationships between the amount of sulfate particles in the air and the incidence of pneumonia and between the amount of airborne tar and the incidence of lung cancer.[6]

In the late 1950s, Gorham moved his research to Canada, and, by the early 1960s, he had demonstrated that acid in the rain near industrial regions was connected to industrial emissions and that increases in the acidity of aquatic ecosystems could be traced to acid precipitation.[7]

Although Gorham's work was a pioneering effort in understanding the causes and effects of acid precipitation, he was not the only scientist since Robert Smith to identify discrete parts of the acid rain phenomenon. In 1911, for example, two English scientists demonstrated that the acidity of precipitation decreased with distance from the center of Leeds. They associated the acidity with the combustion of coal and showed that plant growth and seed germination were inhibited by acid rain.[8] In 1919, an Austrian soil scientist noted that substances falling from the atmosphere accelerated the acidification of forest soils. In 1921 and 1927, a Norwegian limnologist reported on the relationship between trout production and the acidity of water in lakes and streams. In 1939, a Swedish scientist demonstrated the relationship between acidity and the toxicity of aluminum to fish. In 1948, Hans Egner, a Swedish soil scientist, established the first large-scale network for measuring precipitation chemistry in Europe. And, by the early 1950s, meteorologists in Sweden, England, and the United States had begun to search data from this network for evidence of atmospheric acidity.[9]

Thus, a substantial amount of work was being done on different symptoms or problems associated with acid rain. Few of these scientists recognized, however, that they were actually doing research on a much more widespread and complex phenomenon. Most of them saw themselves as working within their particular disciplines of limnology, atmospheric chemistry, or soil science. For example, when Gorham started his work in the Lake District, he had no idea that he would actually be studying air pollution; he was initially interested only in characterizing the ecology of certain types of lakes. The discovery that acid rain could escape from industrial areas into rural areas was "just serendipitous—

as so much research is."[10] Not until the late 1960s did anyone integrate knowledge from the different fields of science and recognize the complete problem of acid rain as it is understood today.

In the early 1960s, Svante Oden, a soil scientist at the Agricultural College of Uppsala, Sweden, began to study surface water chemistry, focusing mainly on the disciplines of limnology, soil science, and atmospheric chemistry. In particular, he relied

> **The discovery that acid rain could escape from industrial areas into rural areas was "just serendipitous—as so much research is."**

heavily on the records of precipitation chemistry collected by Hans Egner, whose network had gradually expanded throughout Scandinavia and most of Europe and came to be called the European Air Chemistry Network. As the first large-scale and long-term study on the changing chemistry of precipitation, the network provided Oden with nearly 20 years of data.

Oden also drew heavily upon the work of two Swedish scientists, Carl Gustav Rossby and Erik Eriksson, for data on the trajectories of air masses. Rossby and Eriksson had founded the science of atmospheric chemistry in the early 1950s and were convinced that atmospheric transport and deposition were major mechanisms for the dispersal and chemical transformation of many substances. They, too, had used Egner's data to test their ideas and to learn about the trajectory of air masses. Oden concluded from their work that invisible plumes of pollution could be transported long distances and could cause chemical changes in ecosystems far from cities and industry.[11]

In 1967, Oden was the first to publish a complete theory of acid rain. His initial conclusions were that

- acid precipitation was a large-scale regional phenomenon with well-defined source and sink regions;
- rain, lakes, and seawater were becoming increasingly acidic;
- air pollutants containing sulfur and nitrogen were being transported by winds over distances of 100 to 2,000 kilometers through several nations of Europe;
- the most likely cause of acid deposition in Scandinavia was airborne sulfur blown in from Great Britain and East and West Germany; and
- the probable ecological consequences would be changes in the chemistry of lakes, decreased fish populations, leaching of toxic metals from soils into lakes and streams, decreased forest growth, increased plant diseases, and accelerated damage to materials.[12]

Oden published his conclusions in two different media. In 1967, he outlined his concept of an insidious "chemical war" among the nations of Europe in Stockholm's prestigious newspaper *Dagens Nyheter*. This article captured the attention of the press and prompted more stories on the subject. In 1968, his article in *Ecology Committee Bulletin* stimulated interest in the scientific community.[13] Scientists from the disciplines of limnology, soil science, and atmospheric chemistry began to argue about his unconventional ideas. If Oden was correct, no longer could the solution to pollution be the dilution of an offending substance. Many scientists were inspired (or provoked) to design experiments to prove or disprove Oden's ideas. Multidisciplinary discussions and international conferences ensued throughout Europe and around the world.

The Swedish government sponsored a study in response to this growing scientific and public concern to assess more comprehensively the acid rain situation. Sweden presented the study's results at the 1972 UN Conference on the Human Environment. This was the first time that acid rain was raised as a

specific international air pollution problem during a major international conference. The paper focused world attention on the general problem of pollution drifting across national borders and on acid rain in particular.[14]

Although North American awareness of acid precipitation did not fully develop until the early 1970s, Canadian scientists had long recognized the potential problem because of the effects of sulfur dioxide emissions and acid precipitation in the vicinity of metal smelters. As early as 1939, a group of scientists had reported on the acidification of soils near a lead-zinc smelter near Trail, British Columbia. And in 1960 and 1963, Eville Gorham and his colleagues reported on the ecological effects of pollution from the smelting facility near Sudbury, Ontario.[15]

Studies in the United States date back to 1963, when two ecologists, Gene E. Likens of Dartmouth College and F. Herbert Bormann of Yale University, started an interdisciplinary study of a small watershed in New Hampshire. Their study included the chemistry of rainwater; and in 1972, they reported that the rain in the region was highly acidic, despite its distance from sources of pollution.[16]

Just prior to this report, Svante Oden had promoted scientific interest in the problem with a series of 14 lectures given throughout North America in late 1971. Oden's work helped to convince several key scientists, including Likens and Gorham, of the validity and immensity of the acid rain phenomenon. For example, Likens had

been skeptical of his own findings until he learned that Oden had found the same things in another hemisphere. Similarly, Gorham had believed that acid rain was only a local problem until he learned from Oden that pollutants could be transported over thousands of kilometers.[17]

In 1974, Likens and Bormann reported that rain in the eastern United States was 100 to 1,000 times more acidic than normal and that the probable cause was emissions of sulfur and

If Svante Oden was correct, no longer could the solution to pollution be the dilution of an offending substance.

nitrogen oxides from industry and electric power plants. This report sounded the alarm for U.S. scientists and the public, and the story was soon published in the popular press. The *New York Times* published an article on acid rain in June 1974.[18] At this point, the phenomenon of acid rain had clearly changed from being a scientific curiosity to being the focus of intense public debate in North America, one that demanded extensive new research activity and a national monitoring program. At congressional hearings in July 1975, Ellis Cowling testified that ignorance in North America about acid rain was largely due to the absence of coordinated research programs and precipitation monitoring networks. Cowling, a biochemist and botanist who had studied in Sweden and in the United States, would later address these problems by chairing a U.S. organization of federal, state, university, and industrial research agencies known as the National Atmospheric Deposition Program.[19]

Stratospheric Ozone Depletion

Unlike the issue of acid rain, which has long been associated with human activities and documented by environmental observations, scientific concern about human-induced destruction of the ozone layer dates only to about 1970. Ozone is a molecule of oxygen (O_3) that absorbs certain wavelengths of biologically damaging ultraviolet light. It is the only gas in the atmosphere that does so and, therefore, is an essential part of the Earth's ecological balance. The evolution of land life is believed to be tied closely to the formation of the protective ozone layer.[20]

Studies of the upper atmosphere started in the late 1800s and stemmed from curiosity about temperature gradients in the air. In 1902, a Frenchman named Teisserenc de Bort reported that the air temperature decreases with altitude up to a height of about 10 kilometers, after which the temperature starts to increase with height. Later studies showed that the temperature continues to rise with altitude up to a height of about 50 kilometers. The region of air between 10 and 50 kilometers of height became known as the stratosphere.

Most of the world's ozone is found in the stratosphere, and its daily creation and destruction is a natural process. The first person to explain the dynamics of ozone was Sidney Chapman, then a scientist at Oxford University. In 1930, he suggested that oxygen molecules (O_2) are split by cosmic rays into two oxygen atoms.[21] Each of these atoms then combines with another oxygen molecule to form ozone. He also suggested, however, that the single oxygen atoms can also destroy ozone by randomly colliding with O_3 molecules and creating two O_2 molecules again. The rate of ozone destruction increases with the amount of ozone present and continues to increase until ozone is being destroyed and created at the same rate.

Chapman was successful in explaining why ozone existed in the stratosphere, but his theory was soon found to be incomplete. Measurements showed that oxygen atoms were destroying ozone at only one-fifth the rate of creation and that the total amount of ozone present was much lower than what Chapman's model would predict. Some unknown process was destroying ozone, and scientists knew that it was not related to any of the other major atmospheric gases (nitrogen, carbon dioxide, and water vapor) because none of them reacts with ozone. The only other possible sinks were atmospheric trace gases, but scientists could not explain how gases of such small concentrations could destroy large amounts of ozone.

The realization that trace gases in the atmosphere could react catalytically with ozone provided an explanation. A catalytic chain is a series of two or more chemical reactions in which one chemical (the catalyst) destroys another chemical without itself being destroyed. The chain can be repeated indefinitely until all of the chemical is destroyed or until the catalyst is removed by some competing process.

In 1950, David R. Bates, an applied mathematician at the University of Belfast, and Marcel Nicolet, an atmospheric scientist at the Institut d'Aeronomie Spatiale de Belgique in Brussels, were the first to suggest that naturally occurring oxides of hydrogen (HO_x) were effective catalysts for ozone destruction.[22] In 1965, John Hampson, a scientist working at the Canadian Armaments Research and Development Establishment in Quebec, was the first to show how HO_x could catalytically destroy ozone in the stratosphere.[23] And in 1966, B. G. Hunt, a scientist with the Australian Weapons Research Establishment, proposed rate constants for Hampson's reactions to obtain agreement between predicted and observed amounts of ozone. (Both Hampson and Hunt were studying the upper atmosphere because their respective institutions were interested in the problem of re-entry of ballistic missiles into the atmosphere).[24]

In 1970, Paul J. Crutzen—an atmospheric chemist at the University of Stockholm who also worked at research institutes in Germany and the United States—was the first to suggest that oxides of nitrogen (NO_x) were another natural catalyst of crucial importance in determining the ozone budget.[25] Crutzen also recognized that oxides of chlorine, sodium, and bromine were able to destroy ozone, but he discounted their roles because, at that time, scientists knew of no significant natural sources of them in the stratosphere.[26]

Scientific concern about human-induced destruction of the ozone layer arose with debate about the environmental impacts of supersonic transports (SSTs). Combustion products of the SST included nitrogen oxides, sulfate particles, and a substantial amount of water vapor. Some environmentalists feared that the water vapor injected into the stratosphere would cause ozone depletion or excess cloud cover. In July 1970, a study group of 70 environmental scientists met at a conference entitled "Study of Critical Environmental Problems" (SCEP) to determine what was known about global environmental problems and what further research was needed. The discussion on the effects of the SST received the most media attention. The group concluded that the amount of ozone depletion caused by water vapor would be insignificant and that any problems associated with nitrogen oxide could be neglected.[27]

In 1971, the SCEP conclusions were attacked by Harold Johnston, a physical chemist at the University of California at Berkeley. Johnston showed that the SCEP report incorrectly discounted NO_x and that the SST did pose a serious threat to stratospheric ozone. Johnston was not a member of the SCEP study group but became involved in the SST/ozone issue by working within a government-sponsored program to study the stratospheric effects of SSTs. This program was a 3-year, $21-million effort known as the Climatic Impact Assessment Program (CIAP) and involved the efforts of more than 1,000 scientists from 10 different countries.[28]

Research attention turned to the role of chlorine in the stratosphere in 1972, when scientists at the U.S. National Aeronautics and Space Administration (NASA) recognized that the space shuttle's solid rocket boosters would inject chlorine directly into the stratosphere. Until that time, scientists thought that volcanic eruptions were the only source of stratospheric chlorine and that that source was negligibly small.[29]

In July 1972, the NASA environmental impact statement for the space shuttle revealed that chlorine in the form of hydrogen chloride (HCl) would be spread as an exhaust product along the shuttle's trajectory and that a large amount would be deposited directly into the stratosphere. Despite this finding, and in the absence of scientific concern about significant amounts of chlorine in the stratosphere, the statement initially concluded that the shuttle was expected to have no negative environmental impacts on the stratosphere.

To check its work, NASA awarded a contract to Richard S. Stolarski and Ralph J. Cicerone, then at the University of Michigan, to examine the statement for weaknesses. Although they were not atmospheric chemists, the two had been working on the dynamics of the ionosphere and saw the contract as a way to break into a new field of research. Stolarski was a physicist by training and Cicerone was an electrical engineer. They thought that the HCl might dissociate to free chlorine, and, by late 1972, they learned of a catalytic chain of ClO_x reactions that could destroy ozone. In the spring of 1973, Stolarski and Cicerone made a formal presentation to NASA that concluded that chlorine compounds from the shuttle could be a significant destroyer of ozone.

Stolarski and Cicerone were not the only scientists to see the shuttle's impact statement and focus on the chlorine problem. After they read the impact statement, Michael E. McElroy and Steven C. Wofsy of Harvard University began to consider chlorine dynamics independently from Stolarski and Cicerone. McElroy was an expert in planetary atmospheres and had studied the chlorine chemistry of Venus as part of NASA's planetary exploration program. He already knew that chlorine could deplete ozone, and he began to apply his pre-

vious findings to the Earth's atmosphere. Within a year, McElroy and Wofsy had designed models to predict how much ozone would be destroyed for a given input of chlorine, regardless of its source.

The activities of the Michigan and Harvard researchers crossed paths for the first time in September 1973 at a meeting of the International Association of Geomagnetism and Aeronomy (IAGA) in Kyoto, Japan. In his pre-

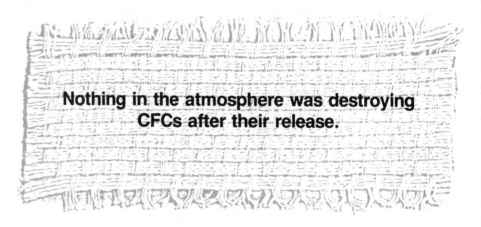

Nothing in the atmosphere was destroying CFCs after their release.

sentation, McElroy focused on the role of NO_x in atmospheric photochemistry. Stolarski, therefore, was the first at the meeting to speak about chlorine chemistry in the Earth's atmosphere. He gave a brief summary of his team's work but did not name the space shuttle as a source of stratospheric chlorine. He spoke instead about volcanic eruptions as a source. After the presentation, McElroy attacked Stolarski's talk, charging that the chemistry was incomplete and that volcanoes were not a significant source of chlorine. Others in the audience agreed that chlorine could indeed destroy ozone but felt that the topic was moot for lack of a major source. They wondered why McElroy and Stolarski were pressing the issue.[30]

All of the papers presented at the IAGA meeting were published in a special issue of the *Canadian Journal of Chemistry* in April 1974. McElroy and Wofsy updated their paper to include their work on chlorine's role in ozone reduction and explicitly stated that they had initiated this work because of concern about the space shut-

tle as a source of stratospheric chlorine. Stolarski and Cicerone also mentioned the space shuttle and showed that oxides of chlorine are even more efficient in destroying ozone than are oxides of nitrogen.[31]

Like the SST, the space shuttle forced scientists to consider the effects of anthropogenic pollutants on the stratosphere. However, this shuttle-induced work on chlorine was just half of the story. Any doubts that oxides of chlo-

rine could be a major destroyer of ozone were dispelled soon after the IAGA meeting when two chemists at the University of California at Irvine, Mario J. Molina and F. Sherwood Rowland, identified chlorofluorocarbons (CFCs) as plentiful sources of stratospheric chlorine. These chemicals are entirely synthetic and have a variety of industrial uses, such as refrigerants, insecticides, and propellants for aerosol cans.

Rowland had become interested in the dynamics of fluorocarbons in 1972, when he learned of the work of James E. Lovelock, an independent scientist working at the University of Reading in England. Out of curiosity, Lovelock had been measuring the concentrations of fluorocarbons in the lower atmosphere. He took measurements in both the Northern and Southern Hemispheres in 1970, 1971, and 1972 and found that the gases were in the air and sea "wherever and whenever they were sought."[32] In his 1973 *Nature* article reporting his findings, Lovelock suggested that the gases could be used as atmospheric tracers of air movements because they were chemically and physically inert. His work was not motivated by any possible link between fluorocarbons and the environment, and, to his later regret, he stated in 1973 that the compounds were of "no conceivable hazard" to the environment.[33]

By 1972, Rowland knew that Lovelock had compared his measurements of atmospheric CFC concentrations with annual amounts of industrial production. Lovelock found that the amount in the atmosphere was very close to the total amount that had ever been produced.[34] In other words, nothing in the atmosphere was destroying CFCs after their release. Rowland, intrigued by Lovelock's findings, began to study CFC dynamics in the summer of 1973. His rationale was that, because scientists were interested in using fluorocarbons as atmospheric tracers, it would be interesting to try to predict their chemical interactions. Molina joined him in October 1973. Like Rowland, Molina was an "outsider" to stratospheric chemistry and chose the issue out of a desire to do something different. Rowland and Molina soon found that CFCs could induce catalytic destruction of ozone and that the current annual amounts of CFC production could cause large rates of ozone destruction. Alarmed about these findings, Rowland met with Harold Johnston in December 1973 to discuss their plausibility. Johnston informed him that the catalytic chlorine chain that Rowland and Molina suggested was, indeed, plausible and had just been discussed at the September IAGA meeting in Kyoto.

In June 1974, Molina and Rowland proposed an alarming hypothesis in *Nature* that the use of chlorofluorocarbons added chlorine to the environment in steadily increasing amounts.[35] They suggested that, once released into the environment, CFCs had lifetimes of between 40 and 150 years and had no obvious sinks other than photodissociation in the stratosphere. Photodissociation would produce chlorine atoms, which would then catalytically destroy ozone. Molina

and Rowland concluded that the stratosphere had only a finite capacity to absorb chlorine atoms and that, even if CFC production were reduced, a lengthy amount of time "of the order of calculated atmospheric lifetimes" would be required for natural moderation. Rowland's and Molina's work suggested that fluorocarbons were indeed an ominous threat to the environment because they were a significant source of stratospheric chlorine and had already been released in sufficient quantities to begin ozone depletion.

News of the fluorocarbon threat hit the popular press in September 1974, when a meeting of the American Chemical Society prompted the *New York Times* to run a front-page story about the work of McElroy and Wofsy.[36] The article stated that the common aerosol can was a major source of fluorocarbons in the environment and that such gases could lead to ozone depletion. The day after the *Times* ran this story, another paper by Stolarski and Cicerone appeared in *Science* and concluded that the chlorine derived from CFCs in the atmosphere would become the dominant factor in ozone depletion.[37] The *Times* article signaled the beginning of public concern over CFCs and their use in aerosol cans and refrigerators. The story was soon picked up by Walter Cronkite of CBS-TV and by the major news magazines.[38] Much new research was initiated, and international collaboration was stimulated.

Global Warming

The greenhouse effect is the process by which heat radiating from the Earth's surface is trapped by atmospheric gases such as carbon dioxide (CO_2) and methane. The trapped heat raises global temperatures, which may significantly alter climate patterns.[39]

Scientific concern about global warming dates nearly to the time of the Industrial Revolution. In 1896,

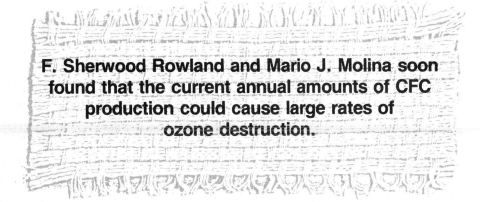

F. Sherwood Rowland and Mario J. Molina soon found that the current annual amounts of CFC production could cause large rates of ozone destruction.

Svante Arrhenius, a Swedish chemist in Stockholm, showed that the amount of coal use at the end of the 19th century was large enough to have an appreciable effect on the atmospheric concentration of carbon dioxide. He provided the first numerical calculations relating CO_2 concentrations to the Earth's surface temperature, and he estimated that a doubling of atmospheric CO_2 would produce a global warming of approximately 4° to 6° C.[40]

Arrhenius was not the first to explore the relationship between atmospheric gases and solar radiation. In 1908, Arrhenius cited French physicist Jean-Baptiste-Joseph Fourier, who had suggested in 1827 that CO_2 in the air helped to keep the Earth warm because the gas trapped heat as if it were a pane of glass. Arrhenius also referred to the work of J. Tyndale, a U.S. scientist, who recognized in 1863 that slight changes in atmospheric composition could bring about climatic variations.[41] And at the same time that Arrhenius was working on his calculations, T. C. Chamberlin, another U.S. scientist, was theorizing that the large variations in Earth's climate, including periodic glaciation, could be attributable to changing CO_2 concentrations.

In the early 1900s, scientists debated the relative importance of different sources of atmospheric CO_2. Arrhenius focused on industrial activities and energy usage and compared the current human activity to "evaporating our coal mines into the air." Others focused on volcanic sources. In 1919, U.S. scientist C. Schuchert noted that "life and its abundance are conditioned by the amount of CO_2 present in the atmosphere."[42] In 1924, Alfred J. Lotka, a physical chemist then at Johns Hopkins University, associated industrial dependence on fossil fuels with increasing amounts of CO_2 in the atmosphere. He noted that industrial activities would double the amount of CO_2 in the atmosphere within a period of 500 years, based on the 1920 rate of coal use.[43] In 1935, V. A. Kostitzin, a French theoretical ecologist, provided a simple model of linear and quadratic differential equations to describe the atmospheric circulation of oxygen, carbon, and nitrogen. His work was one of the first formal attempts to model such cycles.[44]

G. S. Callendar, a British steam technologist, made a direct link between industrial production of CO_2 and global atmospheric temperature in 1938. He reported to the Royal Meteorological Society that he had used temperature records from more than 200 weather stations around the world to calculate that the planet was getting warmer. He also said that this warming would continue indefinitely because it was linked to current human activity. His argument was dismissed, however, because his predicted temperature rise was only one half of one degree by the end of the 22nd century. This result was seen as not very threatening and may even have been viewed as an improvement.[45]

Callendar went on to speculate in 1940 and 1949 that a 10-percent increase in atmospheric CO_2 between 1850 and 1940 could account for the observed warming of northern Europe and North America that had begun in the 1880s. G. Plass, a U.S. scientist in the aeronutronic division of the Ford Motor Company, conducted similar work in the 1950s. He developed a new methodology that yielded the first "modern" estimates of global surface-temperature response to increased CO_2 concentrations.[46]

In 1957, Roger Revelle and H. E. Suess of the Scripps Institute of Oceanography disproved the general

Svante Arrhenius compared the current human activity to "evaporating our coal mines into the air."

scientific assumption that the oceans could harmlessly absorb almost all of the CO_2 emitted by human activities. Instead, they reported that most of the CO_2 produced by the combustion of fossil fuels would stay in the atmosphere and could eventually warm the Earth. Revelle and Suess noted that

human beings are now carrying out a large scale geo-physical experiment of a kind that could not have happened in the past nor be repeated in the future. Within a few centuries we are returning to the atmosphere and oceans the concentrated organic carbon stored in the sedimentary rocks over hundreds of millions of years. This experiment, if adequately documented, may yield a far-reaching insight into the processes of determining weather and climate.[47]

This statement has been quoted many times, but with a change in tone over the years. The original tone was detached and scientific, as if to suggest that humans could make good use of this accidental experiment. Later references to the statement, however, tended to emphasize the first two sentences and engendered an increasing sense of apprehension.

Gordon MacDonald, a scientist at the MITRE Corporation, later pointed out that the collaboration between Revelle and Suess was fortuitous because neither of them had been studying climate. Suess was interested in the cosmic rays that produce the carbon-14 isotope in the atmosphere. Revelle was an expert in marine sediments, which were presumed to be a major sink for atmospheric carbon. Suess had noticed that the carbon-14 content of new tree rings was deficient compared to ones that were 50 years older, which indicated that the ratio of carbon-14 relative to normal carbon-12 had decreased in the atmosphere. Revelle and Suess reasoned that, because the carbon-14 in fossil fuels decays over eons of burial, the CO_2 from fossil fuel combustion would have less carbon-14 than would normal air. In other words, if fossil fuel combustion was diluting the carbon-14 in the atmosphere with carbon-12, the ocean could not be removing CO_2 immediately nor completely from the atmosphere.[48]

In 1958, Charles D. Keeling, a chemist at the Scripps Institute of Oceanography, began to measure atmospheric CO_2 directly at the Mauna Loa Observatory in Hawaii. Keeling found that the total concentration of CO_2 increased annually but that it had a large annual oscillation between spring and fall. This oscillation was caused by green plants fixing CO_2 in the spring and releasing it in the fall when they decayed. Keeling plotted this concentration against time, and the resulting sawtooth diagram soon became a ubiquitous symbol for the greenhouse effect.[49]

In March 1963, the Conservation Foundation held a conference on the implications of the rising CO_2 content of the atmosphere. Participants included Keeling and Plass. The report of the conference concluded that the CO_2 situation might have considerable biological, geographical, and economic consequences; that a doubling of the CO_2 content of the atmosphere would produce an average atmospheric temperature rise of about 3.8°C; and that this increase could cause increased melting of glaciers and immense flooding of low-lying areas of the world's land surface.[50]

More research and assessments of the carbon dioxide problem accumulated in the 1960s and 1970s. In 1965, the President's Science Advisory Committee, a White House panel of distinguished scientists, raised the CO_2 issue as a national concern in a report entitled *Restoring the Quality of Our Environment*.[51] In 1967, Syukuro Manabe and Richard Wetherald of the Geophysical Fluid Dynamics Laboratory in Princeton, New Jersey, used a computer simulation to calculate that the average global temperature might increase by more than 4°C if CO_2 concentrations reached two times their preindustrial level. In Sweden, meteorologist Bert Bolin was undertaking similar studies at the International Meteorological Institute.[52]

Four eminent scientists reported to the U.S. Council on Environmental Quality in 1979 that humans are "setting in motion a series of events that seem certain to cause a significant warming of world climates unless mitigating steps are taken immediately." At about the same time, the U.S. National Academy of Sciences initiated a study of the greenhouse effect at the suggestion of President Jimmy Carter's science advisor, Frank Press. The study panel reported that a doubling of CO_2 in the atmosphere would raise global temperature by 3°C, plus or minus 1.5°.[53]

Despite these reports, not until the 1980s did study groups—such as the Scientific Committee on Problems of the Environment (SCOPE) under the aegis of the International Council of Scientific Unions (ICSU)—begin to treat the CO_2 issue as an international problem that demanded an international perspective in analyses and policy recommendations.[54] In November 1980, ICSU, the United Nations Environment Programme (UNEP), and the World Meteorological Organization (WMO) sponsored a meeting of experts in Villach, Austria. This group reported that CO_2-induced climate

> **Four eminent scientists reported in 1979 that humans are "setting in motion a series of events that seem certain to cause a significant warming of world climates unless mitigating steps are taken immediately."**

change was a major environmental issue but that, because of scientific uncertainties, it was premature to propose limits on CO_2 emissions. The group also emphasized that the CO_2 problem affected both developed and developing nations and suggested a special partnership of effort.[55]

Several major studies from other groups followed in 1982 and in 1983. Some assessments concluded that there was not sufficient scientific evidence to change patterns of fossil fuel consumption. Other studies suggested that immediate changes were necessary. Despite these differences, there was agreement on two basic issues: First, fossil fuel reserves were sufficiently large that, if they were exploited at increasing rates, environmental disturbances would occur; and second, these disturbances would vary from region to region, and the effects of a given change in climate on specific nations could not be predicted.[56]

An important research contribution came in 1985 from V. Ramanathan, a geophysicist at the University of Chicago, and Ralph Cicerone, then at the National Center for Atmospheric Research in Boulder, Colorado. These scientists argued that the trace gases methane, chlorofluorocarbons, and ozone in the lower atmosphere were just as important in causing global climate change as was CO_2.[57]

In late 1985, UNEP, WMO, and ICSU sponsored another study conference in Villach to assess the relative role of CO_2 and other greenhouse gases in climate variations and their associated impacts and to discuss the draft report SCOPE 29, which had been under way since 1980. Scientists from 29 developed and developing countries attended and reached a consensus that an immediate reduction in the rate of carbon emissions was appropriate. This result sharply conflicted with the conclusions of earlier groups because it reflected the consensus of a large and diverse group of experts.[58] Other conclusions of the conference were that past climatic data were no longer a reliable guide for predicting future climate conditions; the greenhouse effect was closely linked with other major environmental issues, such as acid rain and ozone depletion, because all are attributed mostly to human activities that change the composition of the atmosphere; and governmental policies could profoundly affect the rate and degree of future warming of the atmosphere.[59]

Another major result of the two Villach conferences was the work of Bert Bolin at the University of Stockholm.[60] Bolin and three other colleagues edited the SCOPE 29 report, which summarized the states of science and uncertainty as they related to greenhouse-gas concentrations and climate change. SCOPE 29 helped to frame the Villach discussions around three central issues: how to deal with the large uncertainties in the underlying science; the importance of international collaboration in research and resolutions; and the identification of research priorities.[61]

Bolin was also instrumental in creating the Advisory Council on Greenhouse Gases (ACGG). This unit was a small task force designed to help ensure that appropriate agencies and bodies followed up on the recommendations of the 1985 conference in Villach; ascertain that periodic assessments of the state of scientific understanding were conducted; and provide advice on further mechanisms or actions to address this issue. ACGG served until November 1988, when it was superseded by the Intergovernmental Panel on Climate Change (IPCC). IPCC was sponsored by UNEP and WMO, and roughly 30 nations signed on as initial participants.

The popular press seized the issue of global warming in the summer of 1988—one of the hottest summers on record—when James E. Hansen, then director of the NASA Institute for Space Studies, testified before the U.S. Senate:

Number one, the earth was warmer in 1988 than at any time in the history of instrumental measurements. Number two, the global warming is now large enough that we can ascribe with a high degree of confidence a cause and effect relationship to the greenhouse effect. And number three, our computer climate simulations indicate that the greenhouse effect is already large enough to begin to effect the probability of extreme effects such as summer heat waves.[62]

Many in the scientific community thought that Hansen's position was premature and that he made these statements without a proper peer review. Whether or not scientific consensus was near, however, the issue of human-induced global warming was on the public agenda to stay. In 1988, there was a tenfold increase in the annual number of articles that the popular press devoted to the greenhouse issue. Accompanying this media exposure was a flood of congressional activity. In 1989, 32 bills were introduced and 28 days of hearings were held by 9 congressional committees.[63] The issue was clearly a matter of public debate.

Similarities Between the Cases

A number of observations may be drawn from the case studies presented above. First, and most basic, environmental research is international. In each case, the identification and characterization of environmental threats involved the work of scientists from several different countries. Moreover, many of the vital research discoveries that helped identify environmental hazards were not made in the United States. International scientific conferences also helped promote problem identification and characterization. For example, after Svante Oden published his comprehensive work, several conferences were held for multidisciplinary discussions on the causes and effects of acid rain. Many researchers were skeptical about the theory and about their own research findings until they were able to exchange ideas and observations with colleagues from around the world. Such events helped to convince several scientists of the validity and scope of the acid rain phenomenon.[64] Similarly, the international conferences in Villach helped to advance the level of understanding of the greenhouse-gas issue and were valuable mechanisms that allowed a scientific consensus to be built that recommended immediate policy actions.

Secondly, scientific theories that explain environmental problems tend to be inter- and multidisciplinary. The identification of previously unperceived relationships among different disciplines of science led to the holistic understanding of the three environmental problems discussed here. For example, the total concept of acid precipitation could not be understood without integrated knowledge from the fields of limnology, atmospheric chemistry, and soil science. This relationship between disciplinary and interdisciplinary research points to the importance of both in the identification and understanding of environmental problems.

Third, research and development is serendipitous. One cannot predict from where or from whom significant research results will come. Sherwood Rowland and Mario Molina were outsiders to the field of stratospheric chemistry, and yet they discovered an efficient sink for stratospheric ozone. Recognition of these environmental problems percolated and grew from a dynamic body of research that could not have been preplanned, predicted, or prescribed. Contributions came from governmental and nongovernmental sources and from unexpected areas of science.

Finally, environmental monitoring is an essential and undervalued asset. The availability of continuous, long-term data on some elements of an environmental system is critical to monitoring environmental health and to characterizing problems properly. Without some sense of how an ecosystem works, scientists have little chance of understanding its dynamics or of diagnosing a problem. For example, the data provided by the European Air Chemistry Network enabled Svante Oden to advance his theories about acid rain. Similarly, the fluorocarbon measurements made by James Lovelock allowed Rowland and Molina to gauge the magnitude of the CFC problem. In addition, Charles Keeling's long-term measurements of atmospheric CO_2 concentrations allowed the increasing trend and seasonal fixation of CO_2 to become evident.

Analyzing the development of scientific understanding of acid precipitation, stratospheric ozone depletion, and global warming shows that environmental research is international, interdisciplinary, and serendipitous.[65] These similarities in the process of identifying critical environmental problems do not suggest, however, that the generation of knowledge and the steady advance of science led directly to the truth. The science that illuminated these environmental hazards did not grow along a linear path directly toward the identification and resolution of a problem. Rather, the research efforts created a pool of knowledge that expanded in unexpected directions. Scientists were able to tap this pool to learn about Earth processes and to recognize environmental problems associated with human activity. Scientific identification and characterization of environmental issues, therefore, result from individual scientists steadily interpreting a dynamic body of research.

NOTES

1. Gilbert F. White, professor emeritus, Institute of Behavioral Sciences, University of Colorado, Boulder, personal communication with the author, 23 May 1991.

2. E. Cowling, "Acid Precipitation in a Historical Perspective," *Environmental Science and Technology* 16, no. 2 (1982):155A.

3. Ibid., 111A; National Academy of Sciences, *Atmosphere-Biosphere Interactions: Toward a Better Understanding of the Ecological Consequences of Fossil Fuel Combustion* (Washington, D.C.: National Academy Press, 1981), 9–21; and C. C. Park, *Acid Rain: Rhetoric and Reality* (New York: Methuen & Co., 1987), 6.

4. Cowling, note 2 above, page 111A; R. Ostmann, Jr., *Acid Rain: A Plague Upon the Waters* (Minneapolis, Minn.: Dillon Press Inc., 1982), 115; and Park, note 3 above.

5. E. Gorham, "On the Acidity and Salinity of Rain," *Geochimica et Cosmochimica Acta* 7 (1955): 231–39.

6. Ostmann, note 4 above, page 117.

7. See the following series of papers: Gorham, note 5 above; idem, "The Ionic Composition of Some Lowland Lake Waters from Cheshire, England," *Limnology and Oceanography* 2 (1957):22; idem, "Atmospheric Pollution by Hydrochloric Acid," *Quarterly Journal of the Royal Meteorological Society* 84 (1958a): 274–76; idem, "The Influence and Importance of Daily Weather Conditions in the Supply of Chloride, Sulphate, and Other Ions to Fresh Water from Atmospheric Precipitation," *Philosophical Transactions of the Royal Society of London*, Series B, 241, (1958b): 147–78; idem, "Free Acid in British Soils," *Nature* 181 (1958c):106; idem, "Bronchitis and the Acidity of Urban Precipitation," *Lancet* ii (1958d):691; and idem, "Factors Influencing Supply of Major Ions to Inland Waters, with Special Reference to the Atmosphere," *Geological Society American Bulletin* 72 (1961): 1795–840.

8. Cowling, note 2 above, page 111A; and Gorham, note 5 above.

9. Cowling, note 2 above.

10. Ostmann, note 4 above, page 116.

11. Cowling, note 2 above; and Ostmann, note 4 above, page 96.

12. Cowling, note 2 above, pages 114A–115A.

13. S. Oden, "The Acidification of Air and Precipitation and Its Consequences in the Natural Environment," *Ecology Committee Bulletin*, no. 1 (Stockholm: Swedish National Science Research Council, 1968) as cited in Cowling, note 2 above.

14. Cowling, note 2 above.

15. Committee on Energy and Commerce, Subcommittee on Health and the Environment, U.S. House of Representatives, 98th Cong., 2d sess., "Acid Rain: A Survey of Data and Current Analyses" (Washington, D.C.: U.S. Congressional Research Service, 1984), 2; and Cowling, note 2 above, page 117A.

16. G. E. Likens, F. H. Bormann, and N. M. Johnson, "Acid Rain," *Environment*, March 1972, 33–40.

17. Cowling, note 2 above, page 117A.

18. G. E. Likens and F. H. Bormann, "Acid Rain: A Serious Regional Environmental Problem," *Science* 184 (1974):1176–79; and B. Rensberger, "Acid in Rain Found Up Sharply in the East: Smoke Curb Cited," *New York Times*, 13 June 1974, 1.

19. Cowling, note 2 above, page 118A.

20. S. Solomon, "The Earth's Fragile Ozone Shield," in R. S. DeFries and T. Malone, eds., *Global Change and Our Common Future: Papers from a Forum* (Washington, D.C.: National Academy Press, 1989), 73.

21. L. Dotto and H. Schiff, *The Ozone War* (New York: Doubleday, 1978), 33–37.

22. D. R. Bates and M. Nicolet, "The Photochemistry of Atmospheric Water Vapor," *Journal of Geophysical Research* 55 (1950):301.

23. J. Hampson, *Chemiluminescent Emission Observed in the Stratosphere and Mesosphere* (Paris: Presses Universitaires de France, 1965), 393.

24. P. J. Crutzen, "A Review of Upper Atmospheric Photochemistry," *Canadian Journal of Chemistry* 52 (1974):1569; Dotto and Schiff, note 21 above; and B. G. Hunt, "Photochemistry of Ozone in a Moist Atmosphere," *Journal of Geophysical Research* 71 (1966): 1385.

25. P. J. Crutzen, "Influence of Nitrogen Oxides on Atmospheric Ozone Content," *Quarterly Journal of the Royal Meteorological Society* 96 (1970):320.

26. Crutzen, note 24 above.

27. L. J. Carter, "The Global Environment: M.I.T. Study Looks for Danger Signs," *Science* 169 (14 August 1970):660. See, also, A. M. Hammond and T. H. Maugh II, "Stratospheric Pollution: Multiple Threats to Earth's Ozone," *Science* 186 (25 October 1974):335; and the testimony of J. E. McDonald before the U.S. Senate Appropriations Committee on 19 March 1971, *Congressional Record*, 19 March 1971.

28. H. Johnston, "Reductions of Stratospheric Ozone by Nitrogen Oxide Catalysts from Supersonic Transport Exhaust," *Science* 173 (6 August 1971):517; and Dotto and Schiff, note 21 above, pages 39–68.

29. R. S. Stolarski and R. J. Cicerone, "Stratospheric Chlorine: A Possible Sink for Ozone," *Canadian Journal of Chemistry* 52 (1974):1610; and S. C. Wofsy and M. B. McElroy, "HO_X, NO_X, and ClO_X: Their Role in Atmospheric Photochemistry," *Canadian Journal of Chemistry* 52 (1974):1582.

30. Much of this history can be found in Dotto and Schiff, note 21 above, pages 120–44.

31. Wofsy and McElroy, note 29 above, page 1582; and Stolarski and Cicerone, note 29 above, pages 1610–16.

32. J. E. Lovelock, R. J. Maggs, and R. J. Wade, "Halogenated Hydrocarbons In and Over the Atlantic," *Nature* 241 (19 January 1973):195.

33. Ibid.; and Dotto and Schiff, note 21 above, page 9.

34. Dotto and Schiff, note 21 above, page 12.

35. M. J. Molina and F. S. Rowland, "Stratospheric Sink for Chlorofluoromethanes: Chlorine Atom-Catalysed Destruction of Ozone," *Nature* 249 (28 June 1974):810.

36. W. Sullivan, "Tests Show Aerosol Gases May Pose Threat to Earth," *New York Times*, 26 September 1974, A1.

37. R. J. Cicerone, R. S. Stolarski, and S. Walters, "Stratospheric Ozone Destruction by Man-Made Chlorofluoromethanes," *Science* 185 (27 September 1974):1165.

38. Most of this history can be found in Dotto and Schiff, note 21 above, pages 6–26.

39. R. Pomerance, "The Dangers from Climate Warming: A Public Awakening," in D. E. Abrahamson, ed., *The Challenge of Global Warming* (Washington, D.C.: Island Press, 1989), 259.

40. Ibid.; J. H. Ausubel, "Historical Note," in National Research Council, *Changing Climate* (Washington, D.C.: National Academy Press, 1983), annex 2; M. Oppenheimer and R. H. Boyle, *Dead Heat: The Race Against the Greenhouse Effect* (New York: Basic Books, Inc., 1990), 34–42; and H. Ingram, H. B. Milward, and W. Laird, "Scientists and Agenda Setting: Advocacy and Global Warming" (Discussion paper prepared for the Association for Public Policy Analysis and Management Annual Meeting, Bethesda, Md., 24–26 October 1991).

41. J. Firor, *The Changing Atmosphere: A Global Challenge* (New Haven, Conn.: Yale University Press, 1990), 133; and G. MacDonald, "Scientific Basis for the Greenhouse Effect," in Abrahamson, ed., note 39 above, page 125.

42. Ausubel, note 40 above.

43. Ibid.; Oppenheimer and Boyle, note 40 above, page 35; and A. J. Lotka, *Elements of Physical Biology* (New York: Dover, 1956).

44. Ausubel, note 40 above; and see, also, F. M. Scudo and J. R. Ziegler, *The Golden Age of Theoretical Ecology: 1923–1940, A Collection of Works by Voltera, Kostitzin, Lotka, and Kolmogoroff* (New York: Springer-Verlag, 1978).

45. Ausubel, note 40 above.

46. Ibid.

47. Oppenheimer and Boyle, note 40 above, page 36.

48. Ibid., 224.

49. U.S. Environmental Protection Agency, *Policy Options for Stabilizing Global Climate* (Washington, D.C.: U.S. EPA, February 1989), 3.

50. Ausubel, note 40 above.

51. Science Advisory Committee, *Restoring the Quality of Our Environment: Report of the Environmental Pollution Panel* (Washington, D.C., November 1965).

52. Oppenheimer and Boyle, note 40 above.

53. G. M. Woodwell, "Biotic Causes and Effects of the Disruption of the Global Carbon Cycle," in Abrahamson, ed., note 39 above, page 74.

54. B. Bolin, B. R. Döös, J. Jäger, and R. A. Warrick, eds., *The Greenhouse Effect, Climatic Change, and Ecosystems*, SCOPE 29 (New York: John Wiley & Sons, 1986), xx–xxiv.

55. Ibid., 3; and World Climate Programme, *On the Assessment of the Role of CO_2 on Climate Variations and Their Impact: Report of the WMO/UNEP/ICSU Meeting of Experts in Villach, Austria, November 1980* (Geneva: WMO, 1981).

56. Bolin et al., note 54 above, page 7.

57. Ingram, Milward, and Laird, note 40 above, page 15; and V. Ramanathan, R. Cicerone et al., "Trace Gas Trends and Their Potential Role in Climate Change," *Journal of Geophysical Research* 90 (1985):5547–66.

58. Jesse H. Ausubel, personal communication with the author, 19 March 1993.

59. Bolin et al., note 54 above, pages xx–xxiv.

60. Gilbert F. White, personal communication with the author, 13 April 1993.

61. Bolin et al., note 54 above, page xvi.

62. James E. Hansen, Statement before the U.S. Senate Committee on Energy and Natural Resources, *Congressional Record*, 100th Cong., 1st sess., 23 June 1988, 31–80.

63. Ingram, Milward, and Laird, note 40 above, page 14.

64. Cowling, note 2 above, page 117A.

65. An examination of other environmental problems may suggest similar findings. For example, this author also looked at the progression of awareness of the threat caused by polychlorinated biphenyls. See K. P. Shea, "PCB: The Worldwide Pollutant That Nobody Noticed," *Environment*, November 1973, 25–28; and R. H. Boyle, "PCBs: A Case in Point," in Environmental Defense Fund, *Malignant Neglect* (New York: Knopf, 1979), 54–81 and bibliography. Although this subject was not included here, the reader is invited to consider other issues and anecdotes, including the threat to public health from lead poisoning and the danger to environmental quality of decreasing amounts of biodiversity.

Resources: Land, Water, and Air

- **Land (Articles 23 and 24)**
- **Water (Articles 25 and 26)**
- **Air (Articles 27 and 28)**

The worldwide situations regarding scarce energy resources and environmental pollution have received the greatest amount of attention among members of the environmentalist community. But there are a number of other resource issues that demonstrate the interrelated nature of all human activities and the environments in which they occur; these issues may ultimately be of greater significance than whether consumers in modern nations can continue to operate energy-intensive lifestyles. One such issue is that of declining agricultural land. In the developing world, excessive rural populations have forced the overuse of lands and sparked such a shift into marginal areas that, today, the total availability of land is decreasing at an alarming rate of two percent per year. In the developed world, intensive mechanized agriculture has resulted in such a loss of topsoil (millions of tons per year in the United States alone) that some agricultural experts are predicting a decline in food production. Other natural resources, such as minerals and timber, are declining in quantity and quality as well; in some cases they are no longer usable at present levels of technology. The overuse of groundwater reserves has resulted in potential shortages beside which the energy crisis pales in significance. And the very productivity of Earth's environmental systems—their ability to support human and other life—is being threatened by processes that derive at least in part from energy overuse and inefficiency and from pollution. To make matters worse, there is a feeling that both the public and private sectors, including individuals, are continuing to act in a totally irresponsible manner with regard to the natural resources upon which we all depend.

This section of readings begins with a selection that illustrates the relevance of these and other issues and places the concepts of interrelatedness and irresponsibility at the forefront. In "25th Environmental Quality Index: A Year of Crucial Decision," the editors of *National Wildlife* provide their annual report on the environmental crisis. In that report, 1992 is described as a year in which the American electorate moved to a crucial decision about the environment—whether to continue the economic-oriented environmental policies of the past twelve years or to select new policies based on the concept that environmental protection and economic growth were not mutually exclusive. As arguments raged over the true costs of pollution control, the American electorate finally voted to continue to support environmental cleanup by rejecting the environmental policies of the past in favor of still

unclear but promising new strategies. The wisdom of that decision remains to be seen.

Uppermost among the minds of many who think of the environment in terms of an integrated unit is the concept of the threshold or critical limit of human interference with natural systems of land, water, and air. This concept suggests that the environmental systems we occupy have been pushed to the brink of tolerance in terms of stability, and that destabilization of environmental systems has consequences that can only be hinted at, rather than predicted. Although the broad issue of system change and instability, along with the lesser issues such as the quantity of agricultural land, the quality of an iron ore deposit, the sustained yield of forests, or the availability of fresh water seem to be quite diverse, they are all closely tied to a single concept—that of resource marginality. As better, more available resources are used up, it becomes necessary to shift toward the more marginal resources.

In the *Land* section of this unit, two articles deal specifically with the issues of marginality and exploitation. "Beyond the Ark: A New Approach to U.S. Floodplain Management" offers a penetrating analysis of the management of lands that are highly productive and highly marginal—the floodplain environments. Traditionally the site of immensely productive agricultural systems, floodplains and their associated wetlands are also subject to periodic inundations by hurricanes, storm tides, heavy rains, and spring snowmelt. These lowlands adjoining the channels of rivers, streams, and other watercourses and the shorelines of oceans, lakes, and other bodies of water are as marginal as they are productive—a lesson learned by Midwestern farmers during the great spring and summer floods of 1993.

The companion piece to the first article in this subsection deals with the nature of the land and the concept of marginality on a very different level, moving toward the development of scenarios for sustainable agriculture in some of the most marginal of agricultural environments. In "Desktop Farms, Backyard Farms, or No Farms?" Canadian strategist Marc Zwelling addresses the question of how to feed the planet's growing population without destroying the planet's resources by describing a series of scenarios developed with particular application for Canada in the year 2020.

The second subsection of the unit focuses on *Water*, and in this subsection, the two articles deal with water quality on national and regional levels. In the first selec-

tion, the Clean Water Act of 1972 is evaluated by hydrologists Debra Knopman and Richard Smith. In "20 Years of the Clean Water Act," they arrive at a number of sometimes startling conclusions regarding this landmark environmental legislation: in 1993, scientists still cannot reliably answer the most basic questions about national water quality; scientific research on the effectiveness of the act focuses only on statistical summaries of regional conditions and almost no information is or will be available on individual locations; it will probably take more than a decade to amass the kind of data needed to document environmentally and statistically significant trends in water quality.

Moving from the national to the regional scale, the second article in this subsection examines the long and sometimes tragic story of human intervention in the economically marginal environment of the Florida Everglades. In "Redeeming the Everglades," Mark Derr tells the fascinating story of the loss of an environmental system that was, in its natural form, economically marginal; he also relates how, through the combined action of environmental interest groups and public agencies, a process of restoration of this fragile-yet-productive environmental system has begun.

In *Air*, the final subsection in this unit, the first article deals with the most critical of the problems that characterize the global atmosphere—continuing accumulation of greenhouse gases and the concomitant potential for increasing atmospheric heat. In "Global Warming on Trial," environmental scientist Wallace Broecker points out that while scientists such as James Hansen of NASA are convinced that global temperatures are rising as the result of an accumulation of greenhouse gases (mostly carbon dioxide), politicians and the business community remain unconvinced. The stakes in the debate are high, pitting society's short-term well-being against the future of the planet's inhabitants. Broecker notes that the debate is only the opening skirmish in a war of disagreements over the contributing causes to global warming: fossil fuel use, forest clearance, and overpopulation. In the companion article, "Exploring the Links Between Desertification and Climate Change," atmospheric scientists Mike Hulme and Mick Kelly examine the potential for global climate change (of which global warming is the primary component) from the standpoint of a single causation: the process of desertification or the degradation of lands in the marginal environments of drylands.

There are two possible solutions to all these problems posed by the use of increasingly marginal and scarce resources. One is to halt the basic cause of the problem—increasing population and consumption. The other is to provide incentives and techniques for the conservation and management of existing resources and for the discovery of alternative resources to eliminate the demand for more marginal resources.

Looking Ahead: Challenge Questions

What were some of the more crucial environmental issues that influenced the U.S. presidential campaign of 1992? Was there a relationship between public perception of environmental quality and the eventual outcome of the 1992 election?

Why are floodplains considered marginal environments when they are inherently so productive for agriculture?

How can methods of sustainable agriculture be brought into balance with the limitations of marginal agricultural lands?

Why is a good evaluation of the effectiveness of such environmental legislation as the Clear Water Act so difficult to obtain?

How can fragile yet economically marginal environments such as Florida's Everglades be retained as relatively undisturbed ecosystems while still allowing economic use consistent with the needs of contemporary society?

What is the relationship between the process of desertification and global warming? Is there evidence to support the contention that global warming is augmented by dryland degradation rather than the other way around?

25th ENVIRONMENTAL QUALITY INDEX

ILLUSTRATIONS BY SCOTT POLLACK

A Year of Crucial Decision

URING THE STORMY election year of 1992, the state of America's environment and natural resources entered the national politics of the country as seldom before—despite a universal preoccupation with the lingering economic recession. Environmental gains (such as improved air quality in some cities) and persistent problems (such as the continuing degradation of U.S. wetlands, fisheries and the atmospheric ozone layer) often took center stage in the political debate.

The U.S. electorate, besieged by economic worries, was wooed all year by widely divergent views of the relationship between the economy and environmental protection. President Bush, who said in 1988 that he wanted to be the environmental president and that he would counter the greenhouse effect

with "the White House effect," in 1992 ran his administration and his campaign on the premise that the country must choose between economics and ecology—and in a recession must sacrifice the environment.

In the final year of his four-year term, Bush announced his intention to make massive changes to the Endangered Species Act in order, he said, to save jobs for loggers of ancient forests; proposed regulations lifting restrictions of the Clean Air Act in order to make life easier for industries; and undercut an emerging international consensus on the need to reduce emissions of greenhouse gases in order to avoid costs he deemed unacceptable to American business.

His principal challenger, Bill Clinton, sought the White House on an environmental record as governor of Arkansas

that could best be described as spotty. But on April 22, Clinton defined a fundamental difference between himself and President Bush when he declared, "Over the years I've learned something . . . you can't have a healthy economy without a healthy environment, and you don't have to sacrifice environmental protection to get economic growth." The issue of environmental protection was as sharply drawn as it has ever been in a U.S. presidential contest.

Confused by contrasting claims, Americans moved in 1992 toward a crucial decision about the environment, and prepared to make it in the voting booth. On the following pages, *National Wildlife*'s annual Environmental Quality Index recounts many of the key events of the past year which helped form the backdrop for that decision.

WILDLIFE

Throughout last year, the battle lines were sharply drawn over measures to protect imperiled ecosystems and endangered and threatened species—notably regulations protecting the northern spotted owl of the Pacific Northwest. For several years, the bird has been a lightning rod for controversy and 1992 was no different.

To conservationists, studies have established the spotted owl as an indicator species whose plight signaled the ill health of the ancient-forest ecosystem. But for many Bush Administration and timber industry officials, the bird is a symbol of the ill health of an economy supposedly restrained from prosperity by environmental regulation.

This years-long controversy deepened in 1992 with the expiration of the Endangered Species Act. As the debate began last fall over reauthorizing the law, the owl was at center stage (see "Forests." At stake, said National Wildlife Federation President Jay D. Hair, is "the nation's crown jewel environmental law. It is going to be the fight of the century."

The increasing scale of extinctions, along with a few publicized conflicts between species survival and development, has immersed the act in controversy. Housing projects were allegedly held up by the plight of the California gnatcatcher, golf course construction pur-

Authorities refused to extend the use of turtle excluder devices, despite data showing they don't hurt shrimp harvests.

portedly by that of the Oregon silverspot butterfly. A proposed listing of California's delta smelt threatened water supplies of some residents.

Advocates of the law insisted that the

claims of interference were overblown. Studies by the National Wildlife Federation and World Wildlife Fund bear out that perspective. Of more than 120,000 projects found to have the potential to harm endangered species between 1979 and 1991, according to the studies, only 34 were prohibited by federal authorities; the others were modified and allowed to proceed. Yet as reauthorization time approached, said attorney Michael Bean of the Environmental Defense Fund, the law's opponents were trying "to make it appear that there is an endangered species problem in every congressional district in the country."

Such was the case along the Gulf and southeastern U.S. coasts, where shrimpers and their allies have been fighting against the use of turtle excluder devices (TEDs), which essentially are trapdoors inserted in nets to allow endangered sea turtles to escape. The devices have been required during certain times of year since 1989 to prevent loss of turtles,

Controversy heats up as lawmakers begin the debate over renewal of the expired Endangered Species Act

which have died in large numbers in shrimp nets.

Last year, three conservation groups, including the National Wildlife Federation, asked the Bush Administration to expand required use of the devices to all times of year to eliminate continued turtle losses. The groups released results of a two-year study refuting fishing-industry fears that the devices would reduce catches, damage nets or do little to cut turtle mortality.

However, Vice President Quayle's Council on Competitiveness recommended that no action be taken to expand use of TEDs. Then, after Hurricane Andrew hit the region in late summer, the White House issued an order temporarily suspending use of TEDs altogether in the Gulf. "It's easy to blame TEDs when the real source of shrimper problems goes much deeper, to such factors as foreign imports and overfishing," observed NWF attorney Robert Irvin.

While animals like the turtles and

spotted owl dominated the news, ducks continued to serve in 1992, as they have for decades, as a more traditional indicator of the health of the continent's ecosystems. Last spring, the U.S. Fish and Wildlife Service estimated that fewer than 30 million ducks flew from Mexico

A NWF study found that 43 percent of all U.S. endangered and threatened species rely on wetlands during their life cycles.

and the southern United States to Canada—a decline of more than 10 million birds since the early 1970s.

The Fish and Wildlife Service proposed that one duck species—the spectacled eider—be listed as threatened. The bird's numbers have declined by 94 percent since 1971.

Overall, the current duck decline is the worst recorded since the 1930s, when vigorous government intervention, in the form of habitat protection and bag limits, led to recovery of several species. Now the worsening figures have raised calls by scientists for more vigorous protection of wetlands—vital breeding areas for ducks.

Those calls were further buoyed last year by a National Wildlife Federation study, released in September, which found that even though wetlands occupy less than 5 percent of the land area in the Lower 48 states, 43 percent of all plants and animals on the Endangered and Threatened Species lists rely on wetlands at some point in their life cycles. The study also noted that wetlands "are being destroyed at the dizzying rate of 35 acres an hour."

WORSE   BETTER

AIR

LAST YEAR, the Bush Administration handed environmentalists a regulatory defeat after having taken credit for a legislative victory. The subject was the Clean Air Act. Revising the law was a 1988 Bush campaign promise, and the President cited the 1990 reauthorization as a major accomplishment of his administration.

In the legislation, the Environmental Protection Agency (EPA) was required to write regulations for issuing permits to potential polluters. The permits would set specific limits on emissions, and under the law, they could be changed only after government review and public comment. But in mid-1992, the President approved a change in the regulations. New language, proposed by Vice President Quayle's Council on Competitiveness and opposed by EPA Administrator William Reilly, would allow large companies to make "minor" changes in their emissions without prior notice or approval.

By citing a change in production methods, a company would be able to increase pollutant emissions by as much as 245 tons a year. The Competitiveness Council declared that obtaining new permits for such "minor" changes would be expensive. California Congressman Henry A. Waxman accused the administration of "carving the heart out of the new Clean Air Act" with a "knowingly illegal act."

Meanwhile, California continued to present both the worst and best case studies in the struggle for cleaner air. An analysis released at mid-year by the California Air Resources Board showed a 50 percent decline in the past decade in the number of hours per year that locations in the California basin experience hazardous air pollution. During the decade, average peak readings, which pose the gravest threats to human health, dropped by more than 25 percent. However, the lower average was still more than double the amount considered healthful by state standards.

Southern California saw less smog in 1991 than in any year on record. Smog levels, tracked for 15 years, have shown steady decreases since 1979, when the region's smog exceeded U.S. standards on 188 days. (The figure for 1991 was 129 days.) Nor was California the only source of good news; Kansas City became the largest metropolitan area in the nation to meet federal ozone standards, and Fairfield, Connecticut, reported

Smog levels decline in some areas, but the United States remains the leading producer of greenhouse gases

Citing economic harm, the administration refused to consider an international treaty to reduce emissions of greenhouse gases.

smog reductions of about one-third.

To gain further reductions in air pollution, the federal government and some states also deployed new market incentives rather than regulations. The Bush Administration proposed, for example, to offer relief from stricter pollution limits to companies that bought and destroyed old cars, reducing pollution by taking them off the road. The administration said a factory facing a cost of $25,000 to reduce annual emissions by one ton could instead eliminate a ton of emissions from old cars for $3,000.

Trading in pollution credits and debits was authorized by the revised Clean Air Act. The first two sales were made in 1992 by Wisconsin Power and Light, which had already been required by state regulators to clean up its emissions far beyond federal standards. The Wisconsin utility sold another company the right to discharge as many as 25,000 tons of sulfur dioxide.

In much of the country, cars still account for half of all air pollution. To help change this dilemma, the EPA announced stricter vehicle emission inspection programs for 80 major metropolitan areas, and basic testing for 55 cities where it was not required. The agency predicts that 30 percent of cars tested by the more sophisticated apparatus will fail, compared with a current rate of less than 10 percent.

Carbon dioxide emissions, in contrast to the modest gains against smog, remained high in 1992. The United States is the world's leading producer of the principal gas implicated in potential greenhouse warming. The average American used enough energy in 1992 to emit nearly 12 tons of carbon dioxide (while the average Japanese citizen was responsible for 2.5 tons of emissions). The U.S. output of 4.9 billion tons of the gas in 1990 equalled the combined output of China, Japan, Germany and India. But the Bush Administration, insisting that reducing greenhouse gases would harm the economy, refused to consider a proposed international treaty limiting greenhouse gas emissions to 1990 levels.

Worry intensified in 1992 over continuing decline of the ozone layer in the upper atmosphere, where the oxygen molecule screens harmful ultraviolet radiation. Alarmed by new data showing a threat to populations of North America and Europe, the United Nations Environment Program and the Bush Administration moved up to 1995 the date for banning CFCs, the chemicals thought to be doing the most harm to the ozone layer.

WATER

As PART of its latest inventory, released last April, the EPA reported that water pollution has decreased in the United States since passage of the Clean Water Act two decades ago. The study, conducted every two years, found that two-thirds of all surface water in the country meets water-quality standards. However, it also noted that contaminated runoff from farms, streets and lawns has yet to be even partially controlled.

"We've eliminated the gross forms of pollution that cause water bodies to smell and look foul, but those bodies continue to be contaminated by substances we can't see that are equally or more dangerous to our health and the health of our ecosystems," asserted attorney Bob Adler of the Natural Resources Defense Council.

As Congress began considering the rewrite of the 20-year-old law, which mandated the elimination of all water pollution by 1985, it found that the EPA had failed to write and enforce regulations for 80 percent of polluting industrial plants. As a result, said the conservation group American Rivers in its annual survey last year, "Rivers today in America are in worse shape than they've ever been." Supporting evidence was provided by the EPA's own analysis of more than a half-million miles of rivers (about one-third of the national total), which found nearly

The EPA reported that two-thirds of all surface waters now meet U.S. standards, but major contamination problems remain.

50 percent of the river miles too polluted to support their intended uses for recreation, drinking water or fisheries. About 60 percent of the impairment of rivers was traced to agricultural runoff.

The most endangered waterways, according to the American Rivers survey, were the Northwest's Columbia and Snake rivers, where 200 salmon runs (communities of fish identified by the time and place of their spawning) and more than 200 other native fish species are in peril. The problem in this case is not only pollution, but also eight federal dams on the Columbia and lower Snake that provide the area with cheap electricity and irrigation water. The upriver runs of salmon to spawning grounds are still possible because each dam has a fish ladder—a system of ascending pools easily negotiated by mature fish. But the dams take a high toll on young fish headed downriver for the sea.

In marked contrast to attitudes toward the spotted owl in the same region, salmon protection appears to enjoy broad public support. In response, worried by population declines, regulators reduced by one-half the 1992 salmon harvest permitted commercial and sport

U.S. rivers continue to be plagued by polluted runoff, while evidence of coastal contamination mounts

fishermen on the West Coast. The industry was already down 58 percent from its average take in the 1980s.

Meanwhile, there was encouraging news about a river and its salmon from the other U.S. coast. Connecticut's Salmon River was named for Atlantic salmon, which have not been seen in that waterway for two centuries. In 1965, federal and state officials began an ambitious program to restore the fish to the system, despite daunting pollution problems and the presence of 11 large hydro-electric dams and 700 small dams. The effort finally began paying off when, late in 1991, two Atlantic salmon were seen spawning in the Salmon River.

The good news was rare. In February in San Diego, a ruptured pipe designed to carry partially treated city sewage 2 miles out to sea poured 180 million gallons a day of the effluent into water just 3,000 feet from shore. The spill damaged fish populations, closed beaches and brought to public attention the vast

amounts of sewage being dumped into coastal waters. In 1991, there were more than 2,000 beach closings in 14 states because of raw sewage contamination. (An additional 10 coastal states do not test regularly for contamination.)

Florida alone discharges more than

A Florida study found that, instead of dispersing, wastewater can concentrate in coastal areas and possibly kill sea life.

300 million gallons of wastewater into the ocean every day. Researchers there found last year that instead of dispersing from an outfall pipe, as had been assumed, the waste can form stagnant concentrations that are fatal to plankton. According to one scientist on the project, "What we have seen sort of resembles underwater smog."

Coastal pollution is the principal culprit blamed for a spreading epidemic of blooms of harmful algae that now constitute a "major planetary trend," according to oceanographer Theodore Smayna of the University of Rhode Island. The so-called "red tides" are explosive growths of algae that contain trace amounts of toxins which, after concentrating in the tissues of shellfish and other algae-eaters, can become toxic, even fatal, to humans.

Overall, a third of the country's shellfish waters are now closed because of pollution. A NOAA study cited poor water quality, excessive harvesting, disease and the loss of 1.2 million acres of habitat, and predicted current trends could "eventually eliminate the natural harvest of shellfish."

FORESTS

WHILE STUDYING changes in the Earth's surface last year, a NASA research team compared satellite images of the Brazilian rain forest with photos of evergreen forests in the Pacific Northwest of the United States. The scientists found that the destruction of the Brazilian forest has occurred mainly at its edges, leaving vast areas undisturbed. But the U.S. forests are riddled with clearcuts, roads and other development.

Last June, in releasing the images to the public, the NASA team warned that the Northwest forests are on the verge of losing their biological vitality. U.S. Forest Service officials discounted that analysis, saying that they replant all clear-cut areas, but that young trees do not show up on satellite images until they are ten years old. However, biologists contend that this replanting does not eliminate the impact of habitat fragmentation, which causes extinction pressures because of inbreeding of restricted animal populations and increased attacks by predators that live on forest edges.

Meanwhile, the protracted controversy continued in 1992 over remaining ancient forests in the Northwest and the threatened spotted owls that live in them. In February, a U.S. District Court enjoined the federal Bureau of Land Management (BLM) from logging old growth on its land because of the likely

The U.S. Forest Service formalized its intentions to balance timber demands with wildlife habitat and recreation needs.

"adverse effect upon the survival of the northern spotted owl as a species." The decision followed a similar ruling against the Forest Service a year earlier.

In July, the Forest Service responded

to the 1991 injunction by presenting to District Court Judge William Dwyer a plan for meeting its legal obligations for protecting the spotted owl. Judge Dwyer found the plan inadequate, ruling that it failed to consider new evidence of the owl's decline and did not evaluate the plan's impact on other species. The judicial ban on logging old growth was extended, while Forest Service officials protested that it could take them at least two years to get the information Judge Dwyer wants.

The situation was further complicated last year, when a committee of federal cabinet members empowered to overrule the Endangered Species Act in cases of severe economic hardship met for the third time in its 14-year history. The so-called God Squad denied requests by timber interests to open up 31 Oregon old-growth parcels to logging, but voted to overrule the law's provisions on 13 tracts, comprising some 1,700 acres of forest under the jurisdiction of the BLM.

The administration attempts to settle an ancient-forest dispute by empowering the so-called God Squad

The decision had no immediate effect because of several impending legal tangles, including a current lawsuit brought by environmental groups who were challenging the decision.

While announcing the God Squad decision, Interior Department Secretary Manuel Lujan revealed a separate plan of his own. He called it a spotted owl "preservation plan," not a recovery plan, that recommended setting aside only about half of the bird's remaining old-growth habitat in the Northwest. Lujan admitted that his plan violated the Endangered Species Act and would consign the owl to extinction after another century, but said he would ask Congress to vote an exception in order to save logging jobs in the Northwest.

In fact, as another study confirmed in 1992, the forest industry has for some time been abandoning the Pacific Northwest for the South, for reasons unrelated to wildlife protection. Southern states offer a longer growing season that pro-

duces more wood in less time on cheaper land—at a cost of less than three-quarters that of Northwest timber. The study of a 12-year period ending in 1990 showed that the seven largest lumber and plywood companies had reduced their Northwest capacity by over one-third

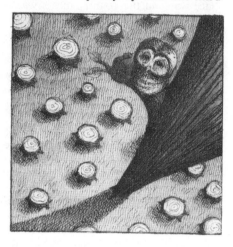

To protect the threatened spotted owl, U.S. courts took action in 1992 to prevent logging of Northwest ancient forests.

while more than doubling production capacity in the South.

The U.S. Forest Service formalized a new direction it has been discussing and testing for three years, by announcing in June that it is embracing the principle of ecosystem management. Forest Service Chief F. Dale Robertson told all regional foresters to implement within 90 days strategies to establish a balance among demands for timber, wildlife habitat, recreational opportunities, and "long-term stability to the ecosystem."

The apparent increased emphasis on recreation and wildlife—and the new commitment to long-term stability—implies a reduction of about 70 percent in the controversial practice of clear-cutting large tracts of forest. Robertson predicted that the new strategies would mean a short-term reduction in the volume of wood cut from the national forests, but no long-term reduction. The 1991 harvest was 8.5 billion board feet. However, many conservationists were skeptical of the announcement, citing continuing pressure on foresters to deliver more timber at the expense of wildlife and recreation.

ENERGY

DURING THE GREAT, lumbering game to determine the nation's energy policy, where the innings go on for months and the number of sides is innumerable, 1992 was the year Congress came to bat. President Bush's side had stepped to the plate in February of 1991 with a highly touted National Energy Strategy designed to boost domestic oil production and the nuclear power industry—and with barely a nod to the need for conservation, efficiency or alternative energy sources. The side was retired with no hits when Congress decisively rejected the President's plan.

The vote signified a rejection of oil exploration in Alaska's Arctic National Wildlife Refuge, and it put Congress squarely behind conservation as a major element of future energy strategy. "This is a watershed," said Nevada Senator Richard Bryan after the final vote. "For decades, our approach to energy problems was just to drill deeper. With this vote, that era is over."

With that, Congress went to work on its own version of an energy policy, determined not only to reduce American reliance on imports, but to promote efficiency and renewable energy. By the time it passed through the process last year, some of the clarity was lost; no attempt was made to raise the tax on gasoline or to increase the efficiency standards of new cars—two moves that could have greatly reduced U.S. dependence on foreign oil supplies.

On the other hand, the final bill which was sent to the White House last fall did contain a number of measures to improve energy efficiency in homes and commercial buildings. The bill mandates new standards for lighting, electric motors, plumbing products and heating and cooling equipment. It imposes new safeguards to protect fish and other wildlife at hydroelectric projects. It also rewards solar, wind and other renewable-energy producers with tax credits, and gives them access to utility-owned transmission lines. And it includes a federal agreement not to tax the progress already being made by the states and elec-

tric utilities. That progress is considerable and growing.

Some states have lost patience with federal dithering about energy policy and have moved ahead with their own strategies. Many electric utilities and their regulators have discovered that reducing demand is far cheaper in the long run

---

While the Congress finally passes an energy bill, the states continue to take positive action

---

With expenditures of some $2.5 billion, U.S. utilities promoted more efficient light bulbs and other conservation programs.

than increasing their electricity-generating supplies.

When regulators cooperate, electric utilities that give away high-efficiency light bulbs, help their customers weatherize homes and offer rebates for increased efficiency can make more money while their individual customers pay lower rates (and the environment suffers far less) than would be the case if they simply kept building generating capacity. Spending on what the utilities prefer to call "demand-side management" passed $2.5 billion in 1992.

As a result, New York City, for exam-

ple, has cut its projected growth in electricity demand by 80 percent and expects its peak summertime demand to be less in the summer of 2008 than it was in 1991. Confidence in the predictions was such that New York State cancelled a $6 billion hydroelectric purchase from Canada, dealing a blow to a massive dam project in Quebec.

Several states began looking at other ways to move ahead without federal leadership in 1992. A consortium of Northeast states moved toward imposing stringent California emissions standards on cars—a move that would force efficiency improvements along with air pollution reductions, while California considered challenging federal hegemony over fuel-efficiency standards.

However, the U.S. government demonstrated that it intends to preserve its exclusive powers in this area. Last year, when authorities in Maryland increased the state sales tax on gas-guzzling cars and provided a sales-tax rebate to those who bought gas-sippers, federal transportation officials ordered the state to rescind the law. Federal authorities argued that letting the states set different efficiency standards would unduly burden all U.S. automakers.

One quarter of the nation's electricity supply currently goes to lighting. Appropriate use of high-efficiency bulbs could reduce that amount by nearly half.

An EPA program called "Green Lights" has been taking that message to American corporations and government agencies, encouraging them to change light bulbs for their own economic good. By mid-year, more than 500 members had begun upgrades of nearly 3 percent of the nation's work space. Columbia University, for instance, expects to save roughly $2 million per year after replacing its wasteful light bulbs; the reduction in pollution will be equivalent to that from 2,300 cars.

SOIL

AFTER YEARS of heavy topsoil losses, a survey of American farming practices showed that more farmers than ever before were using conservation tillage (also known as "no-till") methods, which in some areas of the country have become the standard practice. Unlike the mechanized, chemically dependent farming that has dominated agribusiness for decades, conservation tillage avoids plowing and retains at least one-third of a field's plant residues in order to reduce erosion and increase soil quality.

In 1991, said the survey, conservation tillage was being applied to more than 28 percent of the nation's 281 million planted acres. In the Northeast and mid-Atlantic areas of the country, it was used on half or more of all planted fields.

Developed as a means of controlling erosion and required in some cases by the 1990 U.S. Farm Bill, conservation tillage has surprised practitioners by yielding crops at costs 25 to 30 percent below the standard methods. Authorities at the U.S. Soil Conservation Service said last year that the change in agriculture practices is unparalleled in this century.

In fact, conservation tillage affects many traditional agricultural practices. In recent years, several states have initiated programs to help farmers cut back on their use of pesticides and fertilizers. In Iowa, for instance, a joint EPA-state program helped farmers reduce their use of nitrogen fertilizers by more than 400 million pounds in 1989 and 1990—without significantly reducing corn yields. The program was designed to help solve serious runoff pollution problems in many of the state's drinking water supplies. Similarly, in Maryland, one farmer in a Department of Agriculture program reported last year that he saved approximately $40,000 in 1991 by heavily reducing his applications of chemical fertilizers.

A United Nations study of global soil conditions released in March found that more than 10 percent of the world's best topsoil—more than three billion acres—has been seriously degraded by human activity since World War II. While acknowledging that the United States has one of the world's best soil conservation programs, the study by the World Resources Institute in Washington, D.C., found that more than a quarter of U.S. cropland is still eroding too rapidly to easily correct. Areas of special concern

> **Many U.S. farmers are now embracing conservation tactics, but grazing is taking a toll on public lands**

To reduce erosion, U.S. farmers used conservation tillage on nearly a third of their 281 million planted acres in 1991.

were the Central Valley of California, the watersheds of the Mississippi and Missouri rivers and the hill farms of Washington State.

The issue of the use—or overuse—of publicly owned western rangelands returned to Congress in 1992 in the form of another attempt to enact higher grazing fees and put the fee receipts into range restoration. As the debate resumed (a similar bill failed in 1991), the General Accounting Office (GAO) weighed in with a study of grazing lands in the Southwest deserts managed by the BLM. The GAO concluded that BLM practices were degrading the land and threatening already endangered species while yielding little money to the treasury. The study, observed Oklahoma Congressman Mike Synar, "is just more evidence of the obvious—that taxpayers are subsidizing fiscal and ecological disaster on public land."

Another study by the Interior and Agriculture departments seemed to support Synar's assertion. Released in April, the report estimated the annual losses of the government grazing program at some $52 million. While ranchers paid private landowners an average of $9.66 to graze one cow for a month, the U.S. government rate to ranchers was only $1.92. However, the government's costs for administering the program averaged out to $3.21 per animal-unit-month.

Meanwhile, cattle ranchers trying to continue widespread grazing on sensitive public lands received a blow last year, when a federal court ruled against their challenge of the Toiyabe National Forest management plan in Nevada. The plan called for a sharp curtailment of grazing in the high-desert forest to protect fragile riparian, or streamside, areas.

Intervening on behalf of the Forest Service, the National Wildlife Federation and one of its affiliates, the Nevada Wildlife Federation, joined forces with the Natural Resources Defense Council to refute the ranchers' allegations in court. The ranchers had claimed that the Toiyabe National Forest management plan infringed on their traditional water rights in the region and also posed a serious economic hardship to them.

Announcing a national campaign to help promote grazing reform, NWF President Jay D. Hair said last summer that overgrazing "is not a sustainable agricultural practice and it is going to change." He predicted that there will be an end to the days of taxpayer-subsidized ranchers in this country.

QUALITY OF LIFE

DURING THE THIRD year of the Decade of the Environment, Americans listened to the often-contradictory opinions of scientists and politicians—and continued to support efforts to clean up the nation's pollution problems.

On the eve of the United Nations environmental summit last June in Brazil, a *USA Today* poll found that a majority of U.S. citizens favored efforts to combat potential global warming, even if such efforts are expensive. But President Bush refused to sign a proposed treaty with timetables for reducing greenhouse emissions on the grounds that it would be too costly to the U.S. economy. That argument—that pollution control costs jobs—was echoed repeatedly last year by Vice President Quayle, whose White House Council on Competitiveness sought to reduce many environmental regulations.

The Vice President argued, for instance, that thousands of auto workers could be thrown out of work if automobile fuel-efficiency standards are raised from the current 27.5 miles per gallon to the proposed 40 mpg to reduce emissions of carbon dioxide. But while most experts agreed that achieving the higher standards would impose substantial costs, they also pointed out that there are potentially immense environmental, economic and human-health benefits to be

Recycling is increasing but new EPA data show Americans are producing twice as much garbage today as they did in 1960.

derived from tougher rules. "The flip side is that some industries are going to benefit," Harvard researcher Dale W. Jorgenson told *The New York Times* last summer. One example: pollution-control

equipment and other related services, a market estimated at more than $100 billion a year.

While arguments raged over the true costs of cleanup, American commuters continued to add to the nation's mounting load of air pollutants by avoiding car pooling and mass transit. According to Census Bureau statistics, released last May, 73 percent of the country's 115 million workers drove alone to their jobs in 1990, compared to 64 percent in 1980. During the same period, mass-transit ridership declined.

"Hopefully, these figures will serve as a wake-up call for the country," observed Federal Transit Administrator Brian W. Clymer, referring to the difficulty of meeting federal mandates to relieve traffic congestion and clean polluted air. A surface transportation bill, passed by Congress and signed by President Bush in 1991, may help remedy the situation. The act enables states and communities, for the first time, to use federal gasoline

---

As arguments rage over the true costs of pollution control, Americans continue to support cleanup

---

taxes to fund mass-transit projects and other alternatives to automobiles.

Meanwhile, Americans continued to face serious contamination problems in 1992. New evidence, for example, verified that the number of children in the United States in danger from exposure to lead is much greater than previously thought, and that lead poisoning poses a health threat to middle-class as well as to low-income youngsters.

Lead is an industrial byproduct found in everything from household dust and water passing through old pipes to soil contaminated by auto pollutants and older lead-based paints. It is particularly toxic to children under the age of six, whose bodies absorb lead more easily than adults. It affects the central nervous system and can cause learning disorders and other problems.

In response to growing medical evidence, the Centers for Disease Control (CDC) lowered the danger point for lead poisoning by more than half. "In some

communities, we can expect a tenfold increase in the number of children at risk," said CDC spokesman Jerry Hershovitz last summer.

In the face of such alarming statistics, some Americans struggled to find a way to make the quality of life for future gen-

New evidence confirmed that the number of children at risk from lead exposure is much greater than previously thought.

erations better. More and more U.S. citizens, a remarkable 70 percent by some counts, embraced household recycling, only to learn that the markets for used newspaper and other products were saturated. "The supply revolution is well under way but the demand revolution is just beginning," observed Phil Bailey of the Buy Recycled Business Alliance.

Experts pointed to the nation's aluminum beverage containers, 62 percent of which were recovered and reused in 1991. "Recycled aluminum consumes 95 percent less energy than smelting new stocks of the metal," observed NWF President Jay D. Hair. "But it took the industry 20 years to reach its current level of recycling efficiency."

Though Americans are recycling more than ever, their trash piles are also growing higher than in the past. The latest EPA figures, released last summer, found that Americans threw out 196 million tons of refuse in 1990, an increase of about 8 percent from 1988 and more than twice as much as in 1960. That amounted to 4.3 pounds of trash a day per person in 1990, up from 2.7 pounds three decades earlier.

WORSE   BETTER

Beyond the Ark

A New Approach to U.S. Floodplain Management

Jon Kusler
Larry Larson

JON KUSLER is the executive director of the Association of State Wetland Managers in Berne, New York. **LARRY LARSON** heads the Floodplain Management and Dam Safety programs for the Wisconsin Department of Natural Resources and is the executive director of the Association of State Floodplain Managers in Madison, Wisconsin.

F loodplains occupy a significant portion of the United States. About 7 percent, or 178 million acres, of all U.S. land is floodplain, and, of course, the percentages are much higher along the coasts and major rivers, where most of the larger cities are located.[1] Floodplains are lands subject to periodic inundation by hurricanes, storm tides, heavy rains, and spring snow melt. They are the lowlands adjoining the channels of rivers, streams, and other watercourses and the shorelines of oceans, lakes, and other bodies of water.

Floodplains are shaped by water-related, dynamic physical and biological processes and include many of the nation's most beautiful landscapes, most productive wetlands, and most fertile soils. They are home to many rare and endangered plants and animals, as well as to sites of archaeological and historical significance. In their natural state, floodplains have substantial value. These complex, dynamic systems contribute to the physical and biological support of water resources, living resources, and cultural resources. They provide natural flood and erosion control, help maintain high water quality, and contribute to sustaining groundwater supplies. Therefore, proper management of floodplains is important to preserve their value and to reduce losses caused by flooding.

The United States is now at a pivotal point in floodplain management. A national status report on floodplain management was released last year by the Federal Interagency Floodplain Management Task Force, and federal agencies responsible for reducing the losses caused by floods are about to begin deliberations on future directions in floodplain management.[2] At the same time, the Clinton administration and Congress wish to reduce spending in light of the $4-trillion national debt. Also, little money is available at state and local levels for flood-loss reduction measures and disaster relief.

The task force's status report, entitled *Floodplain Management in the United States: An Assessment Report*, is the first assessment of the status of the nation's floodplains in 25 years and the most comprehensive assessment and description of floodplain management policies ever undertaken (see the box on the next page). It is an excellent, useful, and even-handed report, but its documentation of all aspects of existing conditions is also its chief weakness because a discussion of existing conditions does not, in itself, adequately set the stage for consideration of possible future directions. Some issues and trends are much more important than others for suggesting informed future directions.

Substantial progress has been made in the last 25 years in U.S. floodplain management. This progress is especially evident in the increased public awareness of flood hazards and the ability of humans to predict potential flooding and to influence risk exposure. But floodplain management in the United States has gone about as far as it can go with its existing approaches. Prime dam sites have been exploited; major floodplains have been mapped; and minimal floodplain regulations have been adopted by more than 18,200 communities.[3] In-

From *Environment*, Vol. 35, No. 5, June 1993, pp. 7-11, 31-34. Reprinted with permission of the Helen Dwight Reid Educational Foundation. Published by Heldref Publications, 1319 Eighteenth St., NW, Washington, DC 20036-1802.

creased funding for existing approaches is not the answer to many of the remaining problems. Instead, a fundamental change is needed. The focus of floodplain management must change from consideration of property losses alone to the consideration of the many purposes of floodplains. Management should be extended to smaller rivers and streams and tailored to watershed conditions. Broad-brush approaches to mapping and regulation that reflect only flood elevations should be replaced, in many contexts, by approaches that also reflect water velocity, sediment regimes, and the changes in runoff that are caused by watershed development. Multiobjective mitigation plans and implementation strategies involving landowners, citizen groups, and local governments should not only improve guidance for future development of floodplain areas but also address the restoration of stream, wetland, and riparian zones. A recent report by the National Academy of Sciences calls for the restoration of 400,000 miles of rivers and streams.[4] It notes that, of the nation's total mileage of rivers and streams, only 2 percent are high-quality, free-flowing segments.[5]

There are many examples of such multiobjective protection and restoration efforts.[6] They have been variously called "greenway," "multiobjective river corridor management," and "environmental corridor management" programs.[7] More than 500 communities have implemented such programs for some or all of their rivers and streams. These programs have been characterized by innovative, problem-solving approaches and broad public involvement (see the boxes on pages 176 and 178).

A number of federal programs encourage such efforts, including the National Park Service's Rivers and Trails Program, the Army Corps of Engineers' floodplain management program, the Tennessee Valley Authority's floodplain management program, and the Federal Emergency Management Agency's community rating system. Some state floodplain, river, wetland, and open space programs also encour-

age such efforts. The California Urban Stream Restoration Program has been particularly successful in encouraging low-cost community stream restoration efforts with broad public involvement through technical assistance and small grants-in-aid to communities. Other examples include the Missouri stream restoration program, the Massachusetts greenway program, and the Maryland greenway program. Despite the success of such projects, no coordinated national legislation, policy, or program exists to support such efforts.[8]

At one time, structural changes—such as dams, levees, channel alterations, and shoreline protection—were the primary approach for addressing

flood losses. Although such structural approaches have reduced flood losses, they often do so at great cost and with great environmental impact. Since 1968, considerable progress has been made in implementing nonstructural loss-reduction measures, such as regulations, warning systems, and evacuation plans. Of the 22,000 flood-prone communities in the United States, more than 18,200, or 82 percent, have adopted floodplain management regulations and participate in the National Flood Insurance Program (NFIP). More than 2.6 million flood insurance policies are presently in force through this program. The Federal Emergency Management Agency

---

EVALUATING THE EFFECTIVENESS OF FLOODPLAIN MANAGEMENT

The Federal Interagency Floodplain Management Task Force's assessment report concluded that it was difficult to evaluate the effectiveness of floodplain management because of a lack of specified goals, baseline data, and monitoring. Nevertheless, the report reached a number of significant observations that were outlined in the summary and executive summary:

• Public recognition of flood hazards is now widespread.

• Judicial support for regulations is also widespread, which increases the liability of landowners who undertake activities that increase flood hazards on other lands.

• Some reduction has occurred in floodplain development.

• A reduction in losses to new development has occurred because of regulatory standards.

• A shift from federal domination toward a more equal federal, state, and local partnership has occurred.

• There is now greater awareness that no single floodplain management strategy is appropriate for every area.

• There has been success in reducing loss of life, but there has been no decline in overall flood losses.

• Floodplain regulations have not arrested deterioration of natural and cultural functions.

• A truly unified floodplain management program is not in place.

The report also outlined a number of opportunities for increasing the effectiveness of floodplain management.

These include

• setting flood-loss reduction goals to be achieved by a certain date;

• improving the database;

• conducting new research;

• integrating strategies for flood-loss reduction and for restoring and preserving natural resources;

• improving coordination and integration in floodplain management;

• increasing cooperation among all persons, programs, and agencies with an interest in reducing flood losses;

• developing methods for better management of high-risk flood hazard areas;

• adopting broader watershed-based and river corridor management approaches;

• improving the incorporation of local conditions in floodplain management approaches;

• helping rural and economically disadvantaged areas;

• improving incentives to encourage the best mix of management measures and better enforcement of floodplain regulations;

• improving awareness and education, including more training and education for government officials; and

• improving techniques for restoring and preserving the natural and cultural resources of floodplains.

SOURCE: Federal Interagency Floodplain Management Task Force, *Floodplain Management in the United States: An Assessment Report*, doc. FIA-17/May 1992 (Washington, D.C.: Federal Emergency Management Agency, 1992), Chapters 15 and 16.

(FEMA) has mapped 18,492 communities, and 2,463 restudies have been completed or are in progress.[9]

Despite these efforts, flood losses continue to increase. Per-capita damages have increased despite measures to reduce such losses, although the rate of increase has slowed. A 1987 study for FEMA estimated that 9.6 million households in 17,466 communities with a total of $390 billion in property value were at risk from flooding. From 1916 to 1985, flood-related deaths averaged 104.4 per year. Per-capita flood-related deaths have decreased, but per-capita flood losses were 2.5 times as great from 1951 to 1985 as from 1916 through 1950, after adjustment for inflation.[10]

In fact, 1992 was the most expensive year in U.S. history for natural disasters, with total estimated losses from floods and hurricanes exceeding $30 billion. In fiscal 1992, 46 presidentially declared disasters and 2 emergencies occurred—the largest number in recent history. Of these, 2 were related to contaminated water supplies and 38, or 83 percent, were flood-related.[11] The paid flood insurance claims for three of these events were huge: $115 million for Hurricane Andrew's flooding in Florida; $30 million for Andrew's flooding in Louisiana; and $30 million for Hurricane Iniki's damage in Hawaii.[12] However, damage caused by wind was much more costly, as is typical of most hurricanes. Hurricane Andrew was the most damaging and powerful hurricane to hit the U.S. mainland this century, and yet it could easily have been worse. However, deaths were exceptionally low: fewer than 70 in Florida and none in Louisiana. This low mortality can be attributed to the effectiveness of the evacuation and warning program, which evacuated an estimated 2.7 million people. Andrew's record-low barometric pressure, sustained winds of 115 to 140 miles per hour, and gusts of 145 to 170 miles per hour resulted in a storm surge of up to 17 feet on Key Biscayne Bay and up to 12 feet in other places.

Although these were unusual storms, they do raise issues such as the cost-ef-

The costs of rebuilding after a flood today are much higher than those of historical floods, such as this one, which closed the Chesapeake & Ohio Canal in 1924.

COURTESY OF U.S. NATIONAL PARK SERVICE

fectiveness of various approaches to floodplain management; how cost-benefit ratios are to be calculated; and to what extent floodplain occupancy should continue to be subsidized by the federal government through NFIP. Soaring costs and deficits require, at a minimum, a careful re-evaluation of the effectiveness of various approaches, particularly those that blatantly subsidize continued occupation of high-risk areas at taxpayers' expense and achieve only single-purpose goals. In light of high deficits, how can federal, state, and local governments best reduce future flood losses and respond to floods and other hazards?

Gaps in Current Programs

Structural and nonstructural efforts to reduce losses have been at least moderately effective in addressing certain situations, but they were not designed to address other situations and do not do so. Several major gaps in existing programs must be addressed.

First, despite the expenditure of $873 million for federal mapping of floodplains, approximately 100 million acres, or one-half of the nation's

floodplains, have been mapped.[13] Unmapped floodplains generally are not subject to regulatory standards by communities or states. Much of the land not subject to management lies along smaller rivers and creeks or along smaller lakes.

In addition, more than 31 percent of flood insurance claims were paid for flood damage outside the mapped "100-year" floodplain.[14] This means that development in these areas is covered by federal insurance but is not subject to regulations to guide new development. To remedy this problem, watershed planning and multiobjective river corridor management for these smaller streams and rivers are needed. This is where development is currently unregulated for flood-hazard reduction purposes and where the greatest changes in hydrology are occurring because of urbanization. It is also where community and local organizations can do the most.

Even in mapped and regulated floodplain areas, however, serious deficiencies exist. Much of the mapped floodplain does not have calculated flood elevations but only a general outline of flood risks. As a result, communi-

Most flood insurance policies will pay for rebuilding flood-damaged houses like this one but will not pay to relocate them out of the flood zone.

ties do not have the tools they need to guide and protect new development. Even where flood studies are under way, it typically takes five years or more to complete a study and adopt regulations. Often, much of the floodplain is developed by then, putting buildings and people at risk.

Another gap in current programs concerns high-risk and unusual hazard areas that have been mapped and regulated but whose maps and regulations inadequately address the special problems. Such areas include alluvial fans; floodplains adjacent to rivers or streams with moveable (erodible) channels; combined flooding/erosion areas, both inland and coastal; areas with long-term fluctuations in ground- and surface-water elevations, such as those adjacent to closed basin lakes; ice jam flooding situations; and subsidence areas. In some parts of the United States, such as the West, most of the flood-risk areas are of such a "special" nature.

Despite the massive NFIP and large expenditures for mapping, only limit-

ed efforts have been made to prepare special maps for or to apply regulatory standards to special high-flood hazard areas. When the mapping program was first developed more than 20 years ago, the high costs caused a uniform national approach to be developed to map and regulate all hazard areas, even high-risk ones. In addition, mapping efforts have generally assumed "existing" watershed conditions, despite broad recognition that urbanization causes dramatic increases in future flood peaks.

The time has come for a shift in mapping philosophy from one overall approach for the whole nation—based primarily upon historical flood events —to much more specific mapping of certain areas to reflect geomorphological factors and particular hazards and to anticipate future development. Such mapping is essential to provide local communities with the tools they need to convince citizens that there is a reasonable, credible, and accurate way of identifying and managing the flood hazards on their properties and

to develop multiobjective local regulatory and management efforts. Such mapping will be expensive, but it need not be carried out on a nationwide basis and could be undertaken on a cost-share basis with states and communities.

Another gap in existing floodplain programs is mitigating the losses to existing structures, starting with structures that have experienced repetitive loss and substantial damage. The assessment report and FEMA data indicate that, in the 1980s, 30,000 structures (2 percent of NFIP-insured structures) filed 2 or more claims of $1,000 or more and thus accounted for about 30 percent of the claims paid, or $747 million. In that same period, about 18,000 buildings suffered damages of more than 50 percent of their value, accounting for more than $438 million in claims paid.[15] Despite hopes that present regulations would lead to a gradual upgrading or elimination of existing substandard structures, only limited progress has been made through the regulation approach alone. A multiobjective mitigation program designed to prevent or reduce future flood damage to these few structures through elevation or relocation has tremendous savings potential.

The lack of financial incentives and landowner assistance to relocate or upgrade substantially damaged structures is illustrated by the situation in Florida's Dade County, which includes Miami. In this area, more than 3,000 homes were substantially damaged by the winds of Hurricane Andrew.[16] Insurance will provide funds to rebuild the homes to their "before-hurricane" conditions. The catch is that these homes are in a flood hazard area. Rebuilding at their former level will not reduce the risk of their being significantly damaged by up to four feet of water in the next flood.

It would cost an additional $30,000 each to elevate the structures. Because the residents' entire savings were lost in the hurricane and their places of employment may also be out of business for an indeterminate time, their ability to pay for or to ar-

SOLDIERS GROVE: A SUCCESS STORY

Soldiers Grove is a small town in Wisconsin beside the Kickapoo River. The river is subject to flash flooding, and, as the hillsides in the watershed have changed from forests to farmed fields, the flood levels have increased. Small towns such as Soldiers Grove were built around mill dams that provided the power to saw logs and grind flour. These small towns are now subject to repeated flooding.

In the 1950s and 1960s, the U.S. Army Corps of Engineers developed a plan to build a dam on the Kickapoo River. Because the dam was miles above Soldiers Grove, it would only provide partial protection to the community. The corps told the village it would still need a levee to keep the flood waters out of the town.

The people of Soldiers Grove were distraught. According to Tom Hirsch, a planner for Soldiers Grove, the townspeople felt that a levee would merely change the village from "a small, economically dying village subject to flooding" into "a small, economically dying village NOT subject to flooding."

Therefore, the village residents pleaded with the corps to look at other alternatives, such as relocating structures to clear out the high-risk flood hazard area, elevating some structures in lower-risk flood areas, and creating open space uses in the cleared high-risk areas. These nonstructural alternatives did not meet the corps' benefit-cost criteria, mainly because the value of the at-risk downtown property, although low because it was subject to flooding, was still greater than the value of the same property with no buildings on it.

Village residents developed their own plan, one that was designed to ad-

dress not only the village's flooding problem, but also other community needs, such as economic development, energy management, housing, recreation, and the quality of life in Soldiers Grove. Through a series of meetings with the townspeople, help with studies from the state, and some small economic grants, the plan evolved. The entire downtown business district would be relocated away from the river, clearing the way for a community recreation area. Homes on the fringe would be floodproofed through elevation. The new downtown buildings would have to have half of their heat supplied through solar energy. Residential housing would be incorporated with commercial structures. Housing for senior citizens would be close enough to the downtown area so that elderly people could walk to the pharmacy and grocery store.

No federal or state agency had a program that could help fund any part of the plan, and Soldiers Grove, while still seeking funding help, was devastated by another flood in 1978. Only then did the Secretary of Housing and Urban Development release some discretionary funds that provided some leverage to implement the village's plan. More than half the cost of the plan, however, was paid by the village and property owners.

Despite these setbacks, this is a success story. The floodplain management plan was carried out. The community has been relocated, and many new structures have been added. The old business district is now a park.

SOURCE: Wisconsin Department of Natural Resources, "Come Rain, Come Shine: A Case Study of a Floodplain Relocation Project at Soldiers Grove, WI."

One reason is that such functions are not mentioned in federal standards for floodplain regulations, and only a relatively small number of communities have incorporated additional provisions into their regulations. In some instances, the availability of subsidized federal flood insurance may have promoted development in highly sensitive or valuable areas, such as wetlands and riparian habitat in the West. Secondly, federal and state floodplain management agencies have generally not promoted the protection of natural and cultural functions because of their narrow flood-loss reduction objectives, a lack of expertise concerning such functions, and a lack of multiobjective approaches for both reducing flood losses and promoting natural and cultural functions. A third reason is that floodplain management has for a long time been narrowly conceived of as managing "excess water" rather than as a part of a broader water resources management that encompasses point and nonpoint source pollution control, storm-water management, water supply, erosion and sediment control, recreation, aquatic habitat protection, and wetland protection and management.

Cost-Effective Management

To address all of these problems, floodplain management should be approached as part of multiobjective watershed (water resources) management with adequate protection for natural and cultural functions. Improving the monitoring and enforcement of existing approaches is necessary, but it will address only a portion of the gaps.

Community, citizen-based efforts are essential to closing these gaps, and yet communities are not being provided with adequate incentives and help. The U.S. floodplain management program needs to be more integrated with other programs that affect floodplains, more localized, more comprehensive, and more unified on a watershed and local government level.

At the federal level, less emphasis

range mortgages for this work is almost nonexistent. Insurance policies, even flood insurance policies, do not provide monies to mitigate against future disasters. Similarly, disaster relief funds can only be used for rebuilding.[17]

It is now more than nine months since Hurricane Andrew, and many of the 3,000 homes remain unrepaired. Modest federal mitigation grants or low-interest loans could help reduce future federal and private outlays when floods do occur.

Present programs also fail to protect the natural and cultural functions of floodplains. The task force's assessment report discusses the natural and cultural functions of floodplains in some depth and concludes that the existing federal, state, and local floodplain management programs do not adequately protect the pollution-control, habitat, flood storage, recreational, and other natural functions. The report does not, however, adequately examine the reasons why the programs neglect these functions.

should be placed on new massive programs and new expenses and more should be placed on shifting the federal role from managing the nation's floodplains to being a facilitator for state and local programs that address the gaps in existing efforts. To address flood losses and better protect natural and cultural functions, the administration, Congress, states, and local governments need to provide a rational, multiobjective framework for watershed-level efforts that are facilitated by federal agencies and states. Federal agencies and states should have continued roles in providing technical assistance, grants in aid, and other assistance to facilitate the accomplishment of federal, state, and local goals, including flood-loss reduction, wetland protection, good water quality, and recreation, while avoiding single-purpose programs. FEMA's community rating system, which provides lower insurance rates for communities with floodplain management efforts that exceed federal standards, is a positive example of such a measure.

Federal subsidies should also be revised to better promote individual responsibility and multiobjective approaches. Nonstructural approaches adopted over the last 25 years—as well as NFIP—tend toward that goal. Nevertheless, this goal is undermined by continued disaster assistance and federal subsidies for beach nourishment, beach erosion control, flood control, and flood insurance for certain high-risk areas where the low insurance rates and regulations do not accurately reflect the risk. These subsidies continue to bias local and state decisionmaking toward structural solutions, even when such decisions are not consistent with individual responsibility, cost-effective use of the floodplain, or achievement of natural and cultural functions.

Federal cost-benefit ratios for flood control and other water resources projects also deserve a hard look, particularly in regard to the calculation of benefits for preventing future flood losses, protecting natural and cultural functions, and providing long-term sustainable use of natural systems.

Using flood-prone land for recreational purposes—such as this park near the Arizona Canal Diversion Channel—is often better than rebuilding flood-damaged structures.

Present procedures are applied in a manner that produces high benefit-cost ratios where intensive development has already occurred or is allowed in the floodplain.

In addition, Congress should provide incentives for multiobjective floodplain and watershed planning and management that anticipate future conditions. Preference should be given to state and local governments, watershed planning, open space, floodplain management, and post-disaster assistance programs that integrate floodplain management into future-oriented environmental programs, such as river management, greenways, trails, point and nonpoint source pollution controls, erosion controls, and wetland management. Such incentives and funding could include adopting a special multiobjective river corridor management act that gives small grants to communities and citizen groups that undertake multiobjective programs and implementing new wetland and watershed management ini-

tiatives for states and local communities as part of the Clean Water Act reauthorization. Support for multiobjective programs should be included in public works budgets and in the new "job corps" and economic stimulus package of the Clinton administration.

Coordination among federal, state, and local governments should also be improved. An interagency mechanism is needed to permit and encourage the federal agencies to play a more active and coordinated role in establishing and implementing multiobjective floodplain and water resources management policies. For all of its faults, the U.S. Water Resources Council served this role, and its reestablishment or the formation of a new coordinating and policy-setting body is needed.

Improved joint training and technical assistance are also needed. Two of the principle barriers to effecting change in national floodplain management policy are compartmental-

MULTIOBJECTIVE RIVER CORRIDOR MANAGEMENT

Many communities have rejected the traditional engineering approaches to reduce flood hazards and have adopted various multiobjective river corridor management programs not only to reduce flood loss but also to protect and restore the natural values of rivers and floodplains and meet recreation needs. A number of these are described in *A Casebook in Managing Rivers for Multiple Uses*, published by the National Park Service in 1991. Profiles of several promising management programs are provided in this publication:

• **Charles River.** The Army Corps of Engineers, the Commonwealth of Massachusetts, and local governments protected 8,500 acres of wetlands along the upper Charles River as part of a "Natural Valley Storage" project. Acquisition costs of wetlands totaled $10 million, compared to potential costs of $100 million for the construction of upstream dams and levees.

• **South Platte River.** The city of Littleton, Colorado, established a 625-acre floodplain park along 2.5 miles of the South Platte River. Old gravel pits were reclaimed to form natural areas and ponds for fish, wildlife, and recreation. Fish habitat in the river was restored. A nature center and recreation trail were built in the park.

• **Wildcat and San Pablo Creeks.** Local citizens of North Richmond, California, prepared a "consensus plan," with alternative designs for flood channels to resemble natural channels. The plan that is now being implemented more fully considers sediment management and involves riparian tree restoration and innovative design of bank and floodplain areas.

• **Boulder Creek.** Boulder, Colorado, has created a 5-mile recreational greenway and a bike path along Boulder Creek. Wetlands have been created or restored to double as storm-water retention and detention areas. Bioengineering has been used for bank stabilization; the trout fishery has been restored; and the greenway has become a central focus of the community.

• **Mingo Creek.** After a series of damaging floods, the city of Tulsa, Oklahoma, developed a greenway plan for Mingo Creek that involves the development of parks and trails linking multipurpose flood control structures along the creek. Restoration of riparian vegetation, a system of lakes, and recreational facilities are incorporated in the plan.

• **Kickapoo River.** After repeated severe flooding of the downtown area, Soldiers Grove, Wisconsin, relocated the entire business district from the Kickapoo floodplain to an upland site. The floodway was converted to a riverside park and recreation area. Assessed property values nearly doubled.

• **Chattahoochee River.** In 1973, the Georgia legislature adopted the Metropolitan River Protection Act, which created a 4,000-foot-wide river corridor (including the width of the river). A river corridor plan was adopted that incorporated standards for land vulnerability, buffer zones, and flood hazards. Local governments are responsible for implementing the plan.

SOURCE: National Park Service, Association of State Wetland Managers, and Association of State Floodplain Managers, *A Casebook in Managing Rivers for Multiple Uses* (Washington, D.C.: NPS, 1991).

ism and lack of training. Federal multidisciplinary teams that include representatives of federal agencies, states, local governments, and nonprofit interest groups would particularly benefit from joint training because it would increase cooperation. Priority topics should include mapping and regulation of high-risk and unique hazard areas, facilitation of local government multiobjective watershed efforts, evaluation and protection of natural and cultural functions, restoration of floodplain systems, and nonstructural ("soft") engineering alternatives.

Data gathering, monitoring, and oversight of federal flood-loss reduction programs should be improved in relation to one another and to other resource management programs. Federal flood-loss reduction programs should be evaluated in terms of the overall objective of reducing flood losses and protecting natural and cultural functions rather than achieving the programs' individual statutory mandates. For example, NFIP must be viewed in the context of larger federal disaster and flood-loss reduction efforts—not as a separate program that balances its own books even though the taxpayers' costs for disaster assistance continue to rise. Post-disaster policies should also be re-evaluated to determine whether they are reducing, rather than simply perpetuating, future losses.

Shifting floodplain management to emphasize community-based multiobjective and watershed-based efforts tailored to local conditions will not be easy. Existing institutions have enormous bureaucratic inertia, and many interest groups as well as agency staff tend to favor the status quo. Uniform, broad-brush national approaches for mapping, regulation, and management are less time-consuming for federal agency staff to administer, and measures that increase the federal workload without increasing federal staff will encounter strong opposition. Some federal agency staff also fear the loss of control. They are accustomed to initiating and implementing flood-loss reduction measures and are reluctant to share responsibility and power with states and local governments. In addition, various federal statutes authorize strong, direct federal roles rather than assistance to local governments.

Floodplain managers also often lack the multidisciplinary expertise needed for the protection and restoration of natural and cultural functions, such as wetland protection. In addition, there are concerns that local government and citizen-based programs will be dominated by local real-estate and other special interests and will not meet regional or national needs and that local, citizen-based efforts will lack the expertise to address highly technical flood, pollution control, and wetland restoration efforts. Community-based watershed approaches involving detailed data gathering, computerized geographic information systems, and other sophisticated analysis techniques can also be very expensive. Various land-use planning efforts for watersheds and other areas have often proved to be of limited value if separated from implementation.

All of these arguments, however, do not justify continuing the status quo. Experience over the last 30 years has shown that, despite the expenditure of huge sums of money, the federal government alone cannot "solve" flood problems.

New, community-based efforts can be practical and implementation-oriented. They need not be prohibitively expensive, as indicated by many successful efforts already implemented across the nation (see the box on page 178). Community efforts can be financed by combining funds from a variety of programs meant to address such activities as nonpoint and point pollution control, storm-water management, outdoor recreation, community redevelopment, and wetland and habitat protection and restoration, as well as flood-loss reduction. Landowners, citizen groups, and local governments must be brought more fully into the process as partners to address not only flood control but also other community economic and environmental needs.

Community-based initiatives need not place huge new demands on federal staff or budgets if coordinated use can be made of the many experts throughout the federal agencies, including the Army Corps of Engineers, FEMA, the Soil Conservation Service, the National Park Service, the Environmental Protection Agency, and the U.S. Geological Survey. Strong, continued federal and state involvement in such community efforts is needed to provide technical assistance and training and to reflect regional perspectives.

It is true that floodplain management must be revised technically to continue to reduce flood losses and meet broader, multiobjective goals. But, more importantly, floodplain management must become a more complete and real local, state, and federal partnership.

NOTES

1. See U.S. Water Resources Council, "Estimated Flood Damages: Appendix B," *Nationwide Analysis Report* (Washington, D.C.: U.S. Government Printing Office, 1977); and Federal Interagency Floodplain Management Task Force, *Floodplain Management in the United States: An Assessment Report*, doc. FIA-17/May 1992 (Washington, D.C.: Federal Emergency Management Agency, 1992), Chapter 1.

2. See discussion and many references contained in Federal Interagency Floodplain Management Task Force, note 1 above, Chapter 3.

3. Information provided by the Federal Insurance Administration, Federal Emergency Management Agency, Washington, D.C.

4. National Research Council, *Restoration of Aquatic Ecosystems* (Washington, D.C.: National Academy Press, 1992).

5. Ibid.

6. See J. Kusler, "Innovation in Local Floodplain Management: A Summary of Community Experience," Special Publication 4 (Boulder, Colo.: Natural Hazards Research and Applications Information Center, 1982); J. Kusler and S. Daly, eds., "Wetlands and River Corridor Management" (Paper presented at the Association of State Wetland Managers, Charleston, S.C., 5–9 July 1990); National Park Service, *A Casebook in Managing Rivers for Multiple Uses* (Washington, D.C.: NPS, 1991); C. E. Little, *Greenways for America* (Baltimore, Md.: Johns Hopkins University Press, 1990); E. Grundfest, ed., "Multiobjective River Corridor Planning" (Proceedings of the Multiobjective Workshops in Knoxville, Tenn., and Colorado Springs, Colo., Association of State Floodplain Managers, 1989); and L. M. Labaree, *How Greenways Work: A Handbook on Ecology* (Washington, D.C.: National Park Service, 1992).

7. Ibid.

8. Congressmen Joseph M. McDade (R-Pa.) and Morris K. Udall (D-Ariz.) introduced a State and Local Multiobjective River Corridor Assistance Act in 1989 (HR 4250). This bill would have provided technical assistance and grants-in-aid to local governments initiating multiobjective efforts. The bill was referred to committee and never adopted.

9. Federal Insurance Administration, note 3 above.

10. See Donnelly Marketing Information Service, *System Update Report* (Washington, D.C.: Federal Emergency Management Agency, 1987) as described in Federal Interagency Floodplain Management Task Force, note 1 above, Chapter 3.

11. Disaster Assistance Program, Federal Emergency Management Agency.

12. Federal Insurance Administration, note 3 above. According to hurricane damage estimates provided by various newspapers, Hurricane Iniki's damages totaled approximately $1.6 billion; in Florida, Hurricane Andrew's total damages came to between $15 billion and $30 billion; and in Louisiana, Andrew's damages amounted to $1 billion.

13. Federal Interagency Floodplain Management Task Force, note 1 above, Chapter 6.

14. Information provided by the Federal Insurance Administration, note 3 above.

15. Federal Interagency Floodplain Management Task Force, note 1 above, Chapter 13.

16. Information provided by FEMA and Dade County, Florida, staff at the Association of State Floodplain Managers conference in Atlanta, Georgia, March 1993.

17. Ibid.

Desktop Farms, Backyard Farms, or No Farms?

Alternative Futures for Sustainable Agriculture

Marc Zwelling

Marc Zwelling is a strategist and president of Vector Public Education, Inc., 101 Gordon Road, North York, Ontario M2P 1E5, Canada. He is also publisher of John Kettle's FutureLetter, *a monthly outlook on technological, economic, and demographic trends. John Kettle provided valuable assistance on the development of this article.*

This article is adapted from Sustainable Farming: Possibilities 1990–2020, *a discussion paper prepared as part of a major study of sustainable agriculture by the Science Council of Canada. For copies, contact: Canada Communications Group/Publishing, Ottawa, Ontario K1A 0S9, Canada; telephone: 819/956-4802.*

How to feed the world's growing population without destroying the planet's resource base is a dilemma faced by policy makers, environmentalists, and food producers everywhere.

The concept of "sustainable agriculture" offers a resolution to the dilemma, but exactly what is it? Many associate the concept with organic farming—foods grown without chemical additives of any kind and with minimal intrusion by heavy machinery. Others look to a form of agriculture that would have

> **"Sustainable farming" has many different meanings, and how we attain it—if we do—will depend on a myriad of environmental, political, and emotional issues.**

virtually no natural inputs such as soil and water; "food" would be constructed from genetic materials in farm factories.

Thus, depending on how sustainability is defined, agriculture of the future could be something everyone participates in from his or her own backyard, or something that is run by robots in factories, completely devoid of human intervention.

To understand the environmental, economic, political, and sociological complexities involved in achieving sustainable agriculture, the Science Council of Canada commissioned these six scenarios to provoke thought about what sustainable farming might look like in Canada by the year 2020.

These scenarios attempt to show what conditions may or may not lead to a more-sustainable form of agriculture in the future; they reflect differences in such factors as levels of government intervention in farming practices, the public's demand for "pure" food, global competition in food production, the impacts of genetic engineering and other technological developments, and the perceived value of farming as an important part of life in Canada.

Scenarios 1A and 1B: Continuing Trends

The first two scenarios share similar features, but differ in the extent of government involvement in farming.

In both scenarios, public concern over pollution rose after 1990. Consumers, especially more-affluent shoppers, demanded "green" products, organic foods, and energy-saving, recyclable goods. Governments grew more concerned about the levels of pesticides, herbicides, and other chemicals applied to food. Market forces, especially powerful major buyers such as the giant retail food chains, prodded farmers into gradually offering more organic products, which reached 30% of the market in the middle of the last decade of the twentieth century.

Farmers began to apply new growth hormones and the like, gaining livestock that grew faster, cows that produced more milk, and grain yields that gradually expanded. In general, chemistry and bioengineering were able to increase farm output through the 1990s and the first decade of the twenty-first century.

Farmers switching to organic and so-called sustainable methods produced smaller harvests and herds than their chemical-intensive competitors. But they often found their costs dropping as they gave up pesticides and growth stimulants. By the year 2020, both old and sustainable farming systems coexisted.

Scenario 1A:
Continuing Trends, Deregulation

Through the 1990s, voters demanded that governments reduce deficits and bring their books into balance. With the rural, agricultural population declining rapidly, governments felt there would be little political fallout from steadily reducing support for agriculture.

Governments were often accused of having no agricultural policies as they withdrew subsidies and other support for farmers in the name of deregulation. In fact, there was a policy: to let the market allocate resources in the food business. The market dictated a slow change from conventional to sustainable agriculture.

Scenario 1B:
Continuing Trends, Re-regulation

Despite pressure to reduce deficits and balance their books, governments did not reduce farm support. Instead, agricultural subsidies grew faster than the economy as a whole throughout the period from 1990 to 2020, as governments tried to help farmers switch to sustainable agriculture.

Government scientists put special efforts into soil restoration and into improving breeds and crop varieties. Farmers were retrained in sustainable methods, including crop rotation, biocontrol of pests, and other methods. Because of government support, which grew yearly through 2020, farming became progressively less expensive.

A food tax was introduced in 2000 to help pay the cost of adapting Canadian agriculture to environmentally friendly practices. In effect, farming was run by governments, which dictated a slow change from conventional to sustainable agriculture.

Scenario 2: All-Import Agriculture

Since 1990, there have been many scares about contamination of the food supply. Whole regions have been panicked by headlines and news reports, such as "Alberta Wheat Warning," "B.C. Fish Alert," and "Ontario Corn Danger."

Public concern over herbicides, pesticides, and other potentially harmful inputs forced farmers to reduce their use of chemicals during the 1990s. But farmers' efforts were too feeble and too late. Environmentally benign substitutes were expensive or unavailable.

The demise of domestic agribusiness was once seen as a national-security issue. However, the danger to food security in 2020 was not the lack of a domestic agribusiness. The danger was relying too much (as Canada did in the past) on domestic sources that were increasingly viewed as tainted. Global sourcing is the only way a nation can be really sure of its vital food supplies. Offshore food of all kinds was not only fresh and appealing but rising in demand. Farming (except for personal enjoyment and use) disappeared by 2020 with almost no public outcry. It was regarded as an old, unsustainable practice that required too much environmental sacrifice to make it worthwhile. Sustainability in agriculture was reached by default. Canadians were happy to be out of a dirty business.

Scenario 3: Farmories and Desktop Agriculture

In the late 1990s, the public came to believe that food from traditional sources was not safe. Producing it threatened the environment; eating it jeopardized health. The environmental movement succeeded by the year 2000 in eliminating demand for vegetables grown with pesticides, livestock raised on growth boosters, and any foods nurtured with water from polluted sources.

A change in social values reduced demand for meat to practically zero. Fattening pigs and cattle on grains was viewed as morally wrong; grains are for people, said "green" consumers. By 2020, many people had never tasted meat.

Greater testing to reassure consumers that their foods were safe increased the price of food and at the same time revealed even worse contamination. Testing advanced to the point that toxic substances in traces as small as one part per quintillion could be detected by 2010.

Science responded to Canadians' fetish for clean, safe food with food substitutes. Once consumers appreciated the nutritional content of these new "non-food foods," they accepted it as real and viewed the old, contaminated food as unreal.

By 2020, almost everything once available from farms and oceans was produced in hygienic "farmories" (farm factories). Farmory food is based on the technology of gene manipulation and computer design. In a real sense, it is "desktop agriculture." Like a sperm bank, a farmory's inventory contained minuscule amounts of original genetic material from many varieties of plants and animals. The feedstock for farmory food was sugar from cellulose. The genetically engineered cells were mixed and developed into food in just days with the help of farmory robots.

Farmory food was completely recycled. Scrap was collected and reused. Unlike old, dirty food, which could not be recycled because of traces of fertilizers, pesticides, and other "poisons," non-food food was absolutely sustainable. The production of non-food food led to the demise of the old idea of sustainable agriculture, whose high costs made it economically unsustainable.

Scenario 4: High-Tech, Export-Based Agriculture

Canada became a world leader in agricultural exports because the country's farmers were the first to learn really high-tech, sustainable agriculture. Other countries were tardy in converting to sustainable farming, giving Canada the opening to world leadership.

Farmers found that they *had* to adjust. Additives, chemicals, and other unnatural ingredients in the food chain were, for all practical purposes, banned by the federal government. Consumers demanded clean food and were prepared to treat food as a cause, buying only what rewarded them with the satisfaction of saving the earth. Government withdrew assistance to farmers who employed any unsustainable practices.

Bioengineering enabled farmers to eliminate pesticides since every crop had been made pestproof and weatherproof. Gene manipulation protected herds from diseases; cattle and other livestock never died of anything but accidents or old age. Organic foods and meats commanded higher prices, giving farmers their first real boost in income in decades; exports for Canadian agricultural products boomed.

Scenario 5: Clean Agriculture

Foreign markets for Canadian produce and livestock virtually dried up as the world turned against Canada's impure agriculture. Offshore environmental groups successfully urged boycotts of Canadian farm products in the late 1990s because Canada was a land abuser, a country whose farmers reduced the fertility of their soil through the punishing use of fertilizers, improper crop rotation, and unsustainable tilling. The quality of Canadian agricultural products was not trusted.

The extinction of Canada's foreign food and livestock markets was considered a national catastrophe. Sustainable agriculture was seen as the

way to rebuild Canada's devastated agribusiness. But only local markets were available. Exporting was unthinkable till Canadians learned how to farm again—the sustainable way. Governments launched expensive programs to convert farming, to cleanse the land, and to purify the food-processing plants.

Yields dropped severely because the clean farming techniques eliminated the inputs that had boosted productivity. Without these inputs available to "dirty" farming, clean farming became more labor intensive and less technology intensive. Farming by hand was back in style, since most heavy equipment was consid-

---

Objections to the Scenarios

Seven experts in agriculture, the environment, and food retailing were invited to review and analyze Marc Zwelling's scenarios for sustainable farming. Here are some highlights of their commentaries.

Scenario 1A: Continuing Trends, Deregulation

◆ This scenario comes closest to describing the future of agriculture in Canada, but farmers will not suddenly eliminate such valuable inputs as chemical pesticides and fertilizers and return to agricultural practices of an earlier era. Farm output would drop, and there would be a disastrous shortage of food.

Albert S. Hester, Technical Insights, Inc., Fort Lee, New Jersey

◆ Scenario 1A leaves Canadian agriculture at a standstill for the next 30 years. A market-driven economy, without regulation, is highly unlikely, given the new political power of advocacy groups such as environmentalists in the industrialized world.

Robert E. Morgan, Saskatchewan Wheat Pool, Saskatoon, Saskatchewan

Scenario 1B: Continuing Trends, Re-regulation

◆ There is no question that the public is becoming increasingly concerned about pollution. However, there is no evidence that this concern will translate into changes in government regulation and movements by consumers toward or-

ganic foods to the degree suggested. The suggestion that governments can run agriculture is impractical. Farming is not a business that can be run effectively by bureaucrats.

WESTARC Group, Inc., Brandon University, Brandon, Manitoba

◆ Scenario 1B contains an oxymoron: "Because of government support, . . . farming became progressively less expensive." I seriously disagree that government involvement will ever result in lower costs. Significant levels of involvement by government will not produce more-effective farming. In fact, the experience in other "controlled state" operations has been just the opposite.

Leonard Kubas, Kubas Consultants, Toronto

Scenario 2: All-Import Agriculture

◆ This "all-import" scenario is limited by the realities of international political tensions and the rise of international terrorism. Canadian policy makers are unlikely to place the security of their national food supply in jeopardy. If Canada is unable to produce "safe food," there is little assurance that other countries can.

Robert E. Hudgens, Winrock International Institute for Agricultural Development, Morrilton, Arkansas

◆ For Canada to be "out of farming" is just too farfetched. It is hard to imagine a time when public

opinion would determine that agriculture is no longer needed in Canada's economy. There are many practical, economic, and emotional reasons why Canada would remain a dominant agricultural country.

Leonard Kubas

Scenario 3: Farmories and Desktop Agriculture

◆ The manipulation of genes is not risk free. The wrong sequencing of genes could produce a very toxic food.

Michael J. Phillips, Office of Technology Assessment, U.S. Congress, Washington, D.C.

◆ The notion of food designed by computer and manufactured in factories is an affront to the timeless human drive for organic connection to the earth and for the regeneration of spirit and flesh made possible by the perfecting of natural systems as opposed to the creation of technological contrivances.

Matt Damsker, Organic Gardening magazine, Emmaus, Pennsylvania

Scenario 4: High-Tech, Export-Based Agriculture

◆ If all chemicals and fertilizers were eliminated from agricultural production, there would not be enough food produced to allow the kind and quality of exports suggested.

WESTARC Group

◆ It is unlikely that plants can be engineered to be pestproof and

ered too dirty to operate near crops and livestock.

Because of labor shortages and because large farms were too vast to work by hand, small farms became commonplace. By 2010, the small family farm had become the way of the future. Agriculture at last was considered clean, and resistance to farming "in my backyard" was being overcome. Using land for lawns was seen as wasteful; soil is for food—not fun—said environmental groups.

Clean farming became popular in cities, with hundreds of thousands of backyard gardens and greenhouses established. Government grants to convert yards to mini-farms were available and widely sought after

---

weatherproof. This is a pipe dream. The performance of current hybrid crops depends very much on the use of fertilizers. It is extremely unlikely that higher-yielding crops can be developed without the need for fertilizer.

Robert E. Morgan

Scenario 5: Clean Agriculture

◆ The first part of the scenario illustrates the danger of banning farm chemicals before alternative methods of farm management are in place. If Canadian agriculture were forced out of operation, it would be impossible to rebuild agriculture in the way described. Farming by hand on small farms would require an enormous farm population. Most of the output of the farms would be needed to support this population, with little left over to supply the non-farm population.

Albert S. Hester

◆ This scenario lacks credibility. It could be that public hysteria caused by some highly inaccurate and exaggerated news stories could devastate Canada's agricultural industry, but people would accept some contamination before they would starve or spend most of their incomes on food.

WESTARC Group

Source: *Sustainable Farming: Possibilities 1990-2020.*

2010. Rooftop farming flourished on apartment, condominium, and skyscraper roofs.

Clean farming made agriculture sustainable but increased the cost of food, requiring new government welfare programs to make sure everyone had a guaranteed annual diet. Agriculture became sustainable in 2020, but was less efficient and productive.

The Most-Likely Scenario

What will sustainable agriculture mean, and how could it happen? Seven experts who were invited to evaluate these six scenarios agree that Scenario 1A (continuing trends with further deregulation) represents the most-likely future.

Technology and consumer opinion will change agriculture so that today's food shortages and concerns over chemical contamination and climate change are resolved under a sustainable future. But by 2020, agricultural practices and products will not be much different from those of today. The power of inertia in consumers and producers alike is the reason. Whatever changes are likely, they will be slower than the scenarios say.

Most of the experts believe that sustainable agriculture will save labor but result in more-expensive food. While science is the tool of sustainable agriculture, it is not the key. Rather, public opinion will matter more. The consumer—not government or technology—will be the catalyst. And consumers may not necessarily demand risk-free food.

A final scenario, "The Least That Will Happen," incorporates the trends and forces in the previous scenarios that at least one of the experts who evaluated the scenarios (see box) believes would unfold:

Public concern over pollution rose steadily through the 1990s, and there were many scares about contaminated food supplies. Public opinion convinced governments that sustainable agriculture meant eliminating many of the inputs and practices that had defined Canadian agriculture, and governments began restricting the use of pesticides, herbicides, and other farm chemicals in the late 1990s, intending to phase them out entirely over the next 20 years.

Although other countries with food

---

surpluses continued to subsidize their agribusinesses heavily, Canada remained one of the world's largest food exporters. But it was no longer regarded as a business with a great future. Farmers who switched to more environmentally friendly, organic, and so-called sustainable methods had smaller harvests, smaller herds, and smaller animals. Their costs rose as they discontinued the use of pesticides and hormone-growth stimulants. As the new century dawned, farmers realized that, to keep food affordable and appetizing while protecting the environment, new technologies would be needed.

Efficiency became the real driving force in the race to sustainability in the first decade of the new century. Productivity improvements in all parts of the food cycle meant that a gradually decreasing share of the labor force was growing, processing, and transporting food. Productivity of plants and animals soared. Genetically altered crops and livestock were becoming more commonplace. Biological products like these meant the government-decreed bans on chemical fertilizers and pesticides were no longer a burden to farmers.

By 2020, the Third World could almost feed itself. As a result, Canadian food exports declined steadily. To remain competitive, Canadian agriculture had to refocus on domestic markets.

Genetically improved livestock and crops required less land. More harvests in a year were possible. Farming had become easier. Urban farming developed, and rooftop and backyard farms flourished. No one could produce a variety of crops, but everyone could produce a surplus, spawning a lively commerce in balcony and backyard vegetables and fruits.

In 2020, food producers and consumers faced the question, Could sustainable agriculture be said to have arrived when food was still based on animals, plants, land, and water? Or would a post-farm agriculture take over, producing nutrition through food substitutes? One path into the future continued the age-old human dependency on the earth. The other snapped the links between earth and health, setting humanity on a road that, if taken, could probably not be retraced.

20 years of the CLEAN WATER ACT

HAS U.S. WATER QUALITY IMPROVED?

Debra S. Knopman
Richard A. Smith

DEBRA S. KNOPMAN and RICHARD A. SMITH are hydrologists in the Branch of Systems Analysis of the Water Resources Division of the U.S. Geological Survey in Reston, Virginia.

Once again, it is time to reauthorize the U.S. Clean Water Act—formally known as the Federal Water Pollution Control Act Amendments of 1972. Just as on every previous occasion that this major environmental law has been considered, Congress and the current administration are short on information about the true state of the nation's water quality and the factors affecting it. Since 1972, taxpayers and the private sector have spent more than $541 billion on water pollution control,[1] nearly all of it on the "end-of-pipe" controls on municipal and industrial discharges mandated by the Clean Water Act. After all this time and money, it would be desirable to know whether the act has worked. Is the water cleaner than it would otherwise have been and have the environmental benefits, however they may be counted, exceeded the costs? The answer to both questions is that no one knows for sure. Only now is research beginning to move beyond anecdotes and to improve substantially the understanding of water quality throughout the nation.

More important than justifying past expenditures on water quality, however, is improving policymakers' ability to choose the most economically efficient pollution control strategies for the future. The recent U.S. presidential election debates demonstrated that there is increasing interest in the relationship between environmental protection and the nation's economy. Simply documenting changes in environmental conditions does not provide enough information for dealing with this issue. Policymakers need to understand cause-and-effect relationships and be able to predict the effect of various control strategies on environmental quality.

Several key issues will influence the upcoming debate on reauthorization of the Clean Water Act, a process that began in 1991 and will continue through 1993 (see box on the next page). During the debate, Congress and the new administration will rely on a patchwork of local, state, regional, and national studies about water quality. Unfortunately, the existing information base is fragmentary at

From *Environment*, Vol. 35, No. 1, January/February 1993, pp. 17-20, 34-41. Reprinted with permission of the Helen Dwight Reid Educational Foundation. Published by Heldref Publications, 1319 Eighteenth St., NW, Washington, DC 20036-1802.

LEGISLATIVE STATUS OF THE U.S. CLEAN WATER ACT

Reauthorization of the Clean Water Act last occurred in 1987 and typically occurs every five years. After a quick start in the late spring and summer of 1991, when more than a dozen hearings were held in both the U.S. House and Senate,[1] progress on the reauthorization of the Clean Water Act slowed. A comprehensive reauthorization bill, the Water Pollution Prevention and Control Act of 1991 (S. 1081), was introduced in the Senate in April 1991 by Senators Max Baucus (D-Mont.), John H. Chafee (R–R.I.), and others and was referred to the Senate Committee on Environment and Public Works.[2] A revised version of S. 1081 was circulated at the end of December 1991 as a majority (Democratic) staff draft, and it will probably be reintroduced at the start of the new Congress. The minority (Republican) staff of the Senate committee is working on its own set of revisions to the bill. Majority and minority staff have yet to agree on a number of issues.[3]

Meanwhile, comparable legislation was not introduced in the House in the last Congress even though the chairman of the House Committee on Public Works and Transportation, Congressman Robert Roe (D–N.J.), who retired at the end of the last session, had indicated a strong interest in reporting a Clean Water Act reauthorization bill out of his committee.[4] Although a staff draft of a House bill was being developed, none actually was introduced.

Task groups led by EPA officials began meeting in the spring of 1991 to develop options papers for the Bush administration's positions on individual issues. To prepare for this process, staff from the Office of Management and Budget (OMB) asked EPA and other federal agencies to respond to a set of questions about status and trends in water quality.[5] Extensive information on existing water-quality programs was compiled throughout the federal government. Still, the OMB questions proved difficult to answer in quantitative terms because a picture of national conditions had to be pieced together from a variety of sources, most of which were not designed to provide a representative national view. By the spring of 1992, the administration review process had largely been completed, and a decision was made not to propose its own reauthorization bill in 1992.

The primary reason for the delay in reauthorization can be summed up in one word: wetlands.[6] Because of the Bush administration's proposal to redefine wetlands for purposes of regulation under the Clean Water Act,[7] some members of Congress have been reluctant to advance the reauthorization process in the absence of an agreed definition of the term *wetland*.[8] Federal lawmakers with extensive wetlands in their districts and states have been understandably reticent to offend either real estate developers or citizens with strong environmental concerns.[9]

It remains to be seen whether the wetlands issue will ultimately be separated from the reauthorization of the act to break the congressional impasse in this new session of Congress. A new administration and new chairmen of both authorizing committees—the Senate Committee on Environment and Public Works and the House Committee on Public Works and Transportation—may change the chemistry of the debate, as will the presence of 121 new members of Congress.

1. Hearings Before the Subcommittee on Water Resources of the Committee on Public Works and Transportation, House of Representatives,"Reauthorization of the Federal Water Pollution Control Act," 102-31 and 102-43 (Protection of Wetlands), 102d Cong., 2d Sess. (1992); and Hearings Before the Subcommittee on Environmental Protection of the Committee on Environment and Public Works, United States Senate, "Implementation of Section 404 of the Clean Water Act," 102d Cong., 2d Sess. (1992).

2. U.S. Senate Environment and Public Works Committee, Majority Staff Draft: Amendment in the Nature of a Substitute to S. 1081 Water Pollution Prevention and Control Act of 1991 (Washington, D.C., 31 December 1991).

3. Environmental and Energy Study Institute, "Environmental Energy and Natural Resources Status Report for the 102nd Congress," 14 October 1992, 16–17.

4. Ibid.

5. David M. Gibbons, deputy associate director of the Natural Resources Division, Office of Management and Budget, memorandum to Lajuana S. Wilcher, assistant administrator for water, Environmental Protection Agency, 4 April 1991. The questions were: What are the remaining problems preventing the achievement of water quality goals? What other important goals (e.g. habitat preservation) are affected by water quality problems? Why should these goals be addressed in the context of the Clean Water Act? Quantify each problem by the best measures available and quantify the cost to society posed by each problem whenever possible; How will the severity of each problem change over the next 10 to 20 years, assuming implementation of existing laws and no Clean Water Act amendments? What are the sources of the pollutants posing the remaining problems and what level of control has already been imposed on these sources? Define a range of options for addressing each remaining problem that takes into account expected water quality changes in the absence of any Clean Water Amendment; What are the incremental costs and benefits associated with each option identified in question 6? and Quantify to the maximum extent possible.

6. J. Kusler, "Wetlands Delineation: An Issue of Science or Politics?" Environment, March 1992; S. Nicholas, "The War over Wetlands," Issues in Science and Technology, Summer 1992, 35–41.

7. White House press release, "President's Plan for Wetlands" (Washington, D.C., 9 August 1991).

8. Environmental and Energy Study Institute, note 3 above.

9. Ibid.

best, but new federal programs that are now being implemented may improve the quality of data. However, an attempt to apply the data that are currently available is necessary in order to begin answering the fundamental question: Has the Clean Water Act worked?

The Information Gap

Some of the most contentious political issues in U.S. environmental legislation have been more symbolic than substantive. In the coming debate on the Clean Water Act, however, resolution of such major issues as effective municipal wastewater treatment, nonpoint-source pollution controls, control of toxic chemicals, and wetlands protection could have broad impacts on human and ecological health. Unfortunately, these issues are being debated largely in the absence of conclusive information about existing water quality as well as about the influence of current control measures and of various land-use and basin characteristics on water quality.

This lack of conclusive information is nothing new. In the early 1970s, when Congress was first debating the Clean Water Act, there was a widespread perception that water quality was bad and getting worse. Rachel Carson's 1962 best seller, *Silent Spring*, mobilized a generation of citizens to try to stop the ecological damage wrought by unregulated use of chemicals in the environment.[2] Attitudes about water quality were based mainly on anecdotes and images of the Cuyahoga River on fire and raw sewage spewing into San Francisco Bay. Despite having little reliable information on national water quality, Congress responded to public demands to clean up the nation's surface waters by passing the Clean Water Act.[3] The act was passed despite a veto by then-President Richard M. Nixon, who described the price tag on the bill—$24 billion over five years—as "extreme and needless overspending."[4]

According to the Senate report accompanying the passage of the Clean Water Act in 1972, "much of the information on which the present water quality program is based is inadequate and incomplete. The fact that a clearly defined relationship between effluent discharge and water quality has not been established is evidence of that information gap. . . . The fact that many industrial pollutants continue to be discharged in ignorance of their effect on the water environment is evidence of the information gap."[5]

Scientifically defensible assessments of water quality in the United States, or in any nation for that matter, are far more complicated than Congress realized in 1972. In fact, an information gap was first pointed out in 1971, when M. Gordon Wolman wrote in *Science* that "few observational programs combine the necessary hydrology with measurements of water quality, river characteristics, and biology. While some long-term observations exist, the lack of coordinate observations makes long-term comparisons [of river quality] virtually impossible."[6] This was still true in 1981, when the General Accounting Office released a report that concluded that the federal government had no reliable method of measuring the environmental effect of spending $30 billion to construct municipal sewage treatment plants, which were the cornerstone of the Clean Water Act.[7]

In 1993, scientists still cannot reliably answer the most basic questions about national water quality: How much of specific pollutants do point and nonpoint sources contribute to

streams and rivers, and how does the point/nonpoint ratio vary geographically and with basin and climate characteristics? What fraction of the nation's surface water and groundwater fails to meet water-quality standards for toxic substances and conventional pollutants? What effect have the $70-billion construction grants program and the Department of Agriculture's $8-billion Conservation Reserve Program—implemented in 1986 and designed to reduce erosion of topsoil—had on water quality? Indeed, the Clean Water Act may have generated considerable benefits simply by keeping water quality constant in the face of increases in population and in the gross national product (GNP). To date, however, little progress has been made in reliably measuring the water-quality benefits of the act in any terms. The information gap also affects the assessment of water-quality controls directed at groundwater, lakes, estuaries, and wetlands—all of which can influence the quality of streams and rivers.

One reason for this lack of information is that the mechanism for program evaluation in the original 1972 act is not capable of tracking progress toward water-quality goals. The reporting requirements, section 305(b) of the 1972 act, sound simple enough: States must designate each of their water bodies for such specific uses as drinking-water supply, recreation, and cold water fisheries. Then, each state proceeds with its own monitoring and assessment methods to determine whether these uses are being met. Every two years, the Environmental Protection Agency (EPA) compiles the state results and reports on them to Congress in the National Water Quality Inventory.

In 1972, Congress assumed that the states and EPA would coordinate their data gathering, storage, and retrieval functions. Most importantly, Congress supposed that EPA would take the steps necessary to ensure methodological consistency, at least.[8] However, until very recently, this was not the case; EPA's implementation of the reporting requirements in sec-

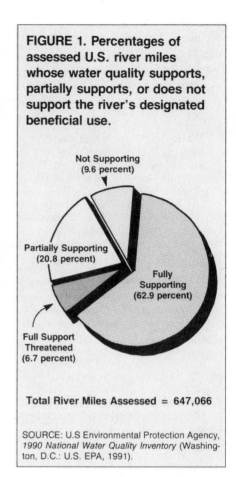

FIGURE 1. Percentages of assessed U.S. river miles whose water quality supports, partially supports, or does not support the river's designated beneficial use.

Not Supporting (9.6 percent)

Partially Supporting (20.8 percent)

Fully Supporting (62.9 percent)

Full Support Threatened (6.7 percent)

Total River Miles Assessed = 647,066

SOURCE: U.S Environmental Protection Agency, *1990 National Water Quality Inventory* (Washington, D.C.: U.S. EPA, 1991).

tion 305(b) has not produced the kind of baseline and comparative database needed to evaluate past water-quality trends and to develop future strategies. For example, according to the latest National Water Quality Inventory, in 1990, 51 U.S. states, territories, and jurisdictions assessed 647,066 river miles. This number reflects 36 percent of the total river miles in the United States, or 53 percent of the total river miles in the states that reported (see Figure 1 on this page). Of the total river miles in these states, 33 percent were judged to support their designated beneficial uses; 20 percent were judged to be either partially supporting, not supporting, or in danger of not supporting; and the remaining 47 percent were not assessed. Interpreting these numbers is problematic at best. No one knows how representative the assessed surface waters were. It is likely that the 64 percent of all U.S. river miles that were unassessed have very different water quality than those that the states targeted for monitoring. In fact, EPA cautions that the numbers should not be compared either state to state or with previous reports because of "inconsistencies among states in how these data were generated."[9] With these limitations, it is difficult to extract much useful national-level information from the data.

Some of the shortcomings of this process may be a consequence of funding. In 1990, EPA estimates that the states spent about $11 million in federal grant money and twice that amount of their own money for ambient monitoring.[10] The magnitude of these expenditures is small when compared to the tens of billions of dollars spent annually by the public and private sectors for water-quality control measures.[11] A comparison of expenditures for information gathering with expenditures for water-quality controls suggests that the information gap is at least partially the result of not enough funds allocated for monitoring.

In addition, there are conceptual and operational problems, both with the 305(b) process itself and with EPA's compilation of the state reports. In general, the state assessments are based on a combination of ambient monitoring networks, local scale studies conducted for regulatory purposes, and professional judgment of state water-quality officials. Although the local studies are useful for local management of water quality, their results are difficult to aggregate nationally because the amount of emphasis on each of these information sources is highly variable from state to state. There is little consistency among state programs in the temporal and spatial scales of data collection, the procedures used to collect and analyze the data, and the types of data collected. As a result of these methodological inconsistencies, comparing results from one biennial report to the next to monitor change is impossible.

The only two long-term, nationally consistent surface-water-quality networks in the United States are operated by the U.S. Geological Survey (USGS). The National Stream-Quality Accounting Network (NASQAN) comprises 420 stations located on large rivers. The stations are located near the outlets of major drainage basins so as to measure collectively a large fraction of the total U.S. runoff.[12] The network also has been used to detect water-quality trends over time, to test for correlative relations to land use, and to measure mass transport from rivers to estuaries, the Great Lakes, and between major river basins. Station locations were not based on the presence or absence of pollution sources in the vicinity nor on specific dominant types of land use in the drainage basin. Also, NASQAN does not monior saline coastal waters, lakes, or reservoirs. In contrast, the USGS Hydrologic Benchmark Network comprises 55 stations located in relatively pristine, mostly headwater, basins. This network was designed to define baseline water quality and has been used to examine the effects of atmospheric deposition on water quality.[13] Both networks measure stream discharge and concentrations of the major dissolved inorganic constituents, nutrients, suspended sediment, fecal bacteria, and some trace metals.[14]

By far the greatest limitation of the existing national networks is that they do not measure toxics and ecological indicators, which have emerged as major water-quality issues in the past several years. Another problem with existing national water-quality networks is that, although station selection was based on hydrologic factors rather than on proximity to pollution sources, the networks were not designed to provide a statistical sample of streams throughout the nation. As a result, the networks can only provide a general overview of national conditions. Moreover, because NASQAN stations represent relatively large drainage basins only, observed trends "are likely to reflect water quality effects resulting from large-scale processes, such as changes in land use, and atmospheric deposition, rather than localized effects such as changes in the amount or quality of point source discharges."[15]

These shortcomings in the national database must be kept in mind when considering existing information on the effects of controlling municipal waste; nonpoint sources of nutrients, pesticides, and heavy metals; and point sources of toxics—all of which are key issues facing Congress in the upcoming reauthorization of the Clean Water Act. Although understanding of water quality has improved over time, inadequate assessment at the national level continues to be a hindrance. Depending on the scale of assessment—from a local river reach or portion of an aquifer, an individual river basin or aquifer, multiple basins and regional aquifer systems, to a national statistical summary—conclusions about water-quality conditions could vary considerably, and, therefore, a variety of monitoring approaches should be applied.

Reducing Municipal Sewage Discharge

The cornerstone of the Clean Water Act was the "construction grant" program, funded predominantly by the federal government, to upgrade municipal wastewater treatment plants to a uniform standard of secondary treatment.[16] In the 1987 amendments to the act, the grant program was converted into a state revolving loan fund financed by direct federal appropriations scheduled to end after fiscal year 1994. However, Congress is under considerable pressure by the states to extend the loan program beyond 1994.[17] Senate bill S. 1081, which must be reintroduced at the start of the new Congress, proposes using the loan program to encourage greater private-sector involvement in financing treatment facilities.

Given the magnitude of public investment, one important issue is the influence of municipal sewage treatment plants on water quality. Understanding the amount and nature of change in water quality, at both the local and regional scales, would provide Congress and the administration

5. RESOURCES: Water

with an indicator of what type of and how much federal investment in additional treatment of municipal sewage is worthwhile.

The preponderance of evidence, based on case studies, subjective opinion surveys, and anecdotal information, seems to show that better municipal waste treatment has improved water quality in certain respects. However, how much water quality has changed and in what ways are less clear. In a review of the effects of treatment upgradings at 13 municipal treatment plants, dissolved oxygen levels clearly had improved downstream at 10 of the plants.[18] The Association of State and Interstate Water Pollution Control Administrators—whose members administer the federal monies for municipal wastewater treatment plant construction—

has periodically polled its membership on how water quality has changed; the consensus is that the act has produced significant gains in water quality.[19]

There is some very good local evidence that improved sewage treatment results in improvements in water quality. For example, when the Blue Plains Treatment Plant in Washington, D.C., was operating with only secondary treatment, dense blue-green algae growth was a normal occurrence in the Potomac estuary, particularly during the summer. Following full implementation of tertiary treatment and denitrification in 1983, phosphorus levels decreased by one-third, ammonia concentrations decreased, and nitrate concentrations increased.[20] The improved wastewater treatment contributed to the return of underwater vegetation, an increase in water clarity,

and increased and balanced biological activity. In a more recent case, statistical analysis of water quality in the White River in Indiana showed clear improvements in dissolved oxygen as a direct result of advanced wastewater treatment.[21] Similar improvements have occurred in the Delaware River, the Neches River in Texas, and the Flint River in Georgia, although heavy metals and organics continue to pose threats to human and ecological health.[22]

It is impossible, however, to take anecdotal information and results of individual case studies and extrapolate them to the nation as a whole in any quantitative way. Statistical analyses of national-level monitoring data can provide a more quantitative, albeit limited, picture of national water-quality changes attributable to sewage treat-

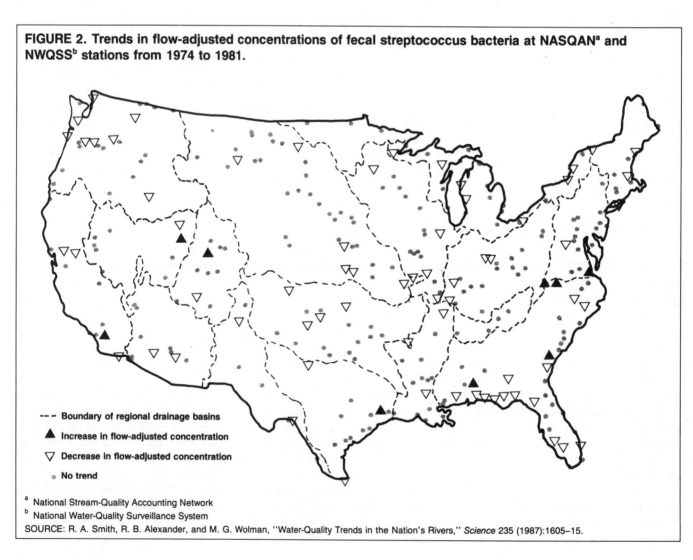

FIGURE 2. Trends in flow-adjusted concentrations of fecal streptococcus bacteria at NASQAN[a] and NWQSS[b] stations from 1974 to 1981.

- - - Boundary of regional drainage basins

▲ Increase in flow-adjusted concentration

▽ Decrease in flow-adjusted concentration

• No trend

[a] National Stream-Quality Accounting Network
[b] National Water-Quality Surveillance System

SOURCE: R. A. Smith, R. B. Alexander, and M. G. Wolman, "Water-Quality Trends in the Nation's Rivers," *Science* 235 (1987):1605–15.

ment plant construction and upgrading. Although not designed to provide a true random sample of surface-water conditions, available water-quality records for a few hundred widely distributed sampling locations provide a generally representative view of conditions in streams and rivers throughout the nation. Two important water-quality measurements collected at these stations—fecal bacteria counts and total phosphorus concentrations—have shown improvement in recent years, which appears to be partly the result of better treatment of municipal effluent (see Figure 2). For example, between 1974 and 1981, the most intensive phase of treatment plant construction, downward (improving) trends were estimated at 23 percent of the 295 stations for which data were available.[23] Results of a recent analysis indicate that these trends, which began in the mid 1970s, continued and even strengthened through 1989.[24]

Perhaps the most noteworthy finding from national-level monitoring is that heavy investment in point-source pollution control has produced no statistically discernible pattern of increases in the water's dissolved oxygen content during the last 15 years.[25] In contrast to case studies showing dissolved oxygen increases in the most oxygen-depleted waters immediately downstream of many sewage outfalls, the data from a 350-station network provide a broader view of stream conditions nationally. The absence of a statistically discernible pattern of increases suggests that the extent of improvement in dissolved oxygen has been limited to a small percentage of the nation's total stream miles. This is notable because the major focus of pollution control expenditures under the act has been on more complete removal of oxygen-demanding wastes from plant effluents.

This lack of visible change may be attributed partly to the probability that much of the investment in point-source pollution controls has simply kept pace with population increases and economic development. Available data indicate that oxygen-de-

manding waste loads declined substantially in the 1970s.[26] However, loads were nearly stable through the 1980s—despite public and private point-source control expenditures of $194 billion—because the population increased by 10 percent and real GNP growth by 30 percent.[27] The nearly constant levels of dissolved oxygen in streams over this period of increasing pollution appear to represent a clear environmental benefit of pollution controls. However, there is still great difficulty in obtaining consistent and comparable pollution-load data on individual point-source discharges, as a recent USGS study of the Upper Illinois River basin clearly shows.[28] Such information might improve the ability to correlate recorded changes in water quality with changes in pollution loads. Here again, although EPA has recently placed a greater emphasis on compiling consistent and comparable point-source data, the lack of reliable historical data makes a meaningful retrospective view of water-quality changes difficult to obtain.

Nonpoint-Source Pollution

A major deficiency in the existing Clean Water Act is the lack of control of nonpoint sources of pollution, such as agricultural lands, feed lots, urban areas, suburban developments, and even silviculture and grazing. The enormous difficulties in identifying, measuring, and controlling nonpoint sources of contamination complicate the allocation of responsibility for its control. In fact, the costs of controlling certain nonpoint sources could greatly exceed the costs of point-source controls.[29] The current debate focuses on the choice between a "command-and-control" regulatory strategy like that used for point sources and voluntary or economic incentives.[30] (Also see "Dealing with Pollution: Market-Based Incentives for Environmental Protection," by Robert N. Stavins and Bradley W. Whitehead in the September 1992 *Environment*.)

Section 319 in the 1987 amendments to the act, which established a

grant program to assist states in developing management plans for nonpoint-source controls, seems to be having little impact.[31] The voluntary approach used in section 319 reflected Congress's unwillingness to encroach on state and local land-use decisions in light of the lack of information both about the actual magnitude of the nonpoint-source problem (although they suspected it was quite large) and about which nonpoint-source control strategies are most effective. In addition, Congress was reluctant to make the federal government vulnerable to a costly operational grant program at a time of mounting budget deficits.[32]

The current Senate majority staff draft revision of S. 1081 (see the box on page 185) proposes to strengthen section 319 by mandating that the states implement a nonpoint-source management program within three years of the reauthorization. States would be required to follow national program guidelines published by EPA that would include, among other elements, "methods to estimate reductions in nonpoint-source pollution loads" and "any necessary monitoring techniques to assess over time the success of management measures."[33] Several options related to nonpoint-source pollution control were discussed within the Bush administration, including economic incentives and adoption of a program similar to one that has been added to the Coastal Zone Management Act, but no specific recommendations were made.

The relative proportion of nonpoint- to point-source loadings of nutrients, suspended sediments, trace metals, agricultural chemicals, and other toxic organic compounds at the national level is unknown. However, an increasing number of studies are beginning to illuminate the scope and magnitude of nonpoint-source pollution in selected areas of the nation. For example, in the Chesapeake Bay, approximately 90 percent of all nitrogen loading is estimated to come from nonpoint sources.[34] Moreover, it could cost from $43 to $325 per pound to reduce nonpoint loadings of phosphorus entering Chesapeake Bay, though the cost of removing phosphorus from point sources is approximately $39 to $71 per pound. One study estimated that between 70 and 94 percent of total annual nitrogen loadings measured at the NASQAN station furthest downstream in four major river basins came from nonpoint sources.[35] Others have cited more equal proportions of point- versus nonpoint-source contributions of nutrients, however.[36]

A study of herbicides in surface waters of the midwestern United States found that large concentrations of herbicides were flushed from cropland and transported through the surface-water system as a result of late spring and early summer rainfall.[37] Median concentrations of atrazine and other herbicides increased by an order of magnitude during this period and then decreased nearly to pre-planting levels by the time of the fall harvest. Water samples from more than 50 percent of the sites exceeded the promulgated maximum contaminant levels for atrazine in drinking water (3 micrograms per liter) during the spring. Some of the herbicides persist from year to year, and degradation products of these herbicides were found to be both persistent and mobile. In fact, the pesticide DDT and its breakdown products have been detected in the agricultural soils of the Yakima River basin in Washington State even though the use of DDT was banned almost 20 years ago.[38] In the San Joaquin Valley of California, irrigation drainage practices have been mobilizing selenium in the soils, thereby causing widespread damage to waterfowl and degradation of surface-water quality when the high-selenium waters drain into reservoirs and wetlands.[39] Knowledge of soils and other hydrologic and geologic conditions, however, has produced a sophisticated and integrated understanding of the causes of selenium contamination and has led to improved management practices.

Nonpoint-source pollution is not limited to agricultural areas, however. Urban areas can also be significant sources. According to data from NASQAN stations located in coastal areas, median loadings of phosphorus in predominantly urban basins are higher than median loadings in basins dominated by croplands or forests. Furthermore, it is becoming increasingly clear that atmospheric deposition has become a major source of nitrogen in surface water.[40]

Wetlands are also a significant part of the nonpoint-source control debate. Wetlands are not only good for wildlife; they play a critical role in local and regional water quality, as well. (For an excellent discussion of wetlands issues, see Jon Kusler's article in the March 1992 *Environment*.) Protection of wetlands brings the debate between the protection of ecological resources and the rights of property owners to the most local level. A clash between the definition of water resources as public goods and the exercise of private property rights, the wetlands debate touches on such difficult policy areas as federal involvement in state and local land-use decisions and the appropriate process for establishing a scientifically defensible definition of wetlands (see box on page 185).

Section 404 is the *de facto* wetlands protection program of the Clean Water Act. The Army Corps of Engineers issues permits for any dredge-and-fill activities in the "navigable waters" of the United States only after administrative procedures are followed to allow for public comment on a draft environmental impact statement (EIS). The final EIS is subsequently issued by the corps with the concurrence of other federal agencies. Wetlands proposals, introduced in the House and Senate as separate bills to amend the Clean Water Act, range from broadening the jurisdiction of the section 404 permit process to limiting its applicability to strictly "navigable" waters. Senate bill S. 1081 includes no new provisions for regulation of wetlands. Complicating resolution of the issue is the difficulty in delineating wetland resources and their change over time. This difficulty makes the assessment of proposed policy changes a highly subjective process.[41]

The complexity of the nonpoint-source problem is hardly academic. What is the appropriate strategy and level of federal, state, local, and private investment in nonpoint-source controls under such varying conditions? To properly address this question for the nation as a whole, regional-scale hydrologic, chemical, and biological processes must be better understood. Simultaneously, consistent and comparable national-level monitoring data that are representative of the varying land uses contributing to nonpoint-source problems are needed to describe and compare conditions and trends among regions.

Toxics in Surface Waters and Sediments

The toxics issue presents EPA with large technical and financial difficulties. The number of potentially toxic pollutants is large, measurement is expensive, and analytical methods are not always sensitive enough to detect biologically significant concentrations. Furthermore, there is no nationally consistent database to rely on for guidance on where control efforts should be focused. EPA has regulated only about one-fifth of the industrial plants that dump toxic substances into rivers, lakes, and estuaries.[42] The existing regulatory structure of the Clean Water Act is almost entirely built around the control of point pollution sources through the pretreatment program (Section 307); nonpoint sources of toxic pollutants are presently unregulated.

Senate bill S. 1081 includes several sections that would increase control of toxic point sources. The bill sets strict deadlines for EPA to issue rules to control existing point sources of toxics. Also, each publicly owned treatment plant would be required to implement a "toxic reduction action program" directed toward significantly reducing specific sources of toxics flowing into the plant. And finally, EPA would be required to submit to Congress an annual "environmental toxics release assessment'' that summarized the industrial sources of toxics, their geographic distribution, and approximate discharges.

Little reliable information at the national or even regional scale can be reported for either inorganic or organic toxic chemicals in natural waters. Because of the vast array of toxic organic chemicals in the environment, it is highly impractical to monitor any but a relatively small number of toxic chemicals on a regular basis. A far more efficient and meaningful approach would be to use bioindicators and chemical measures concurrently, yet the United States presently has no nationally consistent biological database that could be used for this purpose.

The country also lacks a national database on trace metals, some of which are toxic to aquatic life even at very low concentrations, such as less than one part per billion. Indeed, recent evidence suggests that very low ambient concentrations of trace metals are actually dissolved in most streams and rivers, but these concentrations are controlled more by stream chemistry than by anthropogenic sources.[43] Anthropogenic sources can significantly raise concentrations of toxics bound to sediment particles (in comparison with the concentrations dissolved in water), which then are ingested by aquatic biota. As with toxic organics, bioindicators may be the most practical means of monitoring the occurrence and trends of trace metals.

Although bioindicators are valuable, few known indicators are suitable for national-level monitoring, and few biological studies have been national in scope. One exception is the National Contaminant Biomonitoring Program of the U.S. Fish and Wildlife Service. Whole fish, collected periodically between 1970 and 1986 from a network of 113 stations located on major rivers and in the Great Lakes, were measured for trace metals and organochlorine contaminant concentrations.[44] Although there are difficulties in interpreting the significance of the results,[45] the trends reported are still of interest. Concen-

trations of cadmium, arsenic, and lead decreased by approximately 50 to 63 percent between 1976 and 1986, and the concentration of mercury remained nearly constant. Fish tissue concentrations of the pesticide Dieldrin, DDT-related compounds, and total polychlorinated biphenyls (PCBs) decreased by more than 60 percent between 1970 and 1986, a reflection of the fact that use of DDT was stopped in 1972 and use of PCBs stopped in 1976. Dieldrin, however, was not banned until 1988.[46]

Closing the Information Gap

EPA is fully aware of the problems with the section 305(b) reporting procedures and is working with the states to reduce methodological inconsistencies in the program.[47] Several other steps are being taken throughout the federal government to close the information gap.

To support future national, regional, and local decisions related to water-quality management, USGS has undertaken the National Water-Quality Assessment (NAWQA) program. The program builds on an existing base of ongoing water-quality activities within USGS, as well as those of other federal, state, and local agencies. The full-scale NAWQA program was implemented starting in fiscal year 1991 following a five-year pilot effort and an extensive evaluation of the program's concept by the National Academy of Sciences.[48] The goals of the NAWQA program are to describe the status of and trends in the water quality of large representative parts of the nation's streams and groundwater and to provide a sound scientific understanding of the natural and human factors that affect the quality of these resources. To meet these goals, the program integrates information about water quality at local, study-unit, regional, and national scales and focuses on water-quality conditions that affect large areas or that recur frequently on the local scale. Because of its regional focus, the NAWQA program is ideally suited to investigate nonpoint-source

contamination and to define on a regional basis the relative contributions of major contamination sources.

The NAWQA program consists of comprehensive assessments in 60 areas of the nation—each of which cover about 20,000 square miles—that account for 60 to 70 percent of the overall water use and population served by public water supply. These 60 study units cover about 50 percent of the coterminous United States and include at least a part of every state.[49] Biological studies in surface water conducted as part of the NAWQA program include analysis of trace elements and trace organic compounds in fish tissues, measurements of sanitary quality, and analysis of ecosystem structure in relation to habitat and water-quality conditions. The chemistry of suspended sediment and bed sediment samples is also determined as part of the program.

This information base will form a national set of data and interpretations that will contribute to a national assessment of critical water-quality issues; aid the states in focusing their monitoring activities on the most important water-quality issues in the most sensitive settings; help the nation and the states define the status of and trends in regional water quality; and provide sound scientific information that can be used by policymakers and resource managers.

Another new initiative is the Environmental Monitoring and Assessment Program (EMAP), which is now beginning to be implemented. Planning for EMAP began in 1989 in response to a recommendation of EPA's Science Advisory Board to establish a statistically sound national survey for monitoring ecological resources. The Science Advisory Board saw this approach as a means for EPA to detect emerging environmental problems and characterize their geographic extent, and persistence over time. EMAP is intended to estimate "the current status, extent, and geographic distribution of ecological resources; the proportion of these resources that is degraded; trends in the conditions of these resources; and the probable causes of adverse effects.''[50]

Biological response indicators will be EMAP's primary measure of ecological conditions. Other chemical and physical indicators will be used to measure stress and chemical exposure of ecological resources. The EMAP sampling framework consists of a systematic hexagonal grid of approximately 12,500 points for the coterminous United States. As currently envisioned, a subset of roughly 3,200 of these points will be randomly selected for surface-water field sampling. Each year, 800 of the 3,200 lake and stream sites will be sampled, giving a four-year resampling cycle. EMAP's results will represent statistical summaries of regional conditions, not descriptions of conditions at individual locations. The surface-water component of EMAP is in the process of development, and thus details are not available at this time. Groundwater is unlikely to be included in EMAP in the foreseeable future.[51] However, work is proceeding on other resources in the EMAP design, such as forests, inland wetlands, agroecosystems, near coastal lands, deserts, and grasslands. Intensive high-altitude aerial photography will precede actual field sampling to characterize as much of the land cover, land use, and extent of resources as possible within each of the sample hexagons.

The 1987 amendments to the Clean Water Act established the National Estuary Program to ''identify nationally significant estuaries, protect and improve their water quality, and enhance their living resources.''[52] The program uses Comprehensive Conservation and Management Plans (CCMPs) developed through the collaborative efforts of management conferences, whose purposes include assessing trends in water quality, identifying environmental stresses, and planning restoration activities. Congress named 12 estuary projects to receive priority consideration, and EPA may select others in the future. Quantitative comparisons of estuarine water quality may be possible if consistent methods are used among

the studies; however, this is not an element of the program.[53]

All of the program initiatives will contribute to the improvement of the U.S. information base on water quality. The recently formed Intergovernmental Task Force on Monitoring Water Quality,[54] under the joint leadership of EPA and USGS, now provides an institutional mechanism to identify systematically gaps in monitoring efforts and particularly focuses on the need to evaluate the effectiveness of federally mandated pollution control strategies. The task force is charged with identifying what kinds and amounts of data are needed, determining what can be obtained by improving the integration and consistency of existing federal and state programs, and recommending extensions of existing programs or new programs to fill gaps in knowledge. In addition, the task force may recommend modifications to existing networks to provide a more statistically representative view of national conditions, but this is not under consideration at this time.

Science and Policymaking

Reauthorization of the Clean Water Act raises several important questions about the value of scientific information in policymaking. In theory, the five-year reauthorization cycle affords both Congress and the administration the opportunity to engage in serious program evaluation. In practice, decision makers usually deal with issues that arise from anecdotes rather than from a formal evaluation of monitoring data collected specifically to identify and quantify the causes and effects of existing and emerging water-quality problems.

The language of Section 305(b) indicates that Congress intended the National Water Quality Inventory to provide the data to evaluate the success of the act. For a variety of reasons, the inventory has not met expectations of national consistency in monitoring and analysis. Also, although the USGS has operated the NASQAN monitoring program since the early 1970s, a significant USGS

program to describe regional and national water quality has only recently emerged with the initiation of NAWQA in the mid 1980s.

Existing U.S. databases present a fragmented picture of national water quality. Almost no information about potentially toxic organic compounds and trace metals is available. Consistent and comparable ancillary information on land use and point-source discharges at the national scale also is lacking. Use of bioindicators is desirable but currently out of reach on a national scale.

Critical steps are now being taken to produce policy-relevant information, but more could be done. The USGS is proceeding with the full implementation of the NAWQA program to provide the scientific basis for understanding cause-and-effect relationships in water quality. And throughout government, academia, and industry, research and development on bioindicators are increasing, which promises to provide meaningful integrative measures of the effects of both toxic and conventional contaminants on ecological resources. These integrative measures are needed in addition to the physical and chemical measures currently used. EMAP could provide the impetus, framework, and funding for more rapid developments in this area.

In addition, EPA, in close consultation and cooperation with other federal and state agencies, is proceeding with plans to improve the 305(b) reporting program by specifying comparable reporting procedures. States also need to report water-quality measures according to some nationally agreed-on ambient standards and designations of uses. The Intergovernmental Task Force on Monitoring could become the institutional vehicle for implementing these changes, the means of integrating the various federal monitoring and assessment programs, and the instrument to encourage more rigorous consistency from the states. Furthermore, modifications to existing networks could provide for a more statistically based description of water-quality conditions and trends on the national scale.

If the steps outlined above are carried through, then many questions about national water-quality conditions could be answered within the next five years. However, it will probably take more than a decade to amass the kind of data needed to document environmentally and statistically significant trends in water quality. In any case, a prerequisite to bringing accurate and meaningful water quality information to the forefront of the policy debate is recognition by decision makers that they do not now have the information they need to make wise decisions for the future. Only then will the effectiveness of the Clean Water Act be known.

NOTES

1. A. Carlin and the Environmental Law Institute, *Environmental Investments: The Cost of a Clean Environment,* EPA-230-11-90-083 (Washington, D.C.: U.S. Environmental Protection Agency, November 1990).

2. R. Carson, *Silent Spring* (Greenwich, Conn.: Fawcett Publications, 1962).

3. Library of Congress Environmental Policy Division, *A Legislative History of the Water Pollution Control Act Amendments of 1972* (Washington, D.C.: U.S. Government Printing Office, 1973).

4. R. M. Nixon, "Message from the President of the United States Returning Without Approval the Bill (S. 2770) Entitled *The Federal Water Pollution Control Act Amendments of 1972,*" in Library of Congress Environmental Policy Division, *A Legislative History of the Water Pollution Control Act Amendments of 1972,* vol. 1 (Washington, D.C.: U.S. Government Printing Office, 1973), 137-39.

5. U.S. Senate Public Works Committee, page 1473, note 3 above.

6. M. G. Wolman, "The Nation's Rivers," *Science* 174 (1971):905-18.

7. Comptroller General of the United States, *Better Monitoring Techniques Are Needed to Assess the Quality of Rivers and Streams* (Washington, D.C.: U.S. General Accounting Office, 1981), 2 volumes; and P. E. Tyler, "After $30 Billion, No One Knows If Nation's Water Is Any Cleaner," *Washington Post,* 14 May 1981, A2.

8. U.S. Senate Public Works Committee, page 797, note 3 above.

9. U.S. Environmental Protection Agency, *National Water Quality Inventory: 1990 Report to Congress* (Washington, D.C.: U.S. EPA Office of Water, 1992).

10. U.S. Environmental Protection Agency, "Review of Environmental Monitoring Costs Related to Environmental Indicators" (Washington, D.C.: U.S. EPA Environmental Results and Forecasting Branch, Office of Policy, Planning and Evaluation, April 1991).

11. Carlin, note 1 above.

12. J. C. Briggs, "Nationwide Surface Water Quality Monitoring Networks of the U.S. Geological Survey" (Paper presented at the American Water Resources Association Symposium, San Francisco, 12-14 June 1978), 49-57.

13. R. A. Smith and R. B. Alexander, "Correlations Between Stream Sulphate and Regional SO_2 Emissions," *Nature* 322, no. 6081 (1986):722-24.

14. During the 1980s, inflation greatly eroded the fixed annual funding of $5 million. As a result, several variables were dropped, including organic carbon and the measurement of trace elements and nutrients in whole water samples. The original design of the networks called for fixed interval sampling on a monthly basis. Now, because of inflation, the frequency of sampling is quarterly or bimonthly in both networks. As a result, total samples collected in NASQAN decreased from more than 6,000 in 1979 to about 2,000 in 1990.

15. D. P. Lettenmaier, E. R. Hooper, C. Wagoner, and K. B. Faris, "Trends in Stream Quality in the Continental United States, 1978-1987," *Water Resources Research* 27, no. 3 (1991):327-39.

16. A. M. Freeman III, "Water Pollution Policy," in P. R. Portney, ed., *Public Policies for Environmental Protection* (Washington, D.C.: Resources for the Future, 1990).

17. Environmental and Energy Study Institute, "1992 Briefing Book on Environmental Legislation" (Washington, D.C., 1992), 13.

18. W. M. Leo, R. V. Thomann, and T. W. Gallagher, *Before and After Case Studies: Comparisons of Water Quality Following Municipal Treatment Plant Improvements,* EPA Report 430/9-007 (Washington, D.C.: U.S. Environmental Protection Agency, 1984).

19. Association of State and Interstate Water Pollution Control Administrators, *America's Clean Water: The States' Evaluation of Progress: 1972-1982* (Washington, D.C.: ASIWPCA, 1984).

20. V. Carter and N. Rybicki, "Resurgence of Submersed Aquatic Macrophytes in the Tidal Potomac River, Maryland, Virginia, and the District of Columbia," *Estuaries* 9, no. 4B (1986):368-75.

21. C. G. Crawford and D. J. Wangsness, "Effects of Advanced Wastewater Treatment on the Quality of White River, Indiana," *Water Resources Bulletin* 27, no. 5 (1991):769-79.

22. R. Patrick, *Surface Water Quality: Have the Laws Been Successful?* (Princeton, N.J.: Princeton University Press, 1992).

23. R. A. Smith, R. B. Alexander, and M. G. Wolman, "Water-Quality Trends in the Nation's Rivers," *Science* 235 (1987): 1605-15; additional results presented in R. A. Smith, R. B. Alexander, and M. G. Wolman, *Analysis and Interpretation of Water-Quality Trends in Major U.S. Rivers, 1974-81,* U.S. Geological Survey Water-Supply Paper 2307 (Reston, Va.: U.S. Geological Survey, 1987).

24. R. A. Smith, R. B. Alexander, and K. J. Lanfear, *A Graphical Summary of National Water-Quality Conditions and Trends,* Open-File Report 92-70 (Reston, Va.: U.S. Geological Survey, 1992).

25. Smith, Alexander, and Wolman, note 23 above; Smith, Alexander, and Lanfear, note 24 above.

26. Smith, Alexander, and Wolman, note 23 above.

27. Carlin, note 1 above.

28. J. S. Zogorski, S. F. Blanchard, R. D. Romack, and F. A. Fitzpatrick, *Availability and Suitability of Municipal Wastewater Information for Use in a National Water-Quality Assessment: A Case Study of the Upper Illinois River Basin, Illinois, Indiana, and Wisconsin,* Open-File Report 90-375 (Urbana, Ill.: U.S. Geological Survey, 1990).

29. S. A. Freudberg and J. P. Lugbill, "Controlling Point and Nonpoint Nutrient/Organic Inputs: A Technical Perspective" (Paper presented at the conference "Cleaning Up Our Coastal Waters: An Unfinished Agenda," U.S. Environmental Protection Agency and Manhattan College, Riverdale, N.Y., 13 March 1990).

30. For some time, economists have questioned the economic efficiency and effectiveness of technology-based regulatory controls on point sources. For background on this issue, see P. R. Portney, ed., *Public Policies for Environmental Protection* (Washington, D.C.: Resources for the Future, 1990); and J. J. Boland, "Enticing the Colossus," *Johns Hopkins Magazine,* June 1992, 43-45.

31. R. W. Adler, "Returning to the Goals of the Clean Water Act," *Update Water Resources,* no. 88, (1992):23-30.

32. Senate debate on overriding veto of H.R. 1, 4 February 1987, in *A Legislative History of the Water Quality Act of 1987* (Public Law 100-4), vol. 1, November 1988, 322-23; Hearings of the House Committee on Public Works and Transportation, Subcommittee on Water Resources, "Reauthorization of the Federal Water Pollution Control Act (Nonpoint Source Pollution)," 24 April 1991, 701-911.

33. U.S. Senate Environment and Public Works Committee, *Majority Staff Draft: Amendment in the Nature of a Substitute to S.1081 Water Pollution and Prevention and Control Act of 1991* (Washington, D.C., 31 December 1991).

34. L. P. Gianessi and H. M. Peskin, *An Overview of RFF Environmental Data Inventory, Methods, Sources, and Preliminary Results* (Washington, D.C.: Resources for the Future, 1984), 111. Estimates of nonpoint-source loadings were derived using a regression procedure developed by T. A. Cohn of the U.S. Geological Survey in 1991; Freudberg and Lugbill, note 29 above.

35. W. G. Wilber, U.S. Geological Survey, personal communication with the authors, 1992.

36. E. L. Tyler, "Reauthorizing the Federal Water Pollution Control Act," *Update Water Resources* no. 88 (1992):7-16.

37. E. M. Thurman, D. A. Goolsby, M. T. Meyer, and D. W. Kolpin, "Herbicides in Surface Waters of the Midwestern United States: The Effect of Spring Flush," *Environmental Science and Technology* 25, no. 10 (1991):1794-96.

38. J. F. Rinella, P. A. Hamilton, and S. W. McKenzie, *Persistence of the DDT Pesticide in the Yakima River Basin,* USGS Circular 1090 (Washington, D.C.: U.S. Geological Survey, in press).

39. R. J. Gilliom et al., *Preliminary Assessment of Sources, Distribution, and Mobility of Selenium in the San Joaquin Valley, California,* Water-Resources Investigation Report 88-4186 (Sacramento, Calif.: U.S. Geological Survey, 1989).

40. D. C. Fisher and M. Oppenheimer, "Atmospheric Nitrogen Deposition and the Chesapeake Bay Estuary," *Ambio* 20, no. 3/4 (1991):102-08.

41. U.S. Fish and Wildlife Service, *Wetlands Status and Trends* (Washington, D.C.: U.S. Department of the Interior, Fish and Wildlife Service, 1991).

42. U.S. Environmental Protection Agency, *Report to Congress: Water Quality Improvement Study* (Washington, D.C.: U.S. EPA, 1988); and M. Weisskopf, "Default on Industrial Effluent," *Washington Post,* 30 December 1991, A11.

43. H. L. Windom, J. T. Byrd, R. G. Smith, Jr., and F. Huan, "Inadequacy of NASQAN Data for Assessing Metal Trends in the Nation's Rivers," *Environmental Science and Technology* 25, no. 6 (1991): 1137-42. For additional information on the difficulty of measuring dissolved concentrations of trace metals, see A. M. Shiller and E. A. Boyle, "Variability of Dissolved Trace Metals in the Mississippi River," *Geochimica et Cosmochimica Acta* 51 (1987):3273-77; and A. R. Flegal and K. Coale, "Discussion: Trends in Lead Concentration in Major U.S. Rivers and Their Relation to Historical Changes in Gasoline-Lead Consumption by R. B. Alexander and R. A. Smith," *Water Resources Bulletin* 25 (1989):1275-77.

44. C. J. Schmitt, U.S. Fish and Wildlife Service, National Contaminant Biomonitoring Program, personal communication with the authors, 1992; and C. J. Schmitt, J. L. Zajicek, and M. A. Ribick, "National Pesticide Monitoring Program: Residues of Organ-

5. RESOURCES: Water

ochlorine Chemicals in Freshwater Fish, 1980–81,'' *Archives of Environmental Contamination and Toxicology* 14 (1985):225–60.

45. Smith, Alexander, and Lanfear, note 24 above.

46. G. W. Ware, *The Pesticide Book* (Fresno, Calif.: Thompson Publications, 1989).

47. A. E. Mayio and G. H. Grubbs, ''Nationwide Water-Quality Reporting to the Congress as Required under Section 305(b) of the Clean Water Act,'' in *National Water Summary 1990–91—Stream Water Quality* (Reston, Va.: U.S. Geological Survey, in press).

48. R. M. Hirsch, W. M. Alley, and W. G. Wilber, *Concepts for a National Water-Quality Assessment Program*, USGS Circular 1021 (Reston, Va.: U.S.

Geological Survey, 1988); and National Research Council, *A Review of the U.S.G.S. National Water Quality Assessment Pilot Program* (Washington, D.C.: National Academy Press, 1990), 153.

49. P. P. Leahy, J. S. Rosenshein, and D. S. Knopman, *Implementation Plan for the National Water-Quality Assessment Program*, Open-file Report 90–174 (Reston, Va.: U.S. Geological Survey, 1990).

50. U.S. Environmental Protection Agency, *Design Report for EMAP*, EPA/600/3-91/053 (Washington, D.C.: U.S. EPA Office of Research and Development, 1990); and idem, *Reducing Risk: Setting Priorities and Strategies for Environmental Protection*, SAB-EC-90-021 (Washington, D.C.: U.S. EPA Science Advisory Board, 1990).

51. D. A. Rickert, U.S. Geological Survey, personal communication with the authors, 1992.

52. U.S. Environmental Protection Agency, *National Estuary Program After Four Years: Report to Congress*, 5039–92–007 (Washington, D.C.: U.S. EPA, April 1992).

53. T. Armitage, U.S. Environmental Protection Agency, personal communication with the authors, 1991.

54. U.S. Office of Management and Budget, ''Coordination of Water Resources Information,'' Memorandum No. M–92–01 (Washington, D.C., 10 December 1991).

Redeeming the Everglades

For almost a century, engineers reconfigured the rivers and wetlands of Florida's Everglades, despoiling a wilderness to benefit industry and agribusiness, condos and beach resorts. The dredges and earthmovers did their work: The water dried up and the creatures thinned out, almost to the point of disappearance. Now, activists and politicians have begun to correct the damage. For the first time in the United States, a public works project is being reversed to right an environmental wrong.

Mark Derr

Mark Derr is based in Miami Beach, Florida, and is the author of Some Kind of Paradise *and* Over Florida. *His latest book is* The Frontiersman, *a biography of Davy Crockett.*

Highway 27 runs northwestward for 60 miles from Miami through remnant Everglades marshes, across the tabletop-flat fields of the 750,000-acre Everglades Agricultural Area, before turning west at the imposing Hoover Dike, which imprisons Lake Okeechobee. Sugar rules here, but there are also fields of vegetables and sod, and a few ranches. The landscape is broken only by roads and the canals that keep the land dry. When sugarcane is burned just prior to cutting—from October to March—the sky turns black with smoke; tongues of orange flame shimmer along the horizon. The sickly smell of sugar being processed in the area's seven huge mills permeates the atmosphere.

A hundred years ago, travelers knew this area as a seemingly endless marsh with saw grass rising 10 feet or more from pristine water. A custard apple swamp buffered the saw grass from the big lake, where at night adventurers would watch mysterious lights—probably swamp gas—floating on the black water. The Everglades was as forbidding as any place they had encountered.

Today it is one of the most imperiled regions in the United States, its rivers channelized, its wetlands drained. Historically, this was a rich mosaic of plants and animals that covered 13,000 square miles, from the headwaters of the Kissimmee River in central Florida through Florida Bay to the Keys. As it exists now, the Everglades is a fragile ecosystem defined by the most extensive plumbing works in the world—a network of canals and pumping stations that control the movement of water through the entire region.

To rescue the area, environmentalists, water managers, and public officials have joined to plan for the Everglades of the future—an imagined place where water would once again flow freely and the historic mix of plants and animals would flourish, albeit on a dramatically reduced scale. The task of making the Everglades of the future real is daunting; the viability of life in south Florida depends on the project's success.

THE GRASSY WATER For 6,000 years, water shaped the topography and defined the plant and animal communities of the Everglades. During the May-to-October rainy season, it sometimes fell in torrents, joining a mass of water that meandered through the twisting channel of the Kissimmee River and parallel creeks into Lake Okeechobee, which regularly sloshed over its banks to join Pa-hay-okee, the Grassy Water, called the river of grass by conservationist Marjory Stoneman Douglas. From there, the water curled toward the east coast before turning south and west in a slow-moving sheet as much as 40 miles wide, crossing the center of the peninsula to Florida Bay, Whitewater Bay, and the Gulf of Mexico. Just beneath the surface, water cut through a huge aquifer reaching under Biscayne Bay, off the east coast, and Florida Bay, to the south. Occasionally it bubbled up in freshwater springs. To the east, surface water crossed the coastal ridge through transverse glades, rivers, and creeks. This was a lush, wet wilderness, hosting hundreds of species of birds, fish, amphibians, reptiles, mammals, and plants found nowhere else in North America. Millions of plume birds filled the sky; herds of green turtles grazed the grass beds of Biscayne Bay; alligators, crocodiles, and otters abounded. Mosquitoes swarmed so thickly they could suffocate animals and make grown men faint.

In the Everglades, before it was altered by humans, subtle shifts in water levels and the duration of inundation were critical in determining which plants would flourish. There were scrub; cabbage palm hammocks; pine flatwoods and palmetto prairies; wet hammocks; prairie grasslands; cypress swamps; marl prairies and freshwater marshes; Everglades marshes and sloughs and tree islands, or keys; tropical hardwood hammocks; mangrove forests and coastal marshes. Saw grass, which constitutes 70 percent of the vegetation in the Everglades marshes and sloughs, is actually a sedge, *Cladium jamaicensis*.

Differences in soil elevation of just a few feet made the difference between a saw grass marsh and an Everglades key, with its dense mix of hardwoods, including mahogany and red-barked gumbo-limbo, and palms. Water levels dictated the rhythm of reproduction for wading birds and the life cycle of plants.

THE DRAINAGE CAMPAIGN No region on earth has been as extensively replumbed or as intensively managed as this one. The Everglades we know today is crisscrossed with 1,400 miles of canals, levees, spillways, and pumping stations. Water managers for the South Florida Water Management District and the U.S. Army Corps of Engineers monitor and control the movement of water through every part of these modern Everglades.

Managing the system involves a constant juggling of the often conflicting interests of agriculture, the natural system, and the urbanized coasts. Engineers store water in conservation areas and in Lake Okeechobee for release during the dry winter periods to serve agriculture and to recharge the Biscayne Aquifer, the source of drinking water for the 4.5 million residents of Florida's Gold Coast, which stretches from Key West to Palm Beach. They lower water levels in the canals to expel runoff and prevent flooding. They also time the release of water to the Everglades National Park, which would not survive without it.

This elaborate water network represents the culmination of a campaign that began in 1881 and ran for the better part of a century to "reclaim" the Everglades for "civilization." Little more than half of the original 4 million acres of Everglades marshes south of Lake Okeechobee remains in anything approaching its natural state, and half of that is locked into water-conservation areas, which, deprived of sheet flow, function more as reservoirs than as marshes.

Even before the area had been surveyed or fully explored, widespread dredging began. Canals in the beginning were little more than mosquito ditches. In 1916 the Tamiami Trail, the first road across the lower peninsula, was completed from Miami to Naples, effectively cutting off the flow of water south. The Everglades marshes began to dry up.

Although they were sufficient to cause environmental damage, the canals were unable to protect people from floods. In 1926 and again in 1928, hurricanes caused a

The Sugar Subsidy

ig Sugar's costly fight against a court-mandated cleanup has confirmed for many of its critics their belief that the industry merely wants to profit from the largess of the federal government while evading its responsibilities.

A federal price support program keeps the domestic price of raw sugar at roughly twice the world price through import quotas. For a variety of reasons that relate to international supply and demand, it is difficult to calculate precisely to what extent U.S. consumers subsidize the industry.

Estimates range from $1.4 billion to $3 billion a year. Big Sugar's share of the subsidy is thus anywhere from $189 million to $400 million a year, with the bulk going to the U.S. Sugar Corporation and Flo-Sun. Figuring in the cost of building and operating a water-control system in the Everglades and advantageous leases on state land, the Wilderness Society has placed the annual state and water-district subsidy at another $17.57 million.

Ironically, according to USDA analysts, Big Sugar, because of its efficient operations and favorable growing conditions, could remain competitive even without the government's price support program—an assertion that industry spokesmen vehemently deny.

Senator Bob Graham backs the sugar subsidy because he believes the industry is ultimately less problematic than alternatives such as row crops, citrus fruits—and suburbanization.

He also believes that if Big Sugar were to feel economically secure it would be less likely to oppose regulation. But critics say that too often the opposite has happened, to the detriment of the entire region—including, ultimately, the farmers. —M.D.

breach in the earthen levee on the south edge of Lake Okeechobee, resulting in extensive property damage and loss of life. State officials implored the federal government to intervene, and engineers began work on the Hoover Dike to keep the lake from flooding the agricultural area then rising on drained Everglades marshes. In 1947 hurricanes flooded thousands of acres on the east coast near Fort Lauderdale and in the Kissimmee area, leading to cries for the federal government to drain and contain the region's waters once and for all.

Between 1948 and 1971, the Army Corps of Engineers created an engineering marvel and an environmental disaster—the Central and Southern Florida Project for Flood Control and Other Purposes. Its mission was to make the coast safe for development, the interior south of Lake Okeechobee safe for farming and, areas like the park secure for plants and animals. To that end, the Everglades was divided into four major "reservoirs"—Lake Okeechobee and three water-conservation areas numbered 1, 2A and 2B, and 3A and 3B—as well as the Everglades Agricultural Area and the Everglades National

Park. Flows through the Big Cypress Swamp National Preserve—west of the Shark River Slough, which runs through the Everglades—were also regulated.

The water-conservation areas wrap around the east and south sides of the agricultural area and are separated from the park by the Tamiami Trail (see map). The northernmost one—Water Conservation Area 1—is managed by the U.S. fish and Wildlife Service as the Arthur R. Marshall Loxahatchee National Wildlife Refuge. Levees and canals mark the boundaries between these areas and control the flow of water into and out of them. Between Palm Beach and Miami, a 70-mile-long perimeter levee cleaves the interior of the river of grass from the east coast, thus hydrologically isolating acres of saw grass marsh and wet prairies.

The Central and Southern Florida Project also turned the meandering, flood-prone, 103-mile-long Kissimmee River into a 56-mile-long canal, known to engineers as the C-38 and to its numerous opponents as the Big Ditch. Deprived of flowing water, the river's oxbows stagnated and wildlife declined. Before the work was even completed, in 1971, friends of the river launched a drive to return it to its natural course.

By that time it was clear to nearly everyone that the entire Everglades region was imperiled. According to some estimates, wading bird populations had fallen to less than 10 percent of their total in the predrainage period, and vast areas were drying out because of a water-management regime that tended to put too much water in the wrong place at the wrong time.

THE URGE TO PRESERVE Almost from the day the first dredges bit into the limestone that underlies south Florida, activists began to challenge the drainage effort, arguing that it would change the climate and turn the region into a desert; that it would alter vegetation; that it would destroy animals and the ambience that had attracted human settlers in the first place. They demanded a park to preserve part of what remained.

In 1947 the Everglades National Park was established, the first major victory in the drive to save the heart and lungs of south Florida. As the dredges continued their work, environmentalists rallied to save one parcel after another of the vanishing natural system. William Robertson Jr., a park biologist for 30 years, once told me that virtually every major battle since the park's creation has been focused on protecting land that had been left outside the original boundaries for political reasons.

Expanding populations on both coasts put developmental pressure on unprotected lands, and a prolonged drought in the 1970s and 1980s amplified the effects of drainage. Fires, algae blooms in Lake Okeechobee that killed millions of fish, hypersalinity in Florida Bay that killed birds and aquatic life, dried-out wetlands—all were symptoms of a system on the verge of collapse.

In 1983 an ambitious state program to make "the Everglades look and function by the year 2000 more as it did at the turn of the century" was launched by Governor Bob Graham. The Save Our Everglades program was intended "to get

out of the every-week crisis and take a long look at what we needed to do," says Graham, now Florida's senior senator.

Saving this vast ecosystem has become a test case of the ability of major human developments to exist in close proximity to a thriving natural system. For that reason, the success or failure of current restoration efforts has broad implications beyond south Florida.

"There are three major threats to the Everglades," says Pete Rhoads, director of the Office of Everglades Restoration at the South Florida Water Management District, which since 1949 has been responsible for overseeing the region's plumbing system. "They are melaleuca, disruption of the hydroperiod—or water quantity, timing, and distribution—and water quality."

The most damaging of the exotic plants and animals that flourish in tropical south Florida, the melaleuca is a fast-growing Australian tree that was introduced in 1906 to help drain the Everglades. It now threatens much of the ecosystem because it drives out all other plants and animals and has no apparent natural enemies—although some newly discovered biological controls hold out a glimmer of hope.

Everglades National Park, which occupies the southern end of the historic Everglades, has suffered greatly not only from melaleuca but also from the drainage project, despite federal mandates that it receive a fair allotment of water and despite attempts to deliver those quotas during the dry season, when they are most needed. "Redesigning the system to

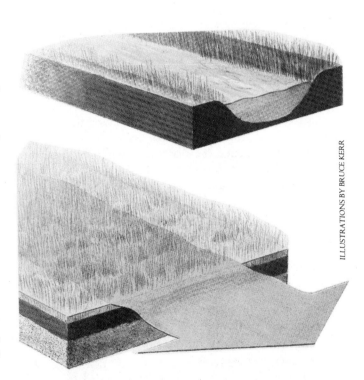

ILLUSTRATIONS BY BRUCE KERR

SHEET FLOW Most rivers flow through narrow channels carved out of the surrounding floodplain (top). When sheet flow is achieved, however, water moves in a slow, steady mass, spread across a wide floodplain whose grasses filter the water as it flows (bottom).

get water in the right place at the right time is essential to the health of the park," says Richard Ring, the park's superintendent. "But you also need to clean up the water from the urban and agricultural areas that comes in here."

Increasingly, ecologists have concluded that the key to nurturing not only the park but the entire remnant Everglades and the urbanized coast as well lies in what Rhoads calls reengineering the system. The Everglades Coalition—founded in 1968 to fight for Everglades National Park water rights and to save the Big Cypress Swamp, reconstituted in 1984 as part of the Graham program, and now comprising 28 national and local environmental groups—has made such a redesign the cornerstone of its agenda for the 21st century. Last fall the National Audubon Society, which is currently chairing the coalition, declared that one of its major campaigns for the 1990s would focus on re-creating something approaching the natural sheet flow of water through the Everglades, and especially through the Shark River and Taylor sloughs. The headwaters of Taylor Slough have been drained for vegetable farms; those of Shark River Slough, for farms and conservation areas.

A COMPLEX TASK Nearly everyone who has considered the possibility of restoring some form of natural flow to the remnant Everglades—the areas lying west of the perimeter levee and south of the Everglades Agricultural Area—recognizes that two steps must be taken: the urbanized east coast, with its 4.5 million residents, must be weaned of its reliance on water from Lake Okeechobee and the water-conservation areas; and the excesses of agriculture must be curbed.

The most ambitious, imaginative, and controversial proposals for recapturing and redirecting the water call for the creation of artificial wetlands. These wetlands would replace the Everglades marshes lost east of the perimeter levee and in the agricultural area and would act as buffers between "civilization" and "nature."

Under one scenario urban storm water would be pumped into interconnected treatment marshes flanking the east side of the levee. Filtering through limestone, the water would seep down to recharge the Biscayne Aquifer, blocking the intrusion of salt water from the coast. It would also establish a hydrological barrier that would prevent water from seeping out of the water-conservation areas. Surface water would flow south into Florida Bay, decreasing its salinity, while treatment marshes would provide valuable wildlife habitat.

More vitally, ecologists are proposing to allow those parts of the agricultural area that go out of cultivation to revert to marshes. Last October a scientific panel convened by the National Audubon Society and chaired by Gordon H. Orians, an ecologist at the University of Washington, recommended establishment of a governor's task force "to find mechanisms for returning this land [the agricultural area] to wetlands over the long term."

The Everglades Coalition has recommended that water managers find a way to allow a sheet flow of water through newly created marshlands to the water-conservation areas, thereby rejoining the upstream system to the Everglades.

Charles Lee, senior vice-president of the Florida Audubon Society, has proposed that those marshes cover as much as 100,000 acres. Several parcels of old farmland—the Holey Land and the Rotenberger Tract—in the southern part of the agricultural area are now being reconverted to Everglades marshes, and four storm-water treatment areas are scheduled for construction for pollution control.

More practical in the short run, if less glamorous, is the examination of ways to break down various internal levees and construct new ones that would allow water to resume a natural, slow movement southward through the water-conservation areas and the park. The Army Corps of Engineers is starting a project that would change levees and pump stations to keep the new 107,600-acre East Everglades addition to the park—part of the Shark River Slough—under water more often (see "Managing for Diversity"). In addition, the South Florida Water Management District is restoring to wetlands 41,000 acres of Southern Golden Gate Estates, an unfinished subdivision east of Naples that has disrupted the hydrology of the Big Cypress Swamp for the past 30 years.

Engineers for the water-management district and the corps have begun a study of ways to reconfigure Canal 111, which drains agricultural fields in southern Dade County into Barnes Sound, on the east coast. That land was once the headwaters of Taylor Slough, the second-largest deepwater channel in the Everglades, which fed Florida Bay. The engineers want to put more fresh water into the bay to try to lower its salinity. At an estimated cost of $100 million, the project will test the ability of water managers to balance the interests of the natural system and those of local farmers.

That balancing act is a political necessity in south Florida and underscores the assertion that dismantling the Everglades' plumbing system is impractical. "The idea of breaking down internal levees alone is not real viable," says Ring. "You will need structures and a plumbing system to deliver the right amount of water at the right time."

Whatever plans are ultimately adopted, those familiar with the region agree that water managers and scientists will have to monitor the various Everglades ecosystems in order to make the constant, subtle adjustments necessary to bring them back to health. "Much of the damage inflicted on the Everglades was done by 'friends' who were well-meaning but did the wrong thing," says Graham, who adds that new projects should be accompanied by a major investment in research "to make sure things are done right."

PROGRESS AND STRIFE Any radical changes in water-control structures could add billions of dollars to the more than $1 billion in state and federal civil-works projects now under way or on the drawing board—including the Kissimmee River restoration and creation of filtration marshes in the agricultural area—forcing difficult political and economic choices. Even sympathetic officials may not have the will or the votes to confront some of the area's agricultural and commercial interests.

Big Sugar—the popular name for the rich and powerful

Managing for Diversity

Designated a world heritage site, international biosphere reserve, and wetland of international importance, Everglades National Park is one of the most unusual—and threatened—regions in the world. Within its boundaries are examples of almost every Everglades ecosystem and several endangered species—including the American crocodile, Florida panther, wood stork, bald eagle, West Indian manatee, red-cockaded woodpecker, and snail kite—and more that are threatened. Its plume birds and alligators attract more than a million visitors a year.

About 1986, researchers like Tom MacVicar, now deputy director of the South Florida Water Management District, began to use computer models to create rainfall plans that would allow delivery of water in something approaching a natural fashion. Those plans required restoration of the eastern part of the great Shark River Slough, which had been left out of the park's original boundaries and which environmentalists had long sought to include. The models helped clinch the case, and in 1989 a 107,600-acre tract of the East Everglades was added. The U.S. Army Corps of Engineers and the National Park Service were ordered to develop a plan for restoring the park's hydrology.

In 1990, however, a U.S. Fish and Wildlife Service field officer issued a "jeopardy opinion" on the grounds that the plan being developed would have an adverse affect on the endangered snail kite, an indigenous raptor that feeds exclusively on apple snails. That report forced a halt to the project and enraged park officials, water managers, environmentalists, and even some officials within the Fish and Wildlife Service,

who criticized the report as narrow and wrong.

To break the impasse, the agencies requested that the National Audubon Society convene a panel to recommend a solution. (See About Audubon, *Audubon*, January-February 1993.) Last October that panel, chaired by Gordon H. Orians, an ecolo-

gist at the University of Washington, concluded that while the proposed plan disrupted some nesting sites, it did not threaten the species over the entire region.

More important, the Orians panel said that for restoration to succeed, the Everglades had to be managed as a whole, instead of on a species-by-species basis.

Observing that bringing water back into one area might negatively affect local populations of certain species even if it opened new habitats for them elsewhere, the scientists urged that all such changes be strictly monitored. In no event, they said, should the Everglades be held hostage to a single species. —*M.D.*

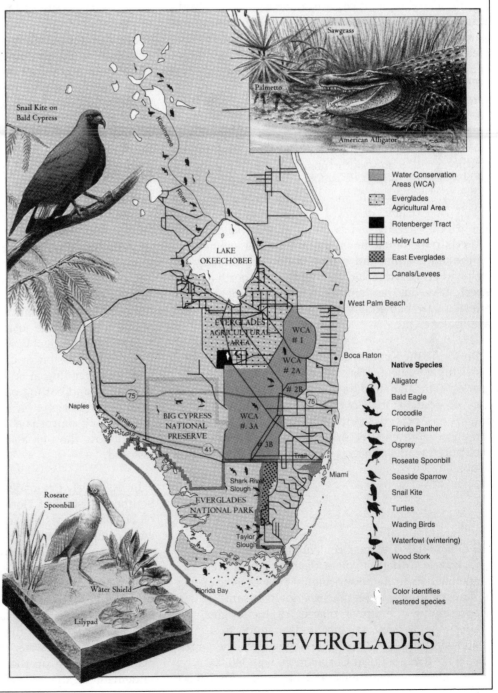

THE EVERGLADES

Water Conservation Areas (WCA)

Everglades Agricultural Area

Rotenberger Tract

Holey Land

East Everglades

Canals/Levees

Native Species

Alligator
Bald Eagle
Crocodile
Florida Panther
Osprey
Roseate Spoonbill
Seaside Sparrow
Snail Kite
Turtles
Wading Birds
Waterfowl (wintering)
Wood Stork

Color identifies restored species

industry that dominates the agricultural area—challenged a plan to clean up water from its fields that is polluting both the Loxahatchee National Wildlife Refuge and other remnant saw grass marshes. Opposition in urban areas can also be expected as the painful realities of realigning an inadequate system become clear: higher taxes to pay for changing the infrastructure and higher prices for water.

"South Florida is the most advanced place of environmental restoration in the country," says Jim Webb, regional director of the Wilderness Society in Florida and a leading proponent of the restructuring. "But political support is like the Everglades—very wide, very shallow." Lee agrees: "The easier things to restore the Everglades have been accomplished: the Big Cypress expansion and the East Everglades addition to the park. These are what could be done with dollars, without stepping on special interests, but the tough questions of how much of the Everglades Agricultural Area has to be converted back to water storage and how much Big Sugar should pay have not been addressed."

The crowning achievement to date of the Save Our Everglades program is the plan to restore the Kissimmee River. Last fall Congress authorized the Corps of Engineers to restore 52 miles in the central stretch of the river, leaving only the northern and southern ends in canal. During the course of the 15-year, $558 million project, which is being funded by both the state and the federal governments, 11 miles of river channel destroyed during construction of the canal will be re-created and 29,000 acres of wetlands will be restored. In the remaining 41 miles, the canal will be filled and the old channel and oxbows reflooded. Initially, the corps had proposed including a section just north of Lake Okeechobee, but residents of two developments there protested so vehemently that water managers decided to leave it alone.

"Restoring the Kissimmee has been *Mission Impossible*," says Estus Whitfield, who helped Graham start his Everglades campaign and now serves as environmental-policy coordinator for Governor Lawton Chiles. But, he adds, it has proved worth the effort. The Kissimmee restoration represents the first time in U.S. history that a major public works project is being reversed for ecological reasons.

THE CLEANUP The most contentious piece of the restoration puzzle to date involves cleaning up nutrient-rich waters that flow from the agricultural area, causing ecological damage in saw grass marshes downstream.

Representing the upper quarter of the original Everglades, the agricultural area today supports 440,000 acres of sugarcane and 60,000 of vegetables. The largest growers are the U.S. Sugar Corporation, with 140,000 acres in cultivation and two mills, and Flo-Sun, with 180,000

acres in cane and three mills. (U.S. Sugar is establishing an 18,000-acre orange grove and a citrus-processing plant west of the agricultural area; a subsidiary farms vegetables.) The agricultural area consumes the bulk of the region's surface water.

It could not exist without the dike that constrains Lake Okeechobee or the area's system of canals and pumps. The water levels must be kept lower than at any time in the past because in the 70 years since the rich organic soil was first exposed to air, as much as six feet of it has oxidized and blown away. It continues to vanish, having dropped 12 to 14 inches between 1977 and 1991—about one inch each year. In the southern part of the agricultural area, some areas are now bare of topsoil.

Farming the muck, which was once submerged wetland, releases phosphorus and nitrogen—and possibly mercury and other elements that are bound to it. Fertilizers also contain phosphorus and nitrogen, although sugar is fertilized less heavily than vegetables. Until 1979 phosphorus-enriched water drained from the fields was pumped into Lake Okeechobee, where it contributed to nutrient overload. That contributed to algae blooms and drove the lake toward eutrophication. To correct the problem, water managers pumped the water directly into the Loxahatchee refuge and Water Conservation Area 2A. The state also imposed controls on dairy farms north of the lake, a major source of the nutrient problem. Water quality in Okeechobee seems to be slowly improving.

Today, says Loxahatchee superintendent Burkett Neely, nutrient overload in the wildlife refuge is as bad as it was in Lake Okeechobee in the mid-1980s, when massive fish kills occurred. The overload has caused changes in periphyton, the algal base of the Everglades marshes, and fueled an explosion of cattails, which cover 34,000 acres and are expanding at an alarming rate, driving out other vegetation and cutting off feeding areas to wading birds. Changes in periphyton have also begun to occur in the park, raising concerns about the quality of the water that would move through a system in which natural flows are restored.

In 1988 Neely persuaded the acting U.S. attorney for the Southern District of Florida, Dexter Lehtinen, to file suit on behalf of the refuge and park against the state Department of Environmental Regulation (DER) and the water-management district, demanding that they enforce Florida's water-quality laws and clean up the polluted discharge from the agricultural area.

In response to that suit, the state legislature passed the Marjory Stoneman Douglas Everglades Protection Act in 1991. It mandates that the water-management district implement a surface-water improvement and management plan for cleaning up the Everglades. In February 1992 a settlement of the federal suit—partly crafted by Carol

Browner, then head of the DER and now of the EPA—was finally approved.

That agreement specified that the water-management district create four storm-water treatment areas, covering a minimum of 35,000 acres in the agricultural area. Further, it required farmers to control runoff from their fields and to take whatever other steps were necessary to reduce phosphorus concentrations in the runoff to levels set for 1997 and 2002.

An Everglades Nutrient Removal Project is under construction on 3,742 acres on the border of the Loxahatchee refuge and is scheduled to begin operation this year. Water managers hope that when fully operational it will help them determine whether larger storm-water treatment areas will work.

The settlement of the federal suit permitted challenges in state administrative hearings. Big Sugar, along with the vegetable growers, has taken full advantage of that provision. Growers challenged the settlement in court, in administrative hearings, and in the press, launching a major public relations campaign to prove that they are not responsible for the problems in the area and should not have to pay for the cleanup. Since there are no other major sources of pollution in the agricultural area, they had a hard time making their case.

The conflict dragged on for months, embroiling a mediator, Gerald W. Cormick; Interior Secretary Bruce Babbitt; the Miccosukee tribe, whose reservation lies downstream from the agricultural area; the water-management district; the park and the wildlife refuge; environmentalists; Governor Chiles; Lieutenant Governor Buddy MacKay; and various state agencies.

Finally, on July 13, Secretary Babbitt announced an agreement in principle among federal and state officials, Big Sugar, and the vegetable growers. At first glance, the complicated plan seems to expand upon the federal court settlement, specifying the construction of filtration marshes, a 25 percent increase in the amount of water moving through the Everglades system, steps by farmers to ensure that runoff from their fields is as clean as technology can make it, and land purchases to restore the headwaters of Taylor Slough.

The agreement also sets a 90-day period during which Big Sugar, the state, and the water-management district will work out technical details and the agricultural groups will move to end their remaining legal challenges.

But the agreement's "statement of principles" has aroused the opposition of environmentalists and the Miccosukee Tribe. Those critics view the Babbitt plan as a retreat from the dictates of the Marjory Stoneman Douglas Act and from the settlement approved in early 1992.

That settlement provided for a two-pronged cleanup. Interim standards were to be met by 1997; the area was to be fully cleaned up by 2002. Critics say that under the Babbitt plan it will not be possible to meet the 1997

standards—much less the 2002 standards. So in effect, they say, the Babbitt plan does not do any more than oblige Big Sugar to meet the interim standards—and it allows growers an extra 10 years to do it. In addition, large portions of the Rotenberger Tract and the Holey Land—areas scheduled for restoration—would be used as storm-water treatment areas, diminishing the amount of new acreage devoted to restoration.

Furthermore, the Babbitt plan calls on Big Sugar and vegetable growers in the agricultural area to pay, over 20 years, as much as $322 million of the cost of the project—which is estimated at $465 million to $700 million. The Everglades Coalition has demanded that the growers be required to pay the full cost of cleaning up their pollution.

In the past, Robert Buker Jr., a senior vice-president of the U.S. Sugar Corporation, has disputed or tended to limit the growers' responsibility and has said that paying the full cost of the cleanup would ruin the industry financially.

The Babbitt settlement also fails to address major issues relating to the restoration of water flows in the remnant Everglades. In short, say its critics, it appears to please no one but polluters and politicians eager to accommodate them.

Florida Audubon's Lee says, "The announcement of the settlement is a grand exercise in unjustified hyperbole." He categorizes the agreement as overly vague, and he, along with Miccosukee tribal chairman Billy Cypress and Dexter Lehtinen, now the tribe's attorney, has protested that the final agreement was negotiated without the participation of the public. In many respects, they say, it represents a victory not for the Everglades but for special interests.

As this article went to press, the Miccosukee and environmental groups were considering what action could be taken to reclaim the momentum for restoration, which they feel has been lost.

THE FUTURE OF AGRICULTURE Although no one can foresee the future of the agricultural area, Big Sugar has repeatedly stated its intention to farm as long as it can. However, if agribusiness deserts the area, its replacement—untrammeled development—might prove even worse for the Everglades.

Lee fears that the scheduled expansion of U.S. Highway 27 to four lanes all the way to Lake Okeechobee will inevitably open the region to suburbanization from fast-growing Fort Lauderdale, especially as the soils become more marginal. It would become, he says, "south Florida's equivalent of the Los Angeles valley," a mess of congestion and pollution. The alternative he prefers is a carefully planned future, in which the state or the water district would purchase land as it moved out of production for restoration to Everglades marshes or for water storage.

Senator Graham adds, "Long-term land use . . . is an

important issue, and it hasn't gathered the attention it needs. We should look at the end of agriculture in a couple of generations and decide what will follow." Observing that the governor could establish a committee to consider the question, he adds, "The idea of some physical corridor to connect the lake to the midpart of the Everglades system is an intriguing one, and using flow ways [long, linear artificial marshes] is part of what a site-specific panel would look at."

By whatever measure, the cost of inaction—to the environment, water supply, tourism, and human health—is greater than that of reengineering an inadequate, flawed system. The quality and fabric of life in south Florida, which in essence is the Everglades—a place people have long praised as an earthly paradise and too often abused for their short-term gain—is at stake. As Allan Milledge, who was chairman of the water-management district when the 1992 settlement was reached and who still serves on its board, says, "We all live off this natural system, and regardless of how we came to it, we have to preserve it."

Global Warming on Trial

*How good is the evidence that the earth is warming,
and where does the burden of proof lie?*

Wallace S. Broecker

*Broecker completed his undergraduate
and graduate studies at Columbia University and is now a professor of geology at
the university's Lamont-Doherty Geological Observatory. For further reading on
climate change within the recent past,
Broecker recommends Jean Grove's* The
Little Ice Age *(London: Routledge, 1990).*

Jim Hansen, a climatologist at NASA's
Goddard Space Institute, is convinced
that the earth's temperature is rising and
places the blame on the buildup of greenhouse gases in the atmosphere. Unconvinced, John Sununu, former White
House chief of staff, doubts that the
warming will be great enough to produce a
serious threat and fears that measures to
reduce the emissions would throw a
wrench into the gears that drive the
United States' troubled economy. During
his three years at the White House,
Sununu's view prevailed, and although his
role in the debate has diminished, others
continue to cast doubt on the reality of
global warming. A new lobbying group
called the Climate Council has been created to do just this.

The stakes in this debate are extremely
high, for it pits society's short-term well-being against the future of all the planet's
inhabitants. Our past transgressions have
altered major portions of the earth's surface, but the effects have been limited.
Now we can foresee the possibility that to
satisfy the energy needs of an expanding
human population, we will rapidly change
the climate of the entire planet, with consequences for even the most remote and
unspoiled regions of the globe.

The notion that certain gases could
warm the planet is not new. In 1896
Svante Arrhenius, a Swedish chemist, resolved the longstanding question of how
the earth's atmosphere could maintain the
planet's relatively warm temperature
when the oxygen and nitrogen that make
up 99 percent of the atmosphere do not
absorb any of the heat escaping as infrared radiation from the earth's surface into
space. He discovered that even the small
amounts of carbon dioxide in the atmosphere could absorb large amounts of
heat. Furthermore, he reasoned that the
burning of coal, oil, and natural gas could
eventually release enough carbon dioxide
to warm the earth.

Hansen and most other climatologists
agree that enough greenhouse gases have
accumulated in the atmosphere to make
Arrhenius's prediction come true. Burning fossil fuels is not the only problem; a
fifth of our emissions of carbon dioxide
now come from clearing and burning forests. Scientists are also tracking a host of
other greenhouse gases that emanate from
a variety of human activities; the warming
effect of methane, chlorofluorocarbons,
and nitrous oxide combined equals that of
carbon dioxide. Although the current
warming from these gases may be difficult to detect against the background
noise of natural climate variation, most
climatologists are certain that as the gases
continue to accumulate, increases in the
earth's temperature will become evident
even to skeptics.

The issue under debate has implications
for our political and social behavior. It
raises the question of whether we should
renew efforts to curb population growth
and reliance on fossil fuels. In other words,
should the age of exponential growth initiated by the Industrial Revolution be
brought to a close?

The battle lines for this particular skirmish are surprisingly well balanced.
Those with concerns about global warming point to the recent report from the
United Nation's Intergovernmental Plan
on Climate Change, which suggests that
with "business as usual," emissions of carbon dioxide by the year 2025 will be 25
percent greater than previously estimated.
On the other side, the George C. Marshall
Institute, a conservative think tank, published a report warning that without
greenhouse gases to warm things up, the
world would become cool in the next century. Stephen Schneider, a leading computer modeler of future climate change,
accused Sununu of "brandishing the
[Marshall] report as if he were holding a
crucifix to repel a vampire."

If the reality of global warming were
put on trial, each side would have trouble
making its case. Jim Hansen's side could
not prove beyond a reasonable doubt that
carbon dioxide and the other greenhouse
gases have warmed the planet. But neither
could John Sununu's side prove beyond a
reasonable doubt that the warming expected from greenhouse gases has not occurred.

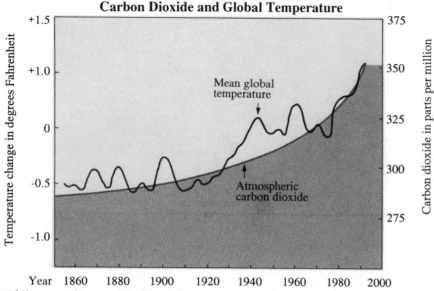

Carbon Dioxide and Global Temperature

While atmospheric carbon dioxide has climbed steadily for more than a century, the earth's temperature shows a more erratic upward trend.

At this point, a wise judge might pose the following question to both sides: "What do we know about the temperature fluctuations that occurred prior to the Industrial Revolution?" The aim of this question would be to determine what course the earth's temperature might have taken if the atmosphere had not been polluted with greenhouse gases. The answer by both sides would have to be that instead of remaining the same as it was in 1850, the planet's temperature would have undergone natural fluctuations, which could have been as large as the changes measured over the last one hundred years. Neither side, however, would be able to supply the judge with an acceptable estimate of what would have happened to the earth's temperature without the release of greenhouse gases.

Perhaps a longer record of the earth's climate would shed light on its natural variability. The climate prior to 1850 can be reconstructed from historical records of changing ice cover on mountaintops

To see why each side would have difficulty proving its case, let us review the arguments that might be presented at such a hearing. The primary evidence would be the temperature records that have been kept by meteorologists since the 1850s. A number of independent analyses of these measurements have reached the same basic conclusions. Over the last century the planet has warmed about one degree. This warming was especially pronounced during the last decade, which had eight of the warmest years on record, with 1990 being the hottest. While Sununu's group might question the adequacy of the geographic coverage of weather stations during the early part of the record and bicker a bit about whether the local warming produced by the growth of cities has biased some of the records, in the end they would concede that this record provides a reasonably good picture of the trend in the earth's temperature. Sununu's advocate would then counter by asking, "Isn't it strange that between about 1940 and 1975 no warming occurred?" The Hansen group would have to admit that there is no widely accepted explanation for this leveling. Sununu's advocate would continue, "Isn't it true that roughly half the warming occurred before 1940, even though almost all the emissions of carbon dioxide and other greenhouse gases have taken place after this date?" Again the Hansen group would have to admit this to be the case.

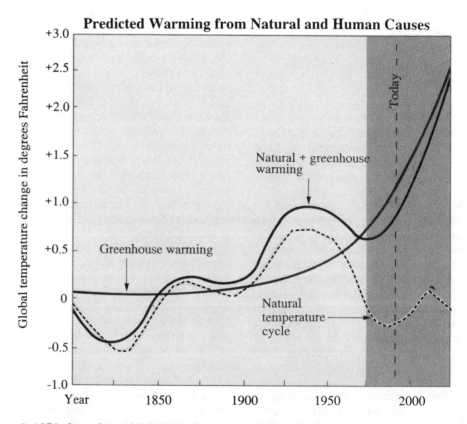

Predicted Warming from Natural and Human Causes

In 1975, the author published this diagram, suggesting that despite a rise in greenhouse gases, the earth cooled somewhat during the 1960s and 1970s because it was at the low point of a natural temperature cycle. He predicted the warming of the 1980s, when the natural cycle again turned upward, no longer masking the effect of the gases.
Graphs by Joe LeMonnier

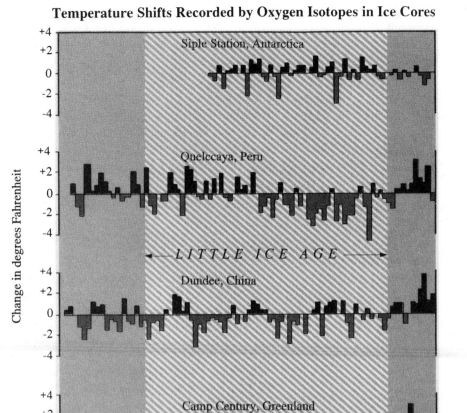

Temperature Shifts Recorded by Oxygen Isotopes in Ice Cores

Siple Station, Antarctica

Quelccaya, Peru

LITTLE ICE AGE

Dundee, China

Camp Century, Greenland

Change in degrees Fahrenheit

Year   1260 1340 1420 1500 1660 1740 1780 1820 1900 1980

*Ice cores from various glaciers around the world show little
agreement on past temperature shifts, reflecting local
conditions that complicate the global record.*

sion of Alpine glaciers heralded the end of the Little Ice Age. Ridges of rock and earth bulldozed into position by the advancing ice still mark the point of maximum glacial progress into the valleys. (The glaciers are still shrinking; less than half of their 1860 volume remains.) The mild conditions that prevailed during the Medieval Warm did not return until this century.

The problem with all this evidence is that it represents only one region of the earth and is, in a sense, anecdotal. An informed judge might also challenge this evidence by pointing out that the northern Atlantic Ocean and its surrounding lands are warmed by powerful ocean currents, collectively known as the Great Conveyor, that transport heat away from the equator (*see* "The Biggest Chill," *Natural History,* October 1987). A temporary shutdown of this circulation 11,000 years ago brought about an 800-year cold period called the Younger Dryas, during which northern Europe was chilled by a whopping 12° F. Could the Little Ice Age have been brought about by a similar weakening of the Great Conveyor? If heat release from the northern Atlantic was the key factor, the Little Ice Age would have been restricted to the surrounding region, and the historical evidence from Iceland and the Alps could not be taken as an index of global temperatures.

Although records of similar duration and quality are not available from other parts of the world, we do have firm evidence that by 1850, mountain glaciers in some regions, such as New Zealand and the Andes, reached down into valleys as far as they had at any time during the last 8,000 years. Furthermore, by 1870 these glaciers had also begun their retreat. This suggests that the Little Ice Age was indeed global in extent.

The global warming that caused the demise of the Little Ice Age confuses attempts to estimate how much of the last century's warming is natural and how much has been caused by pumping greenhouse gases into the atmosphere. The Sununu side would pin as much of the blame as possible on the natural warming trend that ended the Little Ice Age, while Hansen's side would emphasize the role of the greenhouse gases. What is needed to resolve this dispute is a detailed, continuous temperature record that extends back beyond the Medieval Warm to see if cycles could be identified. By extending these cycles into the present century, scientists could estimate the course the

and on the sea. The earliest evidence of this type dates from the end of the tenth century A.D., when Eric the Red first sailed from Iceland to Greenland. Ship logs written between that time and 1190 indicate that sea ice was rarely seen along the Viking sailing routes. The temperature was warm enough that grain could be grown in Iceland. At the end of the twelfth century, however, conditions deteriorated, and sea ice appeared along the Viking sailing routes during the winters. By the midfourteenth century, these routes were forced far to the south because of the ice, and sometime in the late fifteenth century, ships were cut off altogether from Greenland and Iceland because of severe ice conditions. As temperatures dropped, people could no longer grow grain in Ice-

land. The Medieval Warm had given way to the Little Ice Age.

After 1600, records of sea-ice coverage around Iceland and of the extent of mountain glaciers in the Alps improved, giving us an even better idea of recent climate change. The glaciers attracted the attention of seventeenth-century tourists, including artists whose drawings and paintings document the position of a number of major Alpine glaciers. Modern measurements show that the leading edges of these glaciers fluctuated with temperature changes over the last century. Assuming that this correlation held true throughout the Little Ice Age, the historical evidence shows a long interval of glacier expansion, and thus cold climate, lasting until 1860. During the late 1800s, a widespread reces-

earth's temperature would have taken in the absence of the Industrial Revolution.

I made such an attempt in 1975, at a time when the earth's temperature seemed to have remained almost constant since the mid-1940s. Puzzled scientists were asking, "Where's the expected greenhouse warming?" I looked for the answer in the only detailed long-term record then available, which came from a deep hole bored into northern Greenland's icecap at a place called Camp Century. In the 1950s, Willi Dansgaard, a Danish geochemist, had demonstrated that the ratio of heavy to light oxygen isotopes (18 neutrons to 16 neutrons per atom, respectively) in the snow falling in polar regions reflected the air temperature. Dansgaard made measurements of oxygen isotopes in different layers of the ice core; each represented the compressed snowfall of an arctic year. His results served as a proxy for the changes in the mean annual temperature. Dansgaard and his colleagues analyzed the record to see if the temperature fluctuations were cyclic. They found indications of two cycles, one operating on an 80-year time scale and a weaker one operating on a 180-year time scale. (The Milankovitch cycles, caused by changes in the earth's orbit around the sun, operate on a much longer time scale. Ranging from 20,000 years upward, these cycles are thought to control the large swings between glacial and interglacial climates.)

I took Dansgaard's analysis a step further by extending his cyclic pattern into the future. When combined with the expected greenhouse warming, a most interesting result appeared. Temperatures leveled off during the 1940s and 1950s and dropped somewhat during the 1960s and 1970s. Then, in the 1980s, they began to rise sharply. If there is a natural eighty-year cycle and it was acting in conjunction with a greenhouse effect, I would explain the leveling of temperature after 1940 as follows: Dansgaard's eighty-year cycle would have produced a natural warming between 1895 and 1935 and a natural cooling from 1935 to 1975. The cooling in the second half of the cycle might have counterbalanced the fledgling greenhouse warming. After 1975, when the natural cycle turned once again, its warming effect would have been augmented by the ever stronger greenhouse phenomenon, producing a sharp upturn in temperature in the 1980s.

My exercise showed that the lack of warming between 1940 and 1975 could not be used to discount the possibility that the pollution we are pumping into the atmosphere will ultimately warm the globe. We cannot rule out this possibility until that time in the future when the predicted warming is so great that it can no longer be masked by natural temperature fluctuations. My projection suggested that a firm answer will not be available until the first decade of the next century.

While the Camp Century record seemed to provide a good method of determining how natural variations and increasing greenhouse gases were working in concert to produce the measured global temperatures, additional ice core data only created confusion. Oxygen isotope records from ice cores extracted from the Antarctica icecap and mountain glaciers in China and Peru do not follow the Camp Century ice core pattern. Even worse, oxygen isotope records from three additional Greenland ice cores differ significantly from one another and from the original Camp Century record. Perhaps the most disconcerting feature of these ice core records is that the Medieval Warm and Little Ice Age do not even stand out as major features. Local temperature variations could account for these discrepancies, but oxygen isotope ratios also depend on the season the snow falls and the source of the moisture. For these reasons, ice cores may provide good records of large changes, but the smaller ones we are looking for over the last several hundred years are obscured.

At this point, the judge would likely lose his patience and call a halt to this line of argument, saying, "While regional climate changes certainly occurred during the centuries preceding the Industrial Revolution, firm evidence for a coherent global pattern in these natural fluctuations is lacking." The judge might then suggest a different approach to settle the question of whether we are causing the earth to warm. What drives the natural changes? If we could pin down the villain, then perhaps we could say more about how temperature would have changed in the absence of the Industrial Revolution. Witnesses would point to three such mechanisms. First, the sun's energy output may have changed. Second, large volcanic eruptions may have injected enough material into the stratosphere to reflect a substantial amount of solar radiation back into space, cooling the planet. Third, the operation of the ocean-atmosphere system may have changed internally, causing the earth's temperature to wander.

For several centuries astronomers have been observing the cycles of the sun and trying to link them with climate patterns on earth. Sunspots, caused by knots in the sun's magnetic field, undergo cyclic change, alternating between a maximum of spots in the Northern Hemisphere and then a maximum in the Southern Hemisphere. Between these peaks, the number of sunspots drops almost to zero. A complete solar cycle takes twenty-two years. With satellites, astronomers have been able to directly monitor the sun's energy output over the last cycles. Although the energy seems to dip slightly when sunspots disappear, the change seems too small to greatly alter the earth's temperature.

An intriguing proposal was recently made in this regard. Two Danish meteorologists, Eigil Friss-Christensen and Knud Lassen, point out that over the last 130 years for which observations are available, the sunspot cycle has lengthened and shortened with a periodicity of about 80 years, and that these changes closely parallel the earth's temperature. The Danes suggest that during intervals when the sunspot cycle is longer than average, the sun's energy output is a bit lower, and that when the cycle is shorter, the energy output is higher. Could it be that Dansgaard was correct in thinking that the earth's temperature changes on an eighty-year time scale and that these changes are driven by the sun? Most scientists remain skeptical because no physical mechanism has been proposed tying solar output to the length of the sunspot cycle. Others say that the strong similarity between the length of the sunspot cycle and the earth's temperature could be a coincidence.

In addition to the twenty-two-year solar cycle, however, change on a longer time scale has been documented. Between 1660 and 1720, sunspots disappeared altogether. Auroras, which are created when charged particles driven out from the sunspots enter the earth's upper atmosphere, were also absent from the skies during this period. Further, we know from measurements of carbon 14 in tree rings that this radioactive element, produced by cosmic rays bombarding the atmosphere, increased substantially during this time. Normally, charged particles streaming outward from sunspots create a magnetic shield that deflects cosmic rays away from the earth and the inner planets. From 1660 to 1720, this magnetic shield failed, permitting a larger number of cosmic rays to strike our atmosphere and form an unusually large number of radioactive carbon atoms.

Sunspots and Global Temperature

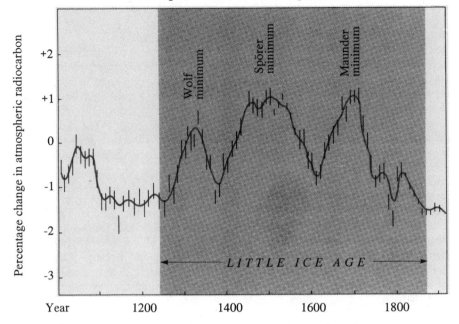

During the Little Ice Age, levels of radiocarbon in tree rings indirectly recorded three periods when sunspots all but disappeared. Sunspot minimums correspond to a slight decrease in the sun's energy output.

From the record of radiocarbon locked up in tree rings, we can identify two even earlier periods of reduced sunspot activity: the Wolf sunspot minimum, from about 1260 to about 1320, and the Spörer sunspot minimum, from about 1400 to 1540. These three periods span a major portion of the Little Ice Age, but the last ended more than a hundred years before the Little Ice Age did—too long a time lag. This mismatch in timing and the small change in the sun's energy output (as measured by satellites over the last solar cycle) make a link between the Little Ice Age and the absence of sunspots unlikely. But the partial match prevents a firm rejection of the sun as a cause of the earth's natural temperature changes.

What about volcanic eruptions? Major volcanic eruptions occur roughly once per decade. Most have little effect on the climate, but occasionally an eruption blasts a large volume of sulfur dioxide high into the stratosphere. Within a month or two, the sulfur dioxide is transformed into droplets of sulfuric acid, which remain aloft in the stratosphere for a year or more. These tiny spheres reflect sunlight away from the earth, cooling the planet. Hansen and his colleagues predict that the recent eruption of Mount Pinatubo in the Philip-

pines (which shot more sulfur dioxide into the upper atmosphere than any other eruption this century) will cool the planet about one degree Fahrenheit over the next two years.

Could the Little Ice Age have been caused by 500 years of intense volcanism releasing copious amounts of sulfur dioxide? This seems implausible, as the world's 100 or so major volcanoes erupt independently of one another and no mechanism exists that could cause them all to erupt with greater frequency. Therefore, the chance is slim that one long interval would be followed by a similar period of lesser activity.

Fortunately, a record is available in ice cores to check this assumption. When the droplets of sulfuric acid from a volcanic eruption drift down from the stratosphere, they are quickly incorporated into raindrops and snowflakes and carried to the earth's surface. So, in the years immediately following a major volcanic eruption, snow layers rich in sulfuric acid are deposited on all the world's icecaps. An ice core taken from the Dye 3 site in southern Greenland reveals that at about the time of the transition from the Medieval Warm to the Little Ice Age the acid content in the ice doubled. On the other hand, low

acidity from 1750 to 1780 (during a time of cold weather) and the relatively high acidity from 1870 to 1920 (when the climate was warming) do not fit the pattern of climate change. Therefore, no strong correlation exists between the trends in volcanic sulfur dioxide and the trend in the earth's temperature.

The last of the three mechanisms that might account for the natural variations in the earth's temperature is a dramatic shift in the way the planet's ocean and wind currents operate. Of the three mechanisms, this one is the hardest to build a case around because we have only a rudimentary understanding of how the interacting elements of the earth's climate system might cause natural fluctuations in temperature. The only well-documented example of such a mechanism is the El Niño cycle, in which winds and ocean currents cause the temperatures of the surface waters of the eastern equatorial Pacific to alternate between warm and cold. The cycle was first noticed because of the severe drops in fish production along the west coast of South America during the warm episodes. Since the timing between these disruptive events ranges from three to seven years, scientists became interested in predicting their arrival. What emerged from these studies is that El Niño cycles are the product of a complex interaction between winds and ocean currents. The importance of this discovery to the global warming debate is that it raises the possibility that cycles involving larger-scale interactions between the atmosphere and oceans—over longer periods—may play an important role. If the earth's temperature is being pushed up and down by such an internal cycle, our chances of determining what would have happened in the absence of the extra greenhouse gases are indeed slim.

Again the judge would become restive and call a halt to this line of evidence as well. At this point he would likely dismiss the case and suggest that the litigants return a decade from now when additional evidence regarding the warming trend has accumulated.

Sununu would deem this decision a victory, for it would provide an excuse to delay actions directed toward reductions in carbon dioxide emissions. On the other hand, Hansen could surely maintain that in the absence of proof that the world is not warming at the rate predicted by computer simulations, we should follow the standard applied to other environmental threats and rule on the side of caution.

Instead of placing the burden of proof on the environmentalists, the proponents of "business as usual" should be obliged to prove that the unfettered release of greenhouse gases will *not* significantly warm the planet. And such proof does not exist; the balance of scientific opinion is that business as usual will alter the climate.

The debate over global warming is merely a small skirmish that marks the beginning of a far broader war. Many of the things that we could do to curb the buildup of greenhouse gases—such as conserving energy, switching to renewable energy sources, or increasing our use of nuclear power—will be stopgap measures if the underlying problem of population growth is not addressed. World population is now 5.5 billion and growing by about 1.8 percent every year. If this rate is not substantially reduced, world population will double by the year 2030. If by that time

the rate of population growth has not been greatly reduced, we run the risk that the population will skyrocket to 20 billion or more before it finally levels off. Each additional person adds to the pressure to increase the use of fossil fuels, pumping ever larger amounts of carbon dioxide into the atmosphere. In countries such as the United States and Canada, where per capita energy consumption is the highest in the world, each person, on average, adds twenty tons of carbon dioxide a year to the atmosphere. In developing countries, where most of the population growth will occur, per capita energy consumption is much smaller, with less than three tons of carbon dioxide emitted per person. But as these countries strive to better the lot of their citizens through industrialization, their energy demands will climb. Most of the increase will be met by burning fossil fuels, particularly coal, which releases

more carbon dioxide per unit of energy produced than oil or gas. Therefore, annual emissions of greenhouse gases are likely to increase.

We are rapidly approaching a limit beyond which we cannot maintain our numbers without long-term damage to our planet's environment and its remaining wildlife and to the quality of life of its human populations. While Sununu may be particularly shortsighted with regard to the effects of greenhouse gases on the climate, most of the world's leaders are shortsighted with regard to the population problem. They seem to ignore it completely. I hope the concern about global warming will force us to develop a broader perspective of our planet's future—one that will include the reality of the population bomb. Only then will we be able to begin the extraordinarily difficult task of defusing it.

Exploring the links between

Desertification and Climate Change

Mike Hulme and Mick Kelly

Mike Hulme is a research climatologist in the Climatic Research Unit at the University of East Anglia in Norwich, England. He specializes in African climate, global climate change, and climate remodeling. Mick Kelly is an atmospheric scientist in the Climatic Research Unit. He is also research director of the Climate and Development Programme for the Centre for Social and Economic Research on the Global Environment at the University of East Anglia and director of the Climate Programme at the International Institute for Environment and Development in London.

More than 100 countries are suffering the consequences of desertification, or land degradation in dryland areas.[1] Loss of produc-tivity and other social, economic, and environmental impacts are directly affecting the perhaps 900 million inhabitants of these nations. There is also concern that the environmental impact of dryland degradation may be felt further afield. Some have suggested that this impact might even be felt worldwide.

The first international effort to address desertification occurred in 1977, when the United Nations Conference on Desertification (UNCOD) recognized that desertification was a major environmental problem with high human, social, and economic costs. The conference adopted the Plan of Action to Combat Desertification (PACD), a 20-year, worldwide program to arrest further dryland degradation. Sixteen years later, after several reviews, PACD has achieved little success.[2] A second phase in the international response to desertification began at the United Nations Conference on Environment and Development (UN-CED) in June 1992. It was agreed then that a Convention to Combat Desertification should be ready for signing and ratification by June 1994 and that an Intergovernmental Negotiating Committee on Desertification(INC-D) should be established to guide this process. It has also been decided that projects to mitigate land degradation in drylands will qualify for allocations from the Global Environment Facility (GEF),[3] but only insofar as the projects pertain to the GEF goals of protecting the global environment by reducing greenhouse-gas emissions, preserving biodiversity, and protecting international waters.

A significant obstacle to the work of INC-D is that desertification is a difficult word to define. In 1991, the UN Environment Programme (UNEP) defined desertification as "land degradation in arid, semi-arid and dry sub-humid areas resulting mainly from adverse human impact."[4] Just

From *Environment*, Vol. 35, No. 6, July/August 1993, pp. 4-11, 39-45. Reprinted with permission of the Helen Dwight Reid Educational Foundation. Published by Heldref Publications, 1319 Eighteenth St., NW, Washington, DC 20036-1802.

one year later, UNCED adopted the definition of ''land degradation in arid, semi-arid and dry sub-humid areas resulting from various factors including climatic variations and human activities.''[5] The different emphasis placed on climate variation in these two definitions is indicative of the disagreement that exists concerning the relative importance of the various causes of dryland degradation. This disagreement may appear to be an impractical, academic issue for the large numbers of people dependent on drylands for their livelihood, but misconceptions and arguments must be resolved for any adequate response to desertification to be made.

The first step in discussing such issues must be to define the most important terms and thereby avoid confusion. The terms *climate change* and *climate variation* are used here to indicate climate variability and trends arising from both natural and anthropogenic causes. The term *global-mean warming* indicates climate change resulting from greenhouse-gas emissions. *Desertification* is taken here to mean land degradation in dryland regions, or the permanent decline in the potential of the land to support biological activity and, hence, human welfare. Desertification should not be confused with drought or desiccation. *Drought* refers to a period of two years or more with below-average rainfall, and *desiccation* is aridification resulting from a dry period lasting a decade or more.[6]

Climate change undoubtedly alters the frequency and severity of drought and can cause desiccation in various regions of the world. It does not necessarily follow, however, that drought and desiccation will, by themselves, induce, or even contribute to, desertification. Whether or not desertification occurs depends upon the nature of resource management in these dryland regions. Identifying the contribution of climate variation to desertification is not a simple matter, and the difficulties are compounded by the possibility that desertification itself may generate climate change.

Does Climate Change Cause Desertification?

The definitions of desertification adopted by UNEP in 1991 and by UNCED in 1992 both implicitly link climate change and the assessment of the extent of desertification. Because arid, semi-arid, and dry subhumid areas are climatically defined,[7] any change in climate that results in an expansion or contraction of these areas is likely to change the formal, measured extent of the problem. For example, when an arid area becomes extremely dry or hyperarid, because of climate change, the area defined as being prone to desertification decreases because hyperarid areas are not included in the accepted definition. Conversely, when a humid area converts to subhumid, the defined area within which desertification is considered possible increases.

That climates do change over the decades has now been established beyond dispute. In the African Sahel, for example, annual rainfall during the most recent three decades has been between 20 and 40 percent less than it was from 1931 to 1960. Table 1 on this page illustrates this change in a different way. Within contiguous Africa, there has been a net shift of land area toward aridity, especially toward hyperaridity, and a consequent net loss of semi-arid and dry subhumid land. Overall, areas prone to desertification have decreased from 52.4 percent of mainland Africa to 51.5 percent between these two 30-year periods—a reduction of 25.3 million hectares. The amount of hy-

A single rain storm like this one in the northern Sahel may constitute a significant percentage of the area's annual rainfall.

TABLE 1 ▬▬▬▬▬
CHANGE IN AREAS OF MOISTURE ZONES IN CONTIGUOUS AFRICA

Moisture zone	Mean land area from 1931 to 1960		Mean land area from 1961 to 1990		Net change between two periods	
	(millions of hectares)	(percentage of total)[a]	(millions of hectares)	(percentage of total)	(millions of of hectares)	(percent)
Hyperarid	450.8	15.1	501.5	16.8	+ 50.7	+ 1.7
Arid	676.9	22.7	680.0	22.8	+ 3.1	+ 0.1
Semi-arid	620.9	20.8	606.9	20.3	– 14.0	– 0.5
Dry subhumid	264.4	8.9	250.0	8.4	– 14.4	– 0.5
Humid	972.4	32.6	947.0	31.7	– 25.4	– 0.9

[a]Percentages total to 100.1 because of rounding.
SOURCE: M. Hulme, R. Marsh, and P. D. Jones, "Global Changes in a Humidity Index Between 1931–60 and 1961–90," *Climate Research* 2 (1992):1–22.

perarid land, however, has increased by more than 50 million hectares.

Determining the precise contribution of climate change to the problem of desertification is not an easy matter. When resource management failure has occurred, there is no doubt that climate variation can aggravate the problem. But separating out the interrelated impacts of climatic and human factors is extremely difficult. Some progress has, however, been made. To cite one example, Compton Tucker and his colleagues at the U.S. National Aeronautics and Space Administration's Space Flight Center in Maryland have used a satellite index of active vegetative cover to determine the extent of the Sahara Desert between 1980 and 1989.[8] Their analysis shows that very substantial interannual variations exist in the extent and quality of surface vegetation in dryland regions. Because much of the vegetation response detected by the satellite is caused by changes in rainfall, much (but not all) of the variability in the extent of the Sahara is due to interannual rainfall variations (see Figure 1). Thus, the index can be used to discriminate between the degradation of vegetation cover caused by rainfall and that which is due to other factors, most notably failures in resource management. Figure 1 relates the estimates of change in the extent of the Sahara to independently derived annual rainfall data for the years from 1980 to 1989. Linear regression analysis indicates that a considerable amount of the year-

to-year variation in areal extent—83 percent—can be explained by the rainfall data. This is a statistically significant relationship. The relationship does, however, leave some residual variability in the extent of the Sahara unexplained, as shown by the lowest curve in Figure 1. This residual component of the variability has tended to increase over the past decade, and statistical analysis suggests that it amounts to an average annual increase in the extent of the Sahara of about 41,000 square kilometers per year. This is equivalent to an average annual areal increase not directly related to annual rainfall variations of almost 0.5 percent from 1980 onwards, which would amount to almost a 5-percent increase in total area over the decade. This trend could be the result of the cumulative impact of a series of dry years on vegetation recovery. For example, a particular rainfall amount in 1989 may generate less vegetation than the same amount in 1980 because of the drought years preceding 1989. Alternatively, the increase in extent may well be due to a deterioration of vegetative cover caused by human activity.

Clearly, the relative contributions of human activity and climate change to desertification will vary from region to region and from time to time. Separating out the relative roles of these factors to identify the most appropriate response in any particular situation—and to accord each factor due weight in the desertification convention—is a pressing challenge.

Can Desertification Change Climate?

Separating cause and effect is rendered more difficult by the fact that desertification may, in turn, affect both local climate and climates further afield. Recently, Bob Balling, Jr., of Arizona State University has suggested that surface air temperature has increased significantly in desertified regions owing to changes in land cover and that this effect has substantially affected global-mean temperature.[9] Desertification is likely to lead to reductions in surface soil moisture, which result in more energy available to heat the air (sensible energy) because less goes to evaporate water (latent energy). While it is conceivable that the warming of desertified areas may have been great enough to produce a measurable increase in global-mean temperature (see the box on page 213), the influence would have been small compared to the potential impact of an enhanced greenhouse effect. The question of whether desertification has had or will have a detectable effect on global climate is, nevertheless, a critical one. If a clear influence could be established, it might be argued that dryland degradation should be classed as a global environmental problem in its own right. At present, though, the evidence of a substantial effect must be considered extremely weak.

There is a better-established, if less direct, link, however, between dryland degradation and global-mean warming through the influence of desertification on the sources and sinks of greenhouse gases. Progressive desertification of drylands in the tropics and elsewhere is likely to reduce a potential carbon sink by reducing the carbon sequestered or stored in these ecosystems. As vegetation dies and soil is disturbed, emissions of carbon dioxide will increase. Desertification may also affect emissions of other greenhouse gases. For example, nitrous oxide emissions might increase because of greater fertilizer use. Methane production may increase in poorly fed cattle. On the other hand, because dry soils are methane sinks, desertification might reduce the gas's atmospheric concentration. There remains a large measure of uncertainty about the relative magnitudes of the various sources and sinks for carbon dioxide and the other greenhouse gases,[10] and such uncertainty adds to the difficulty of quantifying the precise effect of desertification on global-mean warming.

The problem of the "missing" carbon sink illustrates this difficulty: It is impossible to balance the global carbon budget on the basis of current understanding of the major carbon sources and sinks; a certain amount of carbon released into the atmosphere cannot be accounted for. Several possibilities might account for the discrepancy, including a substantial carbon dioxide and/or nitrogen fertilization effect on plant growth; a larger uptake of carbon by the oceans than has previously been thought likely; greater carbon sequestering as a result of recent reforestation programs in northern midlatitudes; and a larger carbon-storing capacity of annual grasses in tropical and subtropical regions. Desertification is clearly relevant to this last possibility as it would alter the effectiveness of a sink that may prove more important than is now estimated.

Although the importance of the net carbon flux associated with desertification is impossible to quantify at this time, a rough order of magnitude can be estimated. Data from the 1980s indicate that atmospheric carbon dioxide accounts for about 55 percent of

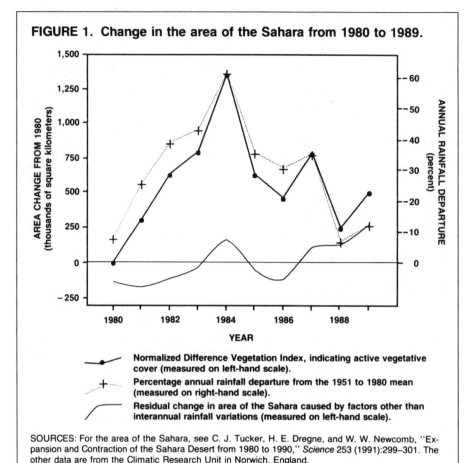

FIGURE 1. Change in the area of the Sahara from 1980 to 1989.

— ● — Normalized Difference Vegetation Index, indicating active vegetative cover (measured on left-hand scale).

--- + --- Percentage annual rainfall departure from the 1951 to 1980 mean (measured on right-hand scale).

⌒ Residual change in area of the Sahara caused by factors other than interannual rainfall variations (measured on left-hand scale).

SOURCES: For the area of the Sahara, see C. J. Tucker, H. E. Dregne, and W. W. Newcomb, "Expansion and Contraction of the Sahara Desert from 1980 to 1990," *Science* 253 (1991):299–301. The other data are from the Climatic Research Unit in Norwich, England.

all greenhouse-gas forcing.[11] Of the global carbon emissions, the net biospheric contribution (that is, the non-industrial component, largely resulting from land-use changes) is variously estimated at between 10 percent and 30 percent (or between 5 percent and 15 percent of total greenhouse forcing). The bulk of these biospheric emissions results from tropical deforestation; land conversion in dryland areas is a minor contributor.[12] When considering the net contribution of dryland regions to greenhouse-gas sources and sinks, a distinction should be made between the contribution arising from nondegrading changes in land use (almost certainly the primary contribution) and that from desertification (the secondary contribution). (An example of a sustainable land-use change is the conversion of a dryland from shrubs to grassland with no subsequent degradation in soil quality.) Therefore, desertification's contribution of carbon

to greenhouse forcing and, hence, to global-mean warming is almost certainly less than a few percent of the global total. It is not possible at this time to estimate net emissions of other greenhouse gases resulting from dryland degradation.

Despite the relative unimportance of desertification as a direct contributor to global warming, there is a clear need to improve understanding of the desertification-related sources and sinks of greenhouse gases. Greater knowledge is important for both scientific and political reasons. The scientific reason is that improved projections of future atmospheric carbon dioxide concentrations and estimates of future rates of global-mean warming depend on better quantification of where and how fast atmospheric carbon is sequestered. The same is true for all the major greenhouse gases. The political reason is that national inventories of greenhouse-gas sources and sinks will be an important element in future negotiations connected with the UN Framework Convention on Climate Change.

By restoring a terrestrial carbon sink and reducing direct emissions of carbon, a reversal of desertification could measurably contribute to reducing global greenhouse forcing but would not by itself lead to a significant reduction in future global-mean warming. The contribution of dryland degradation is too small to have a substantial impact. Nevertheless, because most dryland countries have relatively low industrial carbon emissions, desertification could be a major element of such nations' individual net carbon budgets. In such cases, arresting desertification should be considered a priority action. Only a modest stimulation or protection of carbon sinks in such countries would offset a significant proportion of their other emissions of carbon.

Desiccation in the Sahel

No region has been at the center of the debate over the causal links between climate change, desiccation, and desertification more than the African Sahel. Over the past 25 years, the Sahel has undergone severe desiccation and increasing deterioration of the soil quality and vegetative cover (see box on next page). More than in any other dryland region in the world, it is the simultaneous occurrence of these phenomena in the Sahel that raises questions about the links between climate change and desertification and, in particular, about the cause of the sustained decline in rainfall. Arguments over these questions date back to the mid 1970s, when the meteorology behind the Sahelian crisis of 1972 and 1973 was first discussed.[13] Current ideas about the causes of the desiccation have crystalized around two central themes: internal biogeophysical feedback mechanisms within Africa associated with land-cover changes, such as desertification; and global circulation changes associated with particular patterns of heat distribution in the oceans (ocean temperature departures from the his-

torical mean). With regard to the ultimate cause of the oceanic changes, two possibilities present themselves: They may be a manifestation of quasi-periodic natural fluctuations in ocean circulation, the result of natural climate variability; and/or they may be a response of the ocean system to anthropogenic greenhouse-gas and sulfate aerosol forcing of the climate system. There is some evidence for each of these three causal agents, and each of them has very different impli-

cations in terms of both the appropriate remedial actions and the political repercussions for the INC-D negotiations.

Land-Cover Changes in Africa

The idea that modification of land-cover characteristics in dryland regions might affect regional rainfall was first proposed by Joseph Otterman, an environmental scientist at Tel Aviv University, in 1974 and arose from his empirical work in the

---

HAS DESERTIFICATION "CONTAMINATED" THE GLOBAL-MEAN TEMPERATURE RECORD?

Bob Balling, Jr., of Arizona State University recently suggested that, during the 20th century, desertified areas have warmed by about 0.5° C relative to nondesertified areas.[1] When averaged globally, he argues, this warming significantly "contaminates" the global-mean temperature record and so complicates the search for a warming signal, or component, caused by enhancement of the greenhouse effect.[2] However, Balling's identification of areas that are severely desertified and nondesertified is derived from a map of desertification prepared for the UN Conference on Desertification back in 1977. That map has been superseded by more accurate data collected for the recent assessment by the UN Environment Programme (UNEP).[3] Moreover, Balling's calculated desertification warming signal of 0.5° C per 100 years is based on only a small subset of these "desertified" areas. He extrapolates from this subset to the global scale by suggesting that more than 30 percent of all land (including 90 percent of all drylands) is prone to this desertification warming signal.

UNEP's 1992 global assessment of desertified areas estimated, however, that only 20 percent of dryland regions were seriously degraded. This figure suggests that only 6 percent of all land may exhibit a desertification warming signal, rather than the 30 percent assumed by Balling. The potential bias is, therefore, comparable in magnitude to other biases that may affect the global temperature record. Finally, it should be noted that the differential warming between desertified and neighboring nondesertified areas found by Balling at a resolution

of 5° latitude/longitude may simply indicate that these areas differ in their sensitivity to climate variability and may not be evidence of warming caused by desertification per se. Thus, although Balling may be correct in principle, he has greatly overstated his case, and his analysis provides no convincing evidence that warming caused by desertification has substantially affected the global-mean temperature record.

---

1. R. C. Balling, Jr., "Impact of Desertification on Regional and Global Warming," *Bulletin of the American Meteorological Society* 72 (1991):232-34. The ideas expressed in this article have received a moderate amount of attention in certain circles skeptical of the influence of greenhouse-gas emissions on global climate and have been introduced into discussions of the links between climate change and desertification. See, for example, UN Sudano-Sahelian Office and UN Development Programme, *GEF and Desertification: UNSO/UNDP Workshop, Nairobi, 28-30 October 1992* (New York: UNSO/UNDP, 1992).

2. T. M. L. Wigley, G. I. Pearman, and P. M. Kelly, "Indices and Indicators of Climate Change: Issues of Detection, Validation and Climate Sensitivity," in I. M. Mintzer, ed., *Confronting Climate Change: Risks, Implications and Responses* (Cambridge, England: Cambridge University Press, 1992), 85-96. For a discussion of other potential sources of bias in the global-mean temperature record, see C. K. Folland, T. Karl, and K. Ya. Vinnikov, "Observed Climate Variations and Change," in J. T. Houghton, G. J. Jenkins, and J. J. Ephraums, eds., *Climate Change: The IPCC Scientific Assessment* (Cambridge, England: Cambridge University Press, 1990), 195-238; and C. K. Folland et al., "Observed Climate Variability and Change," in J. T. Houghton, B. A. Callander, and S. K. Varney, eds., *Climate Change 1992: The Supplementary Report to the IPCC Scientific Assessment* (Cambridge, England: Cambridge University Press, 1992), 135-70.

3. UN Environment Programme, *World Atlas of Desertification* (Sevenoaks, U.K.: Edward Arnold, 1992).

Negev Desert.[14] His initial contention was that bared, high-reflecting soils would increase surface albedo (reflectivity), reduce convective processes, and thus decrease rainfall. Around the same time, Jule Charney, a meteorologist at the Massachusetts Institute of Technology, was developing his biogeophysical hypothesis that land-cover changes, primarily around the Sahara, could enhance aridity.[15] Charney's proposed mechanism involved a desertification-induced change in the vertical energy flux in the atmosphere over dryland regions. Charney's mechanism was subsequently criticized because of his omission of the role of soil moisture and the absence of any discussion of latent/sensible heat ratios.[16]

Charney's hypothesis received considerable attention because it, or some variant, would provide an apparent explanation for self-reinforcing drought (that is, desiccation) in dryland regions. According to this hypothesis, an initial change in land-cover characteristics occurs in association with desertification. The initial change may involve a change or removal of vegetation and/or a deterioration in soil quality and moisture-holding capacity. The land-cover change is then amplified as land surface-atmosphere interaction suppresses rainfall, either by reducing surface moisture or by increasing atmospheric subsidence. Lower rainfall, in turn, increases moisture stress on vegetation, lowers soil moisture levels, and further reduces rainfall amounts, thereby closing the feedback loop. The significance of this hypothesis for the present discussion is that, if such land-cover changes can account for the rainfall decline in the Sahel or even for a significant proportion of that decline, then it is the complex matrix of processes leading to desertification in recent decades that is responsible for the Sahel's desiccation.

A substantial amount of effort has been directed over the last 15 years to refining Charney's basic hypothesis and to examining the sensitivity of regional rainfall to large-scale changes

THE DESICCATION OF THE SAHEL

Within recent years, the climate of the Sahel has exhibited a continuing trend toward desiccation. For about 25 years, rainfall has been substantially lower than it was during the first seven decades of the century (see the figure below). This desiccation represents the most substantial and sustained change in rainfall for any region in the world ever recorded by meteorological instruments. Individual years, such as 1984 and 1990, have seen rainfall totals drop to less than 50 percent of those received during the 1930s, 1940s, and 1950s. Contrasting two successive 30-year periods (from 1931 to 1960 and from 1961 to 1990), the rainfall decline over this region has been between 20 and 40 percent. Although this magnitude of desiccation is unprecedented in the instrumental record, it is harder to assess how unusual it is for the longer-term history of the Sahel. Using a combination of lake levels,

landscape descriptions, and historical accounts, Sharon Nicholson at Florida State University has shown that recurrent droughts enduring from one to two decades have been a feature of the Sahelian climate over the last few centuries.[1] Quantifying the severity and precise duration of such droughts, however, is impossible. Examining the historical levels of Lake Chad, one of the great inland lakes of Africa, which lies toward the south of the Sahel, may provide some clues. A comparison of the current decline in the level of Lake Chad to its historical variations suggests that the present desiccation in the Sahel is at least as severe as anything experienced during the last millennium.

1. S. E. Nicholson, "Climatic Variations in the Sahel and Other African Regions During the Past Five Centuries," *Journal of Arid Environments* 1 (1978):3–24.

Annual rainfall departure index for the Sahel from 1901 to 1992, expressed as a percent departure from the mean annual rainfall between 1951 and 1980.

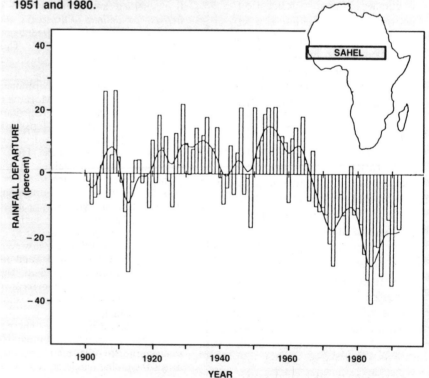

Note: Up to 60 rainfall stations contributed data to this regional series. The smooth curve results from applying a filter that emphasizes variations on a time-scale longer than 10 years.

FIGURE 2. Global field of annual surface air temperature departures associated with drought in the Sahel.

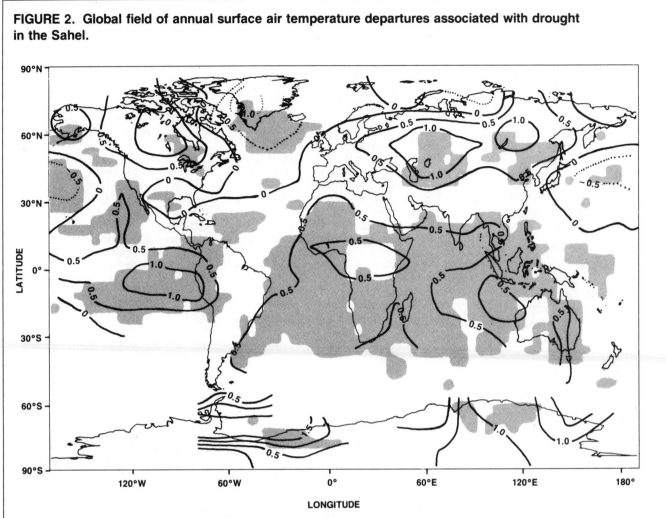

Note: The numbers represent the temperature departures in degrees Celsius, or the differences in temperature between a set of five dry years and a set of five wet years. Shaded areas indicate where the differences are statistically significant.

in land cover through climate modeling experiments. These experiments have been performed for various regions, including the Amazon, the Sahara, and tropical Africa.[17] Such experiments have also addressed a wide range of physical mechanisms for desertification-induced desiccation by modeling interactions among surface albedo, soil moisture and evaporation, and changes in surface roughness and vegetation.[18] These experiments clearly show that large-scale conversion of land-cover characteristics can generate climate change on local and regional scales.

There appears, however, to be a fundamental difficulty in attributing the recent desiccation in the Sahel to land-cover changes on the basis of these model experiments. Observational evidence of the marked, large-scale, sustained changes in surface albedo in dryland regions that are introduced into most model experiments remains weak.[19] (Surface albedo is the measure of land-cover characteristics used in these experiments.) The albedo increases caused by desertification that have been observed are on the order of 25 to 50 percent,[20] and yet a doubling of albedo is used in many model experiments. Moreover, the observed changes have been localized in extent and often short-term, rather than widespread and sustained as assumed in the modeling studies. All of the modeling experiments that have displayed substantial regional rainfall reduction as a response to

land-cover changes have been "sensitivity" rather than "simulation" experiments; rather than imposing observed perturbations to surface vegetation, soil moisture, and so on, they have imposed arbitrarily determined changes, which in all cases have been much larger than those that have actually been observed. Although these experiments are important to understanding how the various physical systems are linked, it is dangerous to draw the conclusion from their results that observed land-cover changes have accounted for observed rainfall changes in the recent past.

A surer way to proceed is to conduct simulation experiments, which impose known perturbations on the model (for example, the observed soil moisture conditions in a given year) and then to examine whether the model reproduces the observed rainfall anomaly of that year. The most impressive set of such simulation experiments has been completed at the British Meteorological Office. The investigators simultaneously perturbed both ocean temperatures and initial soil moisture conditions in a manner consistent with observations.[21] They concluded that ocean temperature forcing appears to dominate the effects of the land surface moisture feedback. Although this work confirms that land surface feedback can play a part in generating self-sustaining drought, the role of this mechanism is secondary to that of variability within the wider climate system.

In light of current empirical and modeling evidence, then, it appears that desertification is not, in itself, a primary cause of the recent desiccation in the Sahel. The degradation of both soil and vegetative cover in dryland regions could well have contributed to the rainfall decline, but this contribution cannot have accounted for anything more than a small fraction of the observed trend. If there is severe and sustained degradation of a substantial dryland area over the next few decades, however, the significance of this internal feedback mechanism may well increase. Over the next 50 years, though, it is more likely that land-cover changes in the humid and subhumid regions of the tropics will lead to substantial changes in regional climate than will those occurring in dryland areas.

Natural Changes in Ocean Circulation

The British Meteorological Office's experiments confirmed the importance of a set of natural mechanisms of climate change that appear to be responsible for the Sahelian desiccation. These mechanisms involve links between Sahelian drought and sea surface temperature (SST) anomalies in the neighboring Atlantic Ocean and other oceans. Research has shown that there is a significant correlation on the

---

WHY IS THE SAHEL DRY?

The Sahel of Africa possesses a monsoonal climate—that is, the climate of the region exhibits a very strong seasonality with a nearly 180° reversal of the prevailing surface wind direction between the wet and dry halves of the year. The winter, or dry, monsoon lasts from October through April and is characterized by northerly or northeasterly surface winds circulating clockwise around the Saharan anticyclone. These winds are extremely dry and lead to no rainfall during these months. The summer, or wet, monsoon commences sometime between April and June and arrives progressively from the south. The moisture in these rain-bearing southerly or southwesterly surface winds originates mostly over the Atlantic Ocean and, to a lesser extent, the Indian Ocean. The sea surface temperatures of these oceans, therefore, exert important control over the rainfall in the Sahel by altering both the moisture characteristics and the vigor of the wet monsoon flow into northern tropical Africa. The wet monsoon varies in duration from five or six months in the southern Sahel (about 10°N) to only a month or two in the far north (about 16°N). This variation in duration creates a very tight gradient in annual rainfall—from less than 100 millimeters north of 16°N to more than 800 millimeters south of about 10°N. Because of this tight gradient, relatively small interannual variations either in the northward penetration or moisture load of the wet monsoon or in the strength of the atmospheric disturbances that lead to the rain outfall result in relatively large variations in the total volume of rainfall received by a locality. Consequently, the rainfall of the Sahel is highly variable from year to year.

---

interannual time-scale between higher-than-normal SSTs south of West Africa and reduced Sahelian rainfall.[22] In the early 1980s, it was argued that a change in atmospheric circulation was affecting both the SST pattern and rainfall over the Sahel. In other words, the SST pattern was just an indicator of the processes affecting Sahelian rainfall rather than the primary cause. It is now thought, however, that the SST pattern may well be the direct cause of the shift in the atmospheric circulation that subsequently affects Sahelian rainfall.

During the mid 1980s, Chris Folland and his colleagues at the Meteorological Office confirmed this statistical correlation between local variations in ocean temperatures and African rainfall on a time-scale of years to decades and found evidence of a broader relationship between Sahelian rainfall and worldwide ocean temperatures.[23] They demonstrated that differences in SST anomalies between the Northern and Southern Hemispheres, most marked in the Atlantic sector, were related to Sahelian rainfall. Higher temperatures south of the equator and lower temperatures north of the equator (see Figure 2) were associated with

lower rainfall over much of northern tropical Africa. Modeled simulations of the effects of these observed SST anomaly patterns confirmed this association. The success of these model experiments in simulating the observed Sahelian rainfall anomalies suggested that the SST anomaly pattern was the direct cause of the rainfall anomalies. This work has since been extended, confirming the original empirical and model results, and the relationship has provided the basis of an experimental seasonal forecasting scheme.[24]

The physical basis of this relationship appears to lie in a disturbance to the meridional, Hadley circulation of the atmosphere over the Atlantic/Africa sector that is induced by the pattern of contrasting hemispheric ocean temperature anomalies. The Hadley circulation exerts a controlling influence on African rainfall patterns. It determines, in part, the position of the Intertropical Convergence Zone—specifically, the extent of its annual north-south migration, which, in turn, affects the strength of the southwesterly airflow originating in the tropical Atlantic that brings the Sahel much of its rain (see the box on

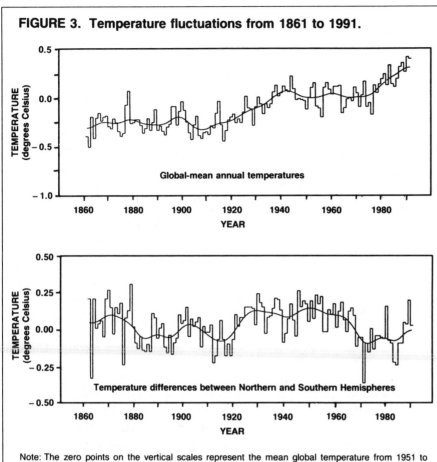

FIGURE 3. Temperature fluctuations from 1861 to 1991.

Note: The zero points on the vertical scales represent the mean global temperature from 1951 to 1980. In the lower graph, positive values indicate that the Northern Hemisphere was warmer than the Southern Hemisphere, and negative values indicate the reverse.

SOURCE: C. K. Folland et al., "Observed Climate Variability and Change," in J. T. Houghton, B. A. Callendar, and S. K. Varney, eds., *Climate Change 1992: The Supplementary Report to the IPCC Scientific Assessment* (Cambridge, England: Cambridge University Press, 1992), 135–70.

page 216).[25] Thus, the link between lower rainfall in the Sahel and a particular pattern of SST anomalies in the oceans has been well established. The initial cause of this SST pattern, however, has yet to be determined. The spatial scale of the phenomenon may provide some evidence of the cause. Although temperatures in the Atlantic Ocean are probably the dominant influence on Sahelian rainfall, the recently observed pattern of SST anomalies in the Atlantic sector is part of a much larger trend in surface air temperatures. Large-scale warming has affected both hemispheres since the late 19th century and has resulted in a net global-mean warming of about 0.5° C (see the top graph in Figure 3). There has been a clear difference, however, in the warming rates of the two hemispheres during recent decades, with the Northern Hemisphere

warming more slowly than the Southern Hemisphere (see the second graph in Figure 3).

This relationship between conditions in the Atlantic/Africa sector and worldwide climatic trends is not confined to temperature; a link also exists between the desiccation of the Sahel in recent decades and global rainfall fluctuations. Although the rainfall deficit in the Sahel is the most striking rainfall change of recent decades, rainfall has been lower in many parts of the northern tropics and subtropics but has increased at higher latitudes (see Figure 4).[26] The Sahel's desiccation could be considered a regional manifestation of the global shift in the climate system that has occurred since the 1950s.

The outstanding question in this line of reasoning concerns the initial cause of the temperature change that

gives rise to the rainfall disturbance. The observed ocean temperature pattern may well be a manifestation of natural climate variability. For example, it could be the result of a reduction in the northward transport of heat in the Atlantic Ocean.[27] More recently, Alayne Street-Perrott and Alan Perrott at the University of Oxford have hypothesized that this reduction in heat transport may be the result of a freshening (reduction of salinity) of the surface waters of the northern North Atlantic.[28] By stabilizing the water column, the freshening reduces deep convection in the North Atlantic and the compensatory surface inflow of water and, hence, heat from the south. Although natural climatic variability is undoubtedly a possible cause of the observed abnormal pattern in ocean temperature, there is equally convincing, albeit equally circumstantial, evidence to suggest that the abnormal pattern may be linked to global-mean warming.

The Link with Global-Mean Warming

Is there a greenhouse-related mechanism that could account for the interhemispheric temperature contrast associated with Sahelian desiccation? Recent research has indicated that there are two alternative—or, more likely, complementary—mechanisms that could induce a temperature difference between the hemispheres.

First, the emission of sulfur compounds as a result of human activity (specifically, fossil fuel combustion) increases the amount of sulfate aerosols in the atmosphere. These aerosols reflect solar radiation, both directly and by altering cloud albedo. Any increase in their concentration in the atmosphere is, therefore, likely to have a cooling effect, offsetting greenhouse warming. Estimates of the scale of this effect vary, but it is considered possible that sulfur dioxide emissions may have reduced the level of warming that rising greenhouse-gas concentrations might have effected by a significant amount.[29] As most sulfur emissions come from the Northern Hemisphere and the sulfate aerosols have a short residence time in the at-

FIGURE 4. Precipitation fluctuations in the Northern Hemisphere.

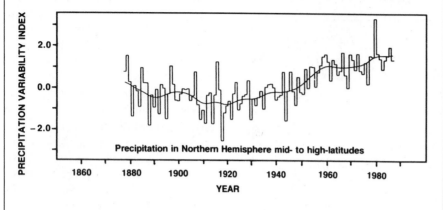

Precipitation in Northern Hemisphere mid- to high-latitudes

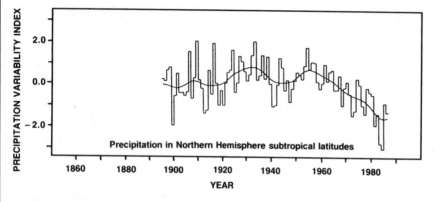

Precipitation in Northern Hemisphere subtropical latitudes

Note: The zero point on the vertical scale of the upper graph represents the mean precipitation from 1877 to 1986. The zero point on the lower graph represents the mean precipitation from 1895 to 1986. The scale is an index of large-scale variations in precipitation, which takes into account the marked changes in mean rainfall that occur from place to place and from month to month.

SOURCE: R. S. Bradley et al., "Precipitation Fluctuations over Northern Hemisphere Land Areas Since the Mid-19th Century," *Science* 237 (1987):171–75.

mosphere, this cooling effect would be largely confined to the Northern Hemisphere. As a result, greenhouse-gas-induced warming would be offset in the Northern Hemisphere but relatively unaffected in the Southern Hemisphere, thereby inducing a differential warming rate and a temperature contrast between the two hemispheres. Estimates of the scale of this effect are not inconsistent with the observed differential between the warming trends in the two hemispheres.[30]

A second greenhouse-related factor exists, particularly in the Atlantic Ocean, that may have caused different rates of warming in the hemispheres. Recent time-dependent ("transient") experiments of enhanced greenhouse-gas forcing using general circulation models (GCMs) that incorporate a dynamic ocean model have suggested that the rate of warming may be retarded in the northern North Atlantic sector and over the Southern Ocean around Antarctica.[31] In these areas, the sinking of dense, saline water masses results in a localized increase in the effective heat capacity and, therefore, in the thermal inertia of the ocean. The increase in thermal inertia slows the warming of the overlying air as heat is drawn down into the ocean. This process induces a meridional gradient in temperature in the Atlantic Ocean north of 60°S, which is not unlike the temperature pattern associated with lower rainfall in the Sahel (see Figure 2). This mechanism may amplify, or even be triggered by, changes in the temperature field and atmospheric circulation that are induced by the effects of sulfate aerosols.

Thus, a physically plausible argu-

ment can be advanced linking the recent desiccation in the Sahel to global-mean warming. However, just as it is impossible to ascribe with any certainty the observed global-mean warming to enhancement of the greenhouse effect,[32] neither can the interhemispheric temperature contrast be attributed with confidence to greenhouse-gas-plus-sulfate forcing or to any other greenhouse-related mechanism. For now, the evidence must be considered circumstantial.

The Future of the Sahel

GCM experiments have recently been conducted to determine whether significant rainfall changes over northern tropical Africa may result from future greenhouse-gas forcing. (Unfortunately, none of these experiments incorporates the sulfate aerosol effect, and only one allows the ocean circulation to respond realistically to greenhouse-gas forcing.) The experiments indicate that global-mean warming should lead to an overall increase in global-mean precipitation because evaporation over the warmer oceans increases the moisture content of the atmosphere.[33] The distribution of rain and snowfall, however, will also be determined by changes in the atmospheric circulation and by other climatic factors.

Figure 5 shows a composite estimate of the percentage change in mean annual rainfall that may accompany each 1° C rise in global-mean temperature induced by greenhouse-gas forcing. The composite is based on a set of seven GCM experiments.[34] An increase in rainfall is apparent in most areas, but annual rainfall decreases over the Mediterranean, North Africa, and a large part of the Sahel, especially the Western Sahel. The effect is most marked over the southwestern margins of the Sahara, in Mauritania and northern Mali and Niger. In the later two areas, annual rainfall decreases by more than 6 percent for every 1° C of global-mean warming. If global-mean warming follows the 1992 projections of the Intergovernmental

Panel on Climate Change,[35] rainfall in these areas decreases by between 6 and 30 percent by 2100. This is, however, a slow rate of decline, as long as it occurs progressively, in comparison with the 20 to 40 percent decline in rainfall experienced by the Sahel during recent decades.

The current performance of climate models in estimating regional rainfall patterns is considered quite weak. Model-to-model differences in predicted rainfall changes over northern Africa are large, and individual models predict a complex spatial pattern of change. Nevertheless, the model composite does show a well-defined reduction in rainfall over much of the Sahel. Compounded by increased temperatures (which can be predicted with greater confidence), lower rainfall would inevitably cause substantial reductions in soil moisture availability.

Negotiating for Survival

Negotiations for the Convention to Combat Desertification will be complicated by the technical and scientific uncertainties underlying many aspects of the desertification issue. It is to be hoped that the negotiators will be assisted by the kind of technical support that was provided by the Intergovernmental Panel on Climate Change during the negotiations for the Framework Convention on Climate Change. Although varying degrees of uncertainty surround the links between climate change, desiccation in dryland regions, and desertification, this brief assessment has shown that there are intrinsic links that should not be ignored (see Figure 6). Even the rather speculative link between global-mean warming and desiccation in the Sahel warrants serious consideration on a precautionary basis because of the serious implications for those living in this dryland area.

The area prone to desertification is, by definition, determined by climatic conditions and, hence, by climate change. It is more difficult to deter-

mine the precise balance between the human and climatic factors that lead to desertification. There is, however, no doubt that, against a background of resource management failure in dryland regions, climate change will aggravate the problem. It is also clear that, by modifying surface characteristics, desertification can induce significant changes in local temperature. But it is far less likely that the global-mean temperature has been affected to any significant extent by dryland degradation.

Progressive desertification of the dryland tropics and other areas is likely to reduce a potential carbon sink by reducing the carbon stored in these ecosystems. Moreover, as vegetation dies and soil is disturbed, desertification increases emissions of carbon dioxide. Desertification may also increase or decrease emissions of other greenhouse gases. Although desertification contributes only a small percentage of all greenhouse-gas emissions, understanding of desertification-related sources and sinks of greenhouse gases should still be improved and desertification rates reduced to enable dryland countries to offset growth in their emissions of other greenhouse gases.

The hypothesized link between the

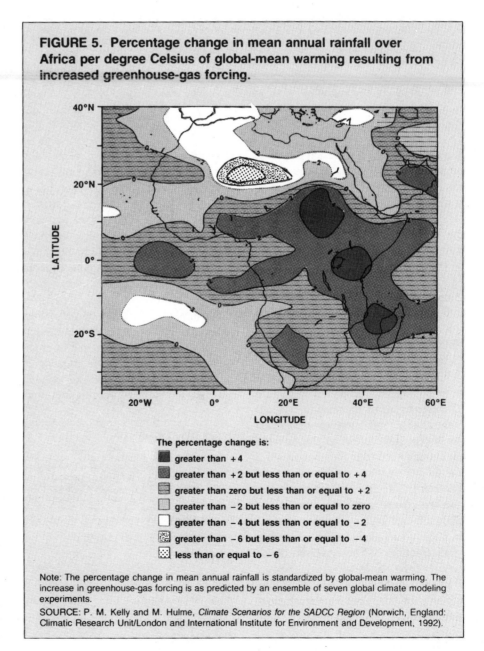

FIGURE 5. Percentage change in mean annual rainfall over Africa per degree Celsius of global-mean warming resulting from increased greenhouse-gas forcing.

The percentage change is:

- greater than + 4
- greater than + 2 but less than or equal to + 4
- greater than zero but less than or equal to + 2
- greater than – 2 but less than or equal to zero
- greater than – 4 but less than or equal to – 2
- greater than – 6 but less than or equal to – 4
- less than or equal to – 6

Note: The percentage change in mean annual rainfall is standardized by global-mean warming. The increase in greenhouse-gas forcing is as predicted by an ensemble of seven global climate modeling experiments.
SOURCE: P. M. Kelly and M. Hulme, *Climate Scenarios for the SADCC Region* (Norwich, England: Climatic Research Unit/London and International Institute for Environment and Development, 1992).

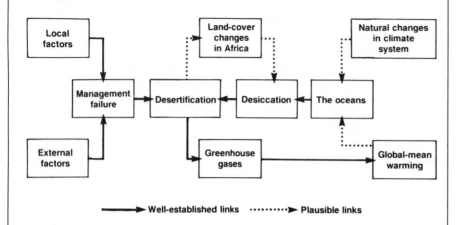

FIGURE 6. The matrix of cause and effect surrounding desertification and the role of climate change.

NOTE: Desertification is the result of resource management failure, the product of both local factors, such as population pressure and inequity, and external factors, such as the state of the global economy, commodity prices, and the burden of debt. Desertification is aggravated by climate change—desiccation—which may be the result of natural mechanisms with the climate system, such as ocean-atmosphere feedback; by desertification itself through, for example, surface-atmosphere interaction; or possibly by global-mean warming. Finally, desertification contributes to global-mean warming through its effect on the sources and sinks of greenhouse gases, such as carbon dioxide. Many uncertainties affect assessment of the relative role of these various factors.

recent desiccation of the Sahel and an interhemispheric contrast in ocean temperature is supported by empirical studies, model simulations, and theoretical argument. The initial cause of the interhemispheric temperature contrast has yet to be determined. It may be the result of natural climate variability or a manifestation of global-mean warming resulting from greenhouse-gas and other anthropogenic emissions. Both possibilities are physically plausible, and neither possibility is contradicted by the available data.

Finally, on the basis of current evidence, it appears likely that the role of desertification in causing (via climate change and land-cover changes) the recent desiccation in the Sahel is very much secondary to that of forcing by ocean temperature anomalies. The rainfall decline estimated to have resulted from past degradation of both soil and vegetative cover in Africa is only sufficient to account for a small fraction of the observed desiccation in the Sahel.

In the future, however, the relative importance of these various factors may change. For example, a sustained degradation of a substantial dryland area may well increase the significance of desertification as a causative agent. On the other hand, climate model experiments suggest that rainfall over the Sahel may decrease even more as global-mean warming develops. In this case, the role of greenhouse-gas emissions in reducing rainfall in the Sahel would become more prominent. Whatever the complexity of these mechanisms, their investigation cannot be considered to be solely of academic interest; as the inhabitants of the Sahel well know, the trend of reduced rainfall continues.

ACKNOWLEDGMENTS

The authors wish to thank Sarah Granich for her contribution to this article and Tim O'Riordan for his helpful comments. The article is based on a technical report prepared for the Overseas Development Administration in London, but it does not necessarily reflect the administration's views.

NOTES

1. UN Environment Programme, *World Atlas of Desertification* (Sevenoaks, U.K.: Edward Arnold, 1992).

2. R. S. Odingo, "Implementation of the Plan of Action to Combat Desertification (PACD) 1978–1991,"

Desertification Control Bulletin 21 (1992):6–14.

3. The Global Environment Facility was established in 1990 to help developing nations respond to global environmental change insofar as this response will reduce the global impact of the problems.

4. UN Environment Programme, *Status of Desertification and Implementation of the UN Plan of Action to Combat Desertification*, UNEP/GCSS.III/3 (Nairobi: UN Environment Programme, 1991).

5. United Nations, "Managing Fragile Ecosystems: Combating Desertification and Drought," chapt. 12 of *Agenda 21* (New York: United Nations, 1992), pt. 2.

6. A. Warren and M. M. Khogali, *Assessment of Desertification and Drought in the Sudano-Sahelian Region: 1985 to 1991* (New York: UN Development Programme and UN Sudano-Sahelian Office, 1992).

7. In the 1992 UNEP desertification assessment, a simple moisture index (the ratio of precipitation to potential evapotranspiration) was used to define these boundaries.

8. The index, known as the Normalized Difference Vegetation Index (NDVI), is derived from the visible and near infrared sensors on polar-orbiting satellites and is an indicator of the photosynthetic vigor of surface biomass. See C. J. Tucker, H. E. Dregne, and W. W. Newcomb, "Expansion and Contraction of the Sahara Desert from 1980 to 1990," *Science* 253 (1991):299–301.

9. R. C. Balling, Jr., "Impact of Desertification on Regional and Global Warming," *Bulletin of the American Meteorological Society* 72 (1991):232–34.

10. R. T. Watson, H. Rodhe, H. Oeschger, and U. Siegenthaler, "Greenhouse Gases and Aerosols," in J. T. Houghton, G. J. Jenkins, and J. J. Ephraums, eds., *Climate Change: The IPCC Scientific Assessment* (Cambridge, England: Cambridge University Press, 1990), 1–40; and R. T. Watson, L. G. Meira Filho, E. Sanhueza, and A. Janetos, "Greenhouse Gases: Sources and Sinks," in J. T. Houghton, B. A. Callander, and S. K. Varney, eds., *Climate Change 1992: The Supplementary Report to the IPCC Scientific Assessment* (Cambridge, England: Cambridge University Press, 1992), 25–46.

11. K. P. Shine, R. G. Derwent, D. J. Wuebbles, and J.-J. Morcrette, "Radiative Forcing of Climate," in Houghton, Jenkins, and Ephraums, eds., note 10 above, pages 41–68. The results of more recent research suggest that the carbon dioxide contribution to the total anthropogenic climate forcing may be higher when the effects of emissions of sulfur compounds and ozone depletion are considered. See Watson, Meira Filho, Sanhueza, and Janetos, note 10 above; and T. M. L. Wigley and S. C. B. Raper, "Implications of Revised IPCC Emissions Scenarios," *Nature* 357 (1992):293–300.

12. A. F. Bouwman, "Land Use Related Sources of Greenhouse Gases," *Land Use Policy*, April 1990, 154–64.

13. M. H. Glantz, "The Value of a Long-Range Weather Forecast for the Sahel," *Bulletin of the American Meteorological Society* 58 (1977):150–58; P. J. Lamb, "Large-Scale Tropical Atlantic Surface Circulation Patterns Associated with Sub-Saharan Weather Anomalies," *Tellus* 30 (1978):240–51; and S. E. Nicholson, "The Nature of Rainfall Fluctuations in Subtropical West Africa," *Monthly Weather Review* 108 (1980):473–87.

14. J. Otterman, "Baring High-Albedo Soils by Overgrazing: A Hypothesised Desertification Mechanism," *Science* 186 (1974):531–33.

15. J. G. Charney, "Dynamics of Deserts and Drought in the Sahel," *Quarterly Journal of the Royal Meteorological Society* 101 (1975):193–202.

16. See, for example, S. B. Idso, "A Note on Some Recently Proposed Mechanisms of Genesis of Deserts," *Quarterly Journal of the Royal Meteorological Society* 103 (1977):369–70.

17. R. E. Dickinson and A. Henderson-Sellers, "Modelling Tropical Deforestation: A Study of GCM Land-Surface Parameterisations," *Quarterly Journal of the Royal Meteorological Society* 114 (1988):439–62; W. M. Cunnington and P. R. Rowntree, "Simulations

of the Saharan Atmosphere: Dependence on Moisture and Albedo," *Quarterly Journal of the Royal Meteorological Society* 112 (1986):971–99; M. F. Mylne and P. R. Rowntree, "Modelling the Effects of Albedo Change Associated with Tropical Deforestation," *Climate Change* 21 (1992):317–43; and Y. C. Sud and M. J. Fenessey, "A Study of the Influence of Surface Albedo on July Circulation in Semi-Arid Regions Using the GLAS GCM," *Journal of Climatology* 2 (1982):105–25.

18. Y. C. Sud and M. J. Fenessey, "Influence of Evaporation in Semi-Arid Regions on the July Circulation: A Numerical Study," *Journal of Climatology* 4 (1984):393–98; and J. Lean and D. A. Warrilow, "Simulation of the Regional Climatic Impact of Amazon Deforestation," *Nature* 342 (1989):411–13.

19. M. Hulme, "Is Environmental Degradation Causing Drought in the Sahel?" *Geography* 74 (1989):38–46.

20. S. I. Rasool, "On Dynamics of Deserts and Climate," in J. T. Houghton, ed., *The Global Climate* (Cambridge, England: Cambridge University Press, 1984), 107–20.

21. D. P. Rowell et al., "Causes and Predictability of Sahel Rainfall Variability," *Geophysical Research Letters* 19 (1992):905–08.

22. Lamb, note 13 above; and J. M. Lough "Atlantic Sea Surface Temperatures and Weather in Africa" (Ph.D. diss., University of East Anglia, 1980).

23. C. K. Folland, D. E. Parker, and F. E. Kates, "Worldwide Marine Temperature Fluctuations, 1856–1981," *Nature* 310 (1984):670–73; and C. K. Folland, T. N. Palmer, and D. E. Parker, "Sahel Rainfall and Worldwide Sea Temperatures, 1901–1985," *Nature* 320 (1986):602–07.

24. C. K. Folland, J. A. Owen, M. N. Ward, and A. W. Colman, "Prediction of Seasonal Rainfall in the Sahel Region Using Empirical and Dynamical Methods," *Journal of Forecasting* 10 (1991):21–56.

25. For a more detailed discussion of the proposed mechanism, see C. K. Folland and J. A. Owen, "GCM Simulation and Prediction of Sahel Rainfall Using Global and Regional Sea Surface Temperatures," in *Modelling the Sensitivity and Variations of the Ocean-Atmosphere System*, WMO/TD no. 254 (Geneva: World Meteorological Organization, 1988), 107–15.

26. R. S. Bradley et al., "Precipitation Fluctuations over Northern Hemisphere Land Areas Since the Mid-19th Century," *Science* 237 (1987):171–75.

27. R. E. Newell and J. Hsiung, "Factors Controlling Free Air and Ocean Temperature of the Last 30 Years and Extrapolation to the Past," in W. H. Berger and L. D. Labeyrie, eds., *Abrupt Climatic Change: Evidence and Implications* (Dordrecht, the Netherlands: Reidel, 1987), 67–87.

28. F. A. Street-Perrott and R. A. Perrott, "Abrupt Climate Fluctuations in the Tropics: The Influence of Atlantic Ocean Circulation," *Nature* 343 (1990):607–12.

29. I. Isaksen, V. Ramaswamy, H. Rodhe, and T. M. L. Wigley, "Radiative Forcing of Climate," in Houghton, Callander, and Varney, eds., note 10 above, pages 47–68; and Wigley and Raper, note 11 above.

30. See, for example, Wigley and Raper, note 11 above.

31. W. L. Gates, P. R. Rowntree, and Q-C. Zeng, "Validation of Climate Models," in Houghton, Jenkins, and Ephraums, eds., note 10 above, pages 93–130; and W. L. Gates et al., "Climate Modelling, Climate Prediction and Model Validation," in Houghton, Callander, and Varney, eds., note 10 above, pages 97–134.

32. T. M. L. Wigley and T. P. Barnett, "Detection of the Greenhouse Effect," in Houghton, Jenkins, and Ephraums, eds., note 10 above, pages 243–55.

33. J. F. B. Mitchell, S. Manabe, V. Meleshko, and T. Tokioka, "Equilibrium Climate Change and Its Implications for the Future," in Houghton, Jenkins, and Ephraums, eds., note 10 above, pages 137–64.

34. Climatic Research Unit, *A Scientific Description of the ESCAPE Model* (Norwich, England: Climatic Research Unit, 1992); and P. M. Kelly and M. Hulme, *Climate Scenarios for the SADCC Region* (Norwich, England: Climatic Research Unit/London and International Institute for Environment and Development, 1992).

35. Houghton, Callander, and Varney, eds., note 10 above; and Wigley and Raper, note 11 above.

Biosphere: Endangered Species

- Plants (Articles 30–32)
- Animals (Articles 33–35)

One of the greatest tragedies of the conservation movement is that it began after it was already too late to save many species of plants and animals from extinction. In fact, even after concern for the biosphere developed among resource managers, their effectiveness in halting the decline of herds and flocks, packs and schools, or groves and grasslands has been limited by the ruthlessness and efficiency of the competition. Plants and animals compete directly with human beings for living space and for other resources such as sunlight, air, water, and soil. As the historical record of this competition in North America and other areas attests, since the seventeenth century settlement has been responsible—either directly or indirectly—for the demise of many plant and wildlife species. In addition to habitat destruction, other factors that have contributed to the decline of animal, bird, and plant populations include pollution of waterways, unnecessary overkills, the inability of indigenous animals to compete with domestic stock, and competition between wild plants and their crop cousins. It should be noted that extinction is a natural process—part of the evolutionary cycle—and not always created by human activity. But human actions have the capacity to accelerate a natural process that might otherwise take millennia.

In the lead article of the unit, consultant Otto Solbrig of the U.N. Man and the Biosphere Program describes this biodiversity from a functional point of view by defining its role in the creation and maintenance of the entire global ecosystem. The tremendous variety in nature has always astonished and delighted humans, Solbrig notes, but the current rate of loss of species diversity seems sure to reduce this immense richness and seriously affect human welfare in the process.

The most important component of any biosphere is the primary production component of living vegetation, and in the three articles in the *Plants* subsection, the issue of human impact on these most important biological systems is central. Certainly the foremost environmental issue related to plants is that of deforestation and each of the articles in this subsection deal with that issue. In the article "Rain Forest Entrepreneurs: Cashing in on Conservation," economists Thomas Carr and Sunder Ramaswamy and mathematician Heather Pedersen explain that most tropical deforestation is the result of economic incentives—cutting forests to extract hardwood timber and fuel and clear-cutting land for crop agriculture and livestock raising. Mounting evidence suggests that these commercial uses of forest lands are not only ecologically devastating but environmentally unsound.

The second article in this subsection also deals with the question of deforestation, focusing on the uncertainty of data regarding the actual rates of deforestation. In their article "Deforestation and Public Policy," Patricia Parisi and Michael Glantz, both of the National Center for Atmospheric Research, note that peer-reviewed estimates of tropical deforestation vary by a factor of six. They assess the impact that such highly variable scientific results have for policymakers who are forced to make decisions on forest management without any consistent time frame for the changes that are taking place. For the most part, discrepancies in data on deforestation allow policymakers to follow their own, often hidden, agendas, whether they be for conservation or full-scale exploitation. Policies based on inaccurate data (or at least facilitated by inaccurate data) do no one any good. No consistent land management strategies in forest regions will work without better and more consistent information. In "Out of the Woods," Jodi Jacobson of the Worldwatch Institute notes that tropical deforestation often produces much more than an altered environment. As development plans for commercial agriculture are implemented following forest clearance, the traditional role of women within the tropical forest society is usually ignored. Under the conditions of shifting subsistence tropical cultivation (the normal economy of preindustrial tropical forest regions), women play the most significant part in cultivation and nurturing of the crops grown within the forest. They also serve as the gathers of natural (wild) plant materials and as the collectors of fuelwood. When they are removed from the forest, as often happens when a livestock grazing economy replaces that of subsistence cultivation, their caretaker role is lost. The result is not only social disruption but the loss of methods of ecologically sustainable methods of forestry, of extensive knowledge of genetic diversity, and of a spirit of nurturing not present either in the commercial agricultural system or among the male hunters of traditional forest economies.

The prospects for the future of wildlife are not any better than that of the world's remaining native forestlands. Land developers destroy animal habitats as cities encroach upon the countryside. Living space for all wild species is destroyed as river valleys are transformed into reservoirs for the generation of hydropower and as forests are removed for construction materials and for paper. Toxic wastes from urban areas work their way into the food chain, annually killing thousands of animals in the United States alone. Rural lands are sprayed with herbicides and pesticides that also kill birdlife and small mammals.

In the first article in the subsection on *Animals*, "Killed by Kindness," journalist Sharon Begley develops a hypothesis on the strength of information from noted wildlife activist and author George Schaller. Begley's (and by extension, Schaller's) hypothesis is that, in order to gain access to areas in which they wish to survey for remaining animals of a species, wildlife groups and zoos are often forced to "buy their way in" through cash donations to local government officials or with the donations of equipment, vehicles, supplies, etc. Such donations support the corrupt governments that allow poaching and other illegal taking of wildlife to begin with.

In "New Species Fever," *Audubon* writer Patrick Huyghe describes the rush among biologists to catalog new (previously unknown) species of animals before they cease to exist as the result of habitat destruction. Although some dramatic discoveries have been made, most new species are small creatures like insects. Calls for their preservation have not engendered the same public sympathy as the battle to save endangered whales or eagles, despite the fact that these small creatures are the foundation of intricate webs that support all life.

A different approach to studying the genetic diversity of animals is taken in the article "Barnyard Biodiversity." Verlyn Klinkenborg writes in *Audubon* about the rich genetic and cultural heritage of agricultural breeds of animals that, like wild species, are vanishing. According to Klinkenborg, the Food and Agriculture Organization of the UN estimates that nearly one-third of the existing breeds of domesticated animals (including fowl) are on the verge of extinction.

The question of plant and animal extinction is much more than one of losing plants and animals that may make our lives richer in less than tangible ways by their very presence. Rather, the question is broader, relating to our own survival as a species. The stability of an ecological system depends upon its diversity. As diversity declines, so does stability. Simplified ecosystems are much more fragile than complex ones. We do not fully understand the role played by plants and animals in all the world's natural systems, but we do know that it must be an important one.

Looking Ahead: Challenge Questions

What is meant by the term "biodiversity," and how does it relate to the welfare of humans?

How can the use of tropical forests be managed so as to maintain the ecological diversity and health of the forest regions and still maintain reasonably high levels of economic productivity?

Why do estimates on the extent of forest loss vary so widely? Are there ways to reconcile conflicting data on deforestation with consistent government policies toward the management of tropical forest regions?

Why do people in developing countries see environmental protection as an obstacle to improving their economies?

What has been the traditional role of women in the management and use of forest resources?

What is the relationship between the desire of conservationist groups to gain influence over wildlife conservation practices and an increase in the very activities (such as poaching) that work against sound conservation and wildlife management?

Why is the discovery of new species of plants and animals of such critical importance to modern biologists?

Why should we be as concerned about the possible extinction of breeds of domestic animals as we are about the disappearance of wild species?

The Origin and Function of Biodiversity

Otto T. Solbrig

OTTO T. SOLBRIG is the Bussey Professor of Biology at Harvard University and a consultant on biodiversity with the Man and the Biosphere Program of the United Nations Educational, Scientific and Cultural Organization.

There is an enormous variety of plants and animals on Earth, and the high degree of species richness in tropical forests and coral reefs is a marvel. But even in environments that are relatively species poor, such as the ocean bottom, hundreds of kinds of microorganisms, plants, and animals flourish. This tremendous prodigality of nature has astonished and delighted humans and is ultimately the source of their sustenance. Yet, in using plants and animals for food and clothing, for building houses and for medicine, humans are endangering this immense richness of species. The current rate of loss of species diversity and the reduction in the genetic variety of crops and wild species could seriously affect human welfare.

People alter the biodiversity of the Earth in both direct and indirect ways. The use of renewable resources, for instance, often directly decreases species diversity. This is especially true in extractive industries such as forestry or fisheries, which tend to overexploit useful species and destroy unwanted ones. Agriculture and animal husbandry also destroy or modify the native biota. In addition, people indirectly change biodiversity by burning fossil fuels and biomass for energy, by altering hydrological patterns, by intentionally or accidentally introducing exotic species that reduce interregional biodiversity; and by destroying hedges, forest fringes, and fallow lands that provide habitat. A new and powerful human threat to species diversity is the release of toxic chemicals, such as lead, mercury, fluorocarbons, and chlorinated pesticides, into the atmosphere, soil, rivers, lakes, and oceans.

The task of describing, cataloguing, and explaining the diversity of living organisms belongs to biologists. But because the number of organisms on the planet is estimated to be 3 to 8 million or possibly higher,[1] and fewer than 1 million of these have been described and catalogued to date, this job is far from complete. The task is so formida-

CENTER FOR MARINE CONSERVATION—MICHAEL WEBER

ble, in fact, that it may never be completed. Nevertheless, it is important to seek patterns in the distribution and abundance of species and to test hypotheses about the function of species diversity in ecosystems.

Today, scientists have only a very rudimentary knowledge of biodiversity. In many temperate and arctic areas, where there are comparatively few species, acceptable catalogues of the vascular plants and vertebrate animals do exist. There are also reasonable estimates of invertebrate animals and nonvascular plants, including fungi. Less known in these areas are soil organisms, bacteria, and viruses. On the other hand, in tropical regions, where there is a high degree of species diversity, the story is very different. There are no detailed lists of the plants and animals of many tropical countries. At best, there are very rough estimates of the sorts of animals, plants, and fungi that live in tropical regions. Most tropical insects, soil organisms, bacteria, and fungi have yet to be collected and described.

Marine species, both temperate and tropical, are probably the least known. The oceans have the highest diversity of animal and plant phyla, yet until recently, the deep sea was believed to be devoid of life. Today, scientists know the ocean floor has an abundant biota with more than 800 known species in more than 100 families and a dozen phyla. Ocean-bottom hydrothermal

From Environment, Vol. 33, No. 5, June 1991, pp. 16-20, 34-38. Reprinted with permission of the Helen Dwight Reid Educational Foundation. Published by Heldref Publications, 1319 Eighteenth St., N.W., Washington, DC 20036-1802.

vents, such as the sulfide chimneys called "black smokers," have been found to contain at least 16 families of invertebrates that were unknown just five years ago.[2] Recently, an entirely new set of unicellular organisms called picoplankton, with cell diameters between 1 and 4 microns, was discovered. The productivity of marine systems may have been underestimated by 50 percent because scientists were ignorant of the role played by picoplankton and had no appropriate methods of measuring them.

Another serious problem is that scientists do not understand exactly how the diversity of genes, genotypes, species, and communities influences ecosystem function. Over the past 100 years, geneticists, taxonomists, evolutionists, and ecologists have accumulated much knowledge about diversity. The information gathered attests to the importance of diversity for the proper functioning of many organisms and ecosystems. However, a comprehensive, rigorous, and general theory of biodiversity is lacking. Because the threat to biological diversity is now great, scientists must learn how living systems are influenced by changes in diversity. Given the rapid pace of landscape transformation worldwide, there is some urgency in obtaining this information. Knowledge of biodiversity is also very important for evaluating the impact of global climatic change.

Fossil records show that drastic environmental change has been a major cause of species extinction.[3] Species losses also derive from mutual interaction, such as competition and predation. Also, because the environment is in a constant state of transformation, some species are always being lost. Some changes in the physical environment are cyclical and repeated, while others are less predictable. (However, even cyclical changes are subject to chance.) In any case, it seems reasonable that genetic diversity provides organisms and ecosystems with the capacity to recuperate after change has occurred. The scientific evidence for this hypothesis is not conclusive, however.[4]

Equilibrium in Nature

An ideal state in which every element

Natural selection probably allowed longer-necked individuals of this African ruminant to survive and reproduce when environmental changes killed others.

is in equilibrium, or "the balance of nature," cannot exist. The weather changes constantly; the diversity of plants and animals fluctuates; mountains erode; and lakes get silted in. Yet, this idea of a balance, of an equilibrium in nature, has persisted. Conservation managers, for example, endeavor to reduce disorder and create undisturbed environments. Those scientists who support the notion of a balance of nature maintain that ecosystems, although not in balance now, are moving constantly toward equilibrium. These scientists argue that ecosystems are prevented from attaining balance because of exterior forces, which are called disturbances. Storms, floods, pests, outbreaks, fires, and human-induced changes are all examples of disturbances that are presumed to keep ecosystems from reaching equilibrium.

The concept of the balance of nature gained credence in the early 18th century when Isaac Newton introduced very successfully the notion that nature could be explained in terms of a few simple laws. According to the Newtonian view, the world is formed by elements that are simple and that respond to regular and deterministic dynamics.[5] Today, physicists believe that nature is complex, not simple, and that the Newtonian view is insufficient as a general explanation of how the universe func-

tions. The universe, physicists are finding, is not in balance. Disturbances and irregularities of all sorts are no longer seen as aberrations but as integral parts of nature. Physicists are also discovering that at the very origin of the cosmos, at the so-called big bang, a singular, irreversible, and complex universe was produced. Similarly, ecologists have been observing and documenting that most, if not all, ecosystems are not in balance.[6]

But why should scientists be concerned with whether an ecosystem is in equilibrium? The answer to this question is crucial for understanding biodiversity. Systems not in equilibrium behave very differently than do systems in equilibrium. Their behavior can even appear strange and mysterious. For example, when the source of a disturbance is removed from a system that is near equilibrium, that system is expected to return to its previous state. When, in an undisturbed forest, a gap is created by a landslide or a falling tree, new seedlings grow in the gap and restore the equilibrium. After a few years, it is almost impossible to tell that there once was a gap. But when the source of change is removed from systems that are not in equilibrium, they do not return to equilibrium. Instead, they adopt a new state. For example, when land used for agriculture in the Amazon forest is abandoned, it grows into a grassland or a savanna and does not return to the original forest. Natural disturbances are a necessary part of the forest ecosystem's function. For example, without gaps creating disturbances, most forests cannot renew themselves. And fires, storms, and hurricanes are and have been a part of life since the beginning of the planet; without them, ecosystems could not function properly.

Knowing about the features of nonequilibrium systems is important for proper ecosystem management. Because a nonequilibrium ecosystem does not necessarily revert to its previous state when the convulsion is removed, it is often impossible for humans to return ecosystems to their original condition. Therefore, one set of management guidelines will not suffice for all situations. The previous history of changes and the present disturbance regime will determine the consequences of any management.

Origin of Diversity

There are two conditions that cause population diversity. First, new genotypes are constantly cropping up in a population through mutation, recombination, and related genetic phenomena and through immigration of individuals, their gametes, or their propagules. Second, diversity in the population is eliminated by natural selection and lost through emigration of individuals. Every genetic variation, from gene mutations to entire species, will disappear eventually. This loss can be a very fast process, or the variants can survive for a long time. Species that have survived for extended periods include horseshoe crabs, which have been around for 200 million years, and cockroaches, which originated even earlier, in the Carboniferous period. The speed at which new variations originate in relation to the rate at which they are eliminated determines the actual diversity of the system.

In the process of gene mutation, all heritable diversity ultimately arises at the molecular level. Gene mutations are chemical changes that take place in the composition of the DNA (deoxyribonucleic acid) molecule, the chemical substance responsible for heredity.

DNA is remarkable in that it shows complex chemical behavior not expected from systems in equilibrium. The most remarkable characteristic of DNA is its ability to regenerate itself, its "autocatalytic" behavior.

The DNA molecule consists of four kinds of repeating units called nucleotides, which can be arranged in any sequential order. DNA has the curious ability to maintain its physicochemical integrity regardless of the order of the four nucleotides. That is, any DNA molecule behaves like any other of the same length, regardless of the nucleotide content. This characteristic of DNA makes life feasible because if only one arrangement were possible, or if the chemical stability of one arrangement were significantly different from that of others, then all DNA molecules would be alike.

The order of the nucleotides uniquely determines the characteristics of the chemical products made by DNA. But the order of the nucleotides in a DNA molecule can change, and these changes are called mutations. The regular appearance of mutations gives life its diversity.[7] Mutated molecules of DNA reproduce their changes and make modified enzymes, which, in turn, make altered cells. Mutated cells result in modified organisms.

Thus, new mutations permit the evolution of new characteristics by changing the structure and/or function of enzymes and other proteins. The characteristics of the enzymes and proteins are determined by those of the DNA molecule, but not the reverse—the "information" flows only in one direction. This flow contrasts with that of an ordinary chemical reaction, in which

CONSERVATION INTERNATIONAL—FULVIO ECCARDI

The continuing destruction of the dense forests between Arizona and Argentina could threaten the survival of the ocelot and, thus, the equilibrium of the entire subtropical ecosystem.

OTTO T. SOLBRIG

The Serengeti plain in Kenya and Tanzania, home to a wide variety of vertebrates, is threatened by rapidly encroaching agricultural activity and growing human populations.

there are forward and backward reactions.

The appearance of mutations is a random process. There is no way to anticipate or predict what mutation is going to occur, only that some will. It is natural selection, however, that determines whether new mutations get established or are eliminated. The processes of mutation and selection can be looked at in two ways. Mutation and selection might be random, independent processes that are not influenced by the characteristics of the system, in which case no active process of diversity maintenance exists. On the other hand, a certain degree of variety may be required for living systems to function properly. If so, there may be system feedbacks that modify the rates of mutation and selection. Below a certain threshold, a system may collapse because of insufficient variety. According to this last view, diversity is actively maintained in a system.

Speciation

Mutation and natural selection generate genetic and morphological variation within a lineage.[8] Yet what is most distinctive in a community is the coexistence of different species, each acting independently and not sharing its DNA or mutations. How, then, does the great variety of species originate?

Speciation is the process that separates genetic variations into distinct units, or species. In speciation, the original population of organisms with similar genes, called a gene pool, is divided into two or more gene pools. Each of these new gene pools then acquires a unique set of characteristics (cellular, tissue, organ, and organismic) through mutation and selection. (The mechanisms that create species have been a focal point of research in the field of organic evolution.[9]) Speciation follows most commonly from the physical division of a gene pool. The separation inhibits interbreeding between individuals in the two populations. In time, one or both populations change enough to prohibit interbreeding.

Because the fate of every species is to become extinct eventually, there must be an influx of new species for life to continue. If, presently, there are about 10 million species, and, on average, each species has a life span of 1 million years, then an average of 10 new species must originate each year. On the other hand, if the life span of a species is only 100,000 years (as many biologists argue), then 100 new species must originate each year. Scientists are often aware of a species' extinction, but the appearance of new species is difficult to observe. Because most species are insects and other small invertebrates, most new species fall into these two poorly studied groups. Thus, there are few well-documented cases of speciation.

Evolution and Resource Limitation

Certain individuals in a population survive and reproduce, while others die without leaving offspring.[10] To understand this phenomenon, scientists must first realize that there usually is not enough of some environmental resources (food, water, minerals) to go around. The availability of resources in the environment limits the rate at which individual organisms can reproduce and their ability to survive. Nucleic acid replication and cell, tissue, and organism growth and reproduction all require energy and materials, of which there are only limited amounts. Individuals that are efficient at harvesting resources have a higher probability of reproducing and surviving than do those that are inefficient. Mutations in DNA provide species with a constant input of physical and chemical variations, some of which improve the efficiency with which the species selects and harvests resources. New mutations may increase the efficiency of protein synthesis at the molecular level, or they may improve the water uptake by a plant at the organism level. Mutations may even increase the efficiency with which a mutualistic association between a plant and its specific pollinator develops at the community level.

DNA molecules that are dissimilar in the order of their nucleotides usually do not have characteristics that directly increase or decrease their own survival or reproductive capacity. Given the appropriate environment and resources, all DNA molecules reproduce at the same rate despite their nucleotide composition. DNA molecules manifest their distinctive survival rates through their effects on the living organisms in which they are embedded. These organisms, be they single cells living in pure culture or multicellular organisms living in complex communities, will inevitably differ in their ability to garner resources. Therefore, they will vary in their survival and reproductive capacities and, consequently, so will the DNA in their cells.

Natural selection is an optimizing process. It is a process that favors efficiency and produces adaptation. It is not a process that involves chance, like the appearance of mutations, but it must operate with the elements that it has. Thus, natural selection never makes a "perfect" organism. Because the environment is constantly changing, a perfect organism, if it could be produced, would soon cease to be perfect as the environment changed. Such systems, in which the optimum state is favored but cannot be attained or maintained, are called frustrated systems.

Diversity is the result of two opposite actions: the processes that produce new genotypes, new varieties, and new species and the processes that eliminate mutations, variants, and species from the system. Natural selection is primarily responsible for the reduction of biodiversity; it acts through differential reproduction and differential mortality. In other words, the probability of reproducing and the probability of dying are not the same for all organisms. Rather, reproduction and death are related to the organism's characteristics as well as to the environment in which that organism lives.

The Ecological Role of Biodiversity

How is species diversity linked to ecosystem structure and function, to the different ways in which organisms and populations interact with each other, and to those community properties that emerge from these interactions? Communities have structures and properties not possessed by the populations within them that are called emergent properties, which include trophic structure, stability, guild structure, and successional stages.

Many scientists believe that species diversity is essential for the proper functioning of communities and for the emergence of community-level properties. Just as many different DNA-encoded enzymes are needed for a complex organism to function properly, so,

scientists believe, are many kinds of species necessary to maintain community structure. But is any diversity sufficient, or are specific mixes of species necessary for the communities and ecosystems to function? This is a very old question in ecology, and two opposing views exist. One view is that a community is formed by the species that happened to arrive first—that the mix of species in a community is a matter of chance: "The vegetation of an area is merely the resultant of two factors, the fluctuating and fortuitous immigration of plants, and an equally fluctuating and variable environment."[11] According to the opposing view, "In any fairly limited area, only a fraction of the forms that could theoretically do so actually form a community at any one time. . . . The community really is an organized community in that it has a 'limited membership.'"[12]

Are these two very different views really irreconcilable? Not necessarily. If a community is defined as the sum of all the plants and animals that grow together in an area, certain patterns can

Oceans have the highest diversity of flora and fauna, and marine species, such as the tube sponge and flower coral, are among the least understood.

be observed: Areas with more than 1,000 millimeters of evenly distributed rainfall always contain a woodland; conifers prevail in areas of extremely low winter temperatures; trees in warm climates have broad leaves; and succulent plants are found in dry climates. The same is true for animals. The mix of species encountered in a community is not a random sample of all plants and animals in the world. Yet, under close scrutiny, differences that are difficult to explain can be found between similar localities. Ecologists cannot predict what species will occur in a given climate and soil type. Also, there have been cases where dominant species, such as the American chestnut, have been lost, with apparently little effect in the overall working of the biological community.

The community behaves in a "frustrated" fashion. Two types of processes are at work and neither is dominant. First, there is natural selection. Species in the community are constantly evolving to increase their ability to withstand the rigors of the environment. This process accounts for cold-hardy species near the poles, for grass-eating ruminants in savannas, and for fruit-eating bats in forests. The individuals that escape their predators and survive other environmental challenges pass on their characteristics to their offspring.

Second, there is chance. When the climate changes, characteristics that were once advantageous can become a burden. Diseases might be introduced, such as Dutch Elm disease, for which local species have no defense. Hurricanes, tornadoes, floods, and droughts are unpredictable and their frequency changes over time. Chance also plays a role in dispersal. For instance, a species might be absent from a community purely by chance. Thus, both natural selection and chance result in the steady coming and going of species through immigration, emigration, extinction, and gene mutation.

A large number of additional unanswered questions exist regarding the role of biodiversity in communities. For example, can more and more species be packed into a community, or are there upper and lower limits? A related question concerns optimal levels of diversity and the factors that control them.[13] Other issues include the role of different individuals within a population, different populations within a species, and so on.

Do the members of the community collaborate in the efficient use of energy and resources? In a variable environment, the existence of individuals with different characteristics may increase the ways in which the population can respond to change.[14] For example, if all plants of a species had similar water requirements, all of them would suffer water stress in any year that was drier than normal, and such periods would result in significantly reduced seed production. But if there is genetic variation, some individuals might perform above average each year within certain limits of environmental variation. Therefore, seed production would be satisfactory in both wet and dry years. Moreover, genetically adaptable organisms should survive in more variable environments than do genetically uniform populations.[15] Experience with crops shows that highly productive but genetically uniform varieties have more restricted environmental requirements than do less productive but more variable varieties. Also, plantations formed by uniform varieties are more susceptible to pest and disease outbreaks.

So, variability within a species seems to be important for long-term survival. Does the same hold for a community, which is formed of species that live together but do not have a common gene pool? Communities with high species diversity may also cope with long-term environmental fluctuations better than do communities with few species. However, the evidence is contradictory. Terrestrial communities in the climatically variable midlatitudes are less diverse than tropical communities in more uniform environments. Also, deep benthic communities are among the most variable communities anywhere though they exist in possibly the most equitable environment on the planet.[16]

Diversity and Niche Structure

According to modern ecological theory, every species in a community occupies a singular ecological niche, or a

Unfortunately, there are no detailed lists of the native plants and animals of many tropical countries.

role in the environment. A niche is thought of as a multidimensional space where each dimension is a characteristic of a species. Because every species has at least one physical or behavioral characteristic that separates it from other species, every species has a unique niche.

In principle, niche theory supplies the theoretical framework to explain the number and types of species in a community. Niche theory predicts that communities that vary in total resources or in the quality of those resources will contain more niches and, therefore, more species. Nevertheless, resources are sometimes partitioned differently in communities with very similar resources. In other words, communities may have large or small niches, which result in a few "generalist" species (those with broad niches) or many "specialist" species (those with small niches). Finally, communities with similar resources and niche sizes may still vary in the number of species because of different degrees of niche overlap. Niche overlap is the degree to which two species exploit the same resource.

Communities differ in niche size and overlap for a variety of reasons, such as differences in climatic stability and predictability, spatial heterogeneity, primary productivity, competition, predation, and degree of disturbance. Thus, niche theory is unable to predict the patterns of species diversity.

Using niche theory to understand why communities vary in number of species has proven very difficult in practice. Many studies have shown very elegantly how species partition various resources in space and in time,[17] but it has proven very hard to show the precise contribution of each factor. A key question is whether two species can have the same niche. In other words, is there species redundancy in a community? Present theory predicts that two species with identical niches cannot coexist, and empirical studies seem to verify this.[18] Still, the question of how similar two coexisting species can be remains to be answered.

Trophic Diversity

Additional unanswered questions concern the role of diversity in food webs. All organisms get their energy from the sun, either directly or indirectly. At the simplest level, cells of the green tissues of plants and some microorganisms capture the sun's energy. Organisms also need specific chemicals from the environment to develop. For instance, all plants obtain the materials for development from soil or water and the air. With these materials and energy from the sun, plants create energy-rich, complex chemical compounds, such as carbohydrates, cellulose, starch, proteins, fats, and wood. Animals ingest and extract the energy from these compounds and incorporate some of them into their bodies. Decomposers reduce them to simple ions that eventually are returned to the soil or water. In a cyclic fashion, then, energy and chemical substances move from the environment to plants, to animals, to decomposing microorganisms, and back to the environment. This trophic diversity, or "food web," is essential for the proper operation of ecosystems. Many studies have shown that the trophic structure of ecosystems is complex and varies from one ecosystem to another. Understanding how food webs function is essential for understanding the place of humans in the ecosystem,[19] because humans also derive their energy and materials from plants and animals. Moreover, interpretation of food webs is necessary to understand the effect of toxic compounds released into the environment and the effect of the introduction of species into new areas.

Are diverse ecosystems with complex food webs more stable than simple ecosystems? Stability, in this instance, refers to the ability of an ecosystem to withstand disturbances, including those produced by people. Scientists used to believe that simple systems are less stable than complex ones. Clearly, there is a threshold of diversity below which most ecosystems cannot function. For example, no ecosystem can function without some plants and decomposers. Without plants, solar energy cannot be captured, and without decomposers, substances are not returned to the environment. But, is there a threshold below which present complex ecosystems lose their stability? How many different sorts of plants and animals are needed? And, if there are many of each sort, does the ecosystem become more stable?

Productivity, or the total quantity of energy captured and biomass produced by an ecosystem in a year, is related to the issue of stability. Are ecosystems that are formed by many species more productive than those with few species? It would appear that simple systems, both natural (*Spartina* marshes and high-latitude marine systems) and artificial (agricultural systems), are more productive than diverse ones. However, some scientists fear that these systems are not stable and have argued that society's increasing reliance on the productivity of these simple systems could eventually put humans in jeopardy.[20]

Humans and Species Diversity

Apart from the ethical and aesthetic reasons, is there any reason to fear that human survival is at risk if biodiversity is not preserved? Can humans exist surrounded only by agricultural fields, planted forests, and the like? This question is not easy to answer. In the short term, natural ecosystems can probably lose species without any great impairment of function. For instance, many planted forests are much simpler than the natural forests they replaced. As long as the environment does not change very much, ecosystems can apparently lose many of their rare species without any visible effects. In some cases, common animal species, such as passenger pigeons, have disappeared or been drastically reduced without endangering human survival.

Yet, environments inevitably change, and sometimes drastically. It is then that the role of obscure species sometimes becomes very important. When dinosaurs became extinct, an obscure group, primitive mammals, suddenly evolved into the dominant role they still play. And now, humans seem on the verge of a drastic change, one of their own making—the result of global warming induced by increased carbon dioxide in the atmosphere.

Predicted climatic changes pose a series of specific questions relating to biodiversity for which there are presently no answers. What is the impact of elevated atmospheric concentrations of carbon dioxide on tree physiology and forest composition? How does elevated carbon dioxide affect soil organisms and soil nutrients? How is climatic change going to affect insect populations, fire frequency, and wind storms? Coastal zones, in particular, are likely to reflect phenomena occurring over a larger geographical area. Already, the North Sea coast is showing the effects of the industrial and agricultural activity of central Europe, which is hundreds of miles away. The effects of industrialization may be catastrophic for many species, which may become extinct. But a few species, and there is no way of predicting which ones, may thrive in the new environment. Therein lies the scientific importance of biodiversity: It increases the likelihood that at least some species will survive and give rise to new lineages that will replenish the Earth's biodiversity.

Possible Research on Biodiversity

A global survey of diversity is essential to understanding the distribution of life forms on the planet.[21] Biogeographical studies have shown that ecosystem diversity varies greatly from one region to another. Tropical forests, coral reefs, and deep benthic communities are very rich in species; deserts and high-latitude areas are species poor. Many studies have shown that species abundance changes naturally along latitudinal and humidity gradients.

An unprecedented transformation of the natural landscapes of the world is now occurring. Yet, different areas are being transformed at different rates. Some regions, such as the Mediterranean basin, have been altered by people for a long time. Other regions have been influenced only moderately by humans. Also, studies show that the response to disturbance varies greatly from one ecosystem to another. Therefore, it is important to determine how different biomes respond to human disturbance and to document the resulting change in biodiversity and ecosystem function.

Classical biogeography concentrated on the study of species ranges and the causes of their distribution. Given scientists' uncertainty about the number of species and the large number of undescribed species, new and more efficient approaches must be developed to complement past studies. These new approaches should take advantage of remote sensing technology and produce a geographic information system with statistical and analytical capabilities. However, because satellites cannot detect species diversity, new measures that are indicative of diversity and are more easily obtained by remote sensing must be used. For example, satellites can detect standing biomass and productivity, and research should help to establish the relationship between biodiversity and these measures.

Other problems to be studied include the effect that habitat fragmentation has on species ranges and on the probability of extinction and speciation; whether fragmented habitats are more prone to invasion; how invasions influence biodiversity; and the effect of increased geographical isolation on populations and species. Also, research should be conducted to develop a methodology for the systematic comparison of biodiversity across regions. To this end, a biogeographical information system for assessing diversity might use satellite information to identify assemblages of species at different geographical scales for long-term monitoring.

Until now, research in biodiversity has been undertaken by taxonomists at such institutions as the Smithsonian Institution, the Harvard University Herbaria and Musea, the Missouri Botanical Garden, the Royal Botanical Gardens in Kew, the Musée d'Histoire Naturelle in Paris, and the Komarov Institute in Leningrad, which are located for the most part in the north temperate zone, while biodiversity is found mostly in tropical countries. Taxonomic institutions in both the north and the south have been in relative decline over the last 50 years. Staffs have not increased or have increased only slightly, and resources allocated to taxonomy have not grown in the same proportion as those devoted to other areas of biology. Consequently, at a time when the conceptual and practical importance of biodiversity is becoming clear, there is a shortage of trained personnel and a lack of necessary funds.

Taxonomists alone cannot answer the questions that have been raised by the massive transformation of the Earth. A collaborative effort by many kinds of biologists, including geneticists, physiologists, and ecologists, is required. As new techniques of genome analysis become routine, even molecular biologists may contribute to identifying genetic variation and its biological role.

Both governmental and nongovernmental institutions should collaborate to advance scientific understanding of

To survive near the poles, modern cold-hardy species have undergone eons of genetic mutation and natural selection.

the significance of biodiversity, and many more resources will have to be made available to fund the collaborative research. The International Union of Biological Sciences, the Scientific Committee on Problems of the Environment, and the Man and the Biosphere Program of the United Nations Educational, Scientific and Cultural Organization are spearheading research in biodiversity. These organizations are in close contact with the International Geosphere-Biosphere Program and the United Nations Environmental Programme, to minimize overlap and to coordinate activities. Many national institutions also have active programs of research in biodiversity, but most of these efforts suffer from inadequate funding.

Biodiversity research and conservation figure prominently in the agenda for the 1992 United Nations Conference on Environment and Development in Rio de Janeiro, Brazil, where conference planners are seeking a sub-

stantial increase in financial support for research in biodiversity.

It is imperative that biologists and the public learn more about the importance of biodiversity and its role in ecosystem function. Only an international collaborative effort that is supported by adequate resources can accomplish this task. Increased public awareness of the serious implications of humanity's depletion of biodiversity before its importance is understood is helping to create a climate that may stimulate governments to support national and international efforts in this field.

NOTES

1.   J. J. Sepkoski, Jr., "A Kinetic Model of Phanerozomic Taxonomic Diversity 1: Analysis of Marine Orders," *Paleobiology* 4 (1978):223–51; R. M. May, *Exploitation of Marine Communities* (Berlin: Springer, 1984); and A. Hoffman, *Arguments on Evolution* (Oxford, England: Oxford University Press, 1989).

2.   J. F. Grassle, "Species Diversity in Deep-Sea Communities," *Trends in Ecology and Evolution* 4 (1989):12–15.

3.   S. M. Stanley, *Macroevolution: Pattern and Process* (San Francisco, Calif.: Freeman Press, 1979); S. M. Stanley, "Rates of Evolution," *Paleobiology* 11 (1985): 13–26; and P. W. Signor, "The Geological History of Diversity," *Annual Review of Ecology and Systematics* 21 (1990):509–39.

4.   The International Union of Biological Sciences, the Scientific Committee on Problems of the Environment (SCOPE), and the Man and the Biosphere Program of the United Nations Educational, Scientific and Cultural Organization (UNESCO) recently joined forces to develop a scientific program to study the role of biodiversity in the function of ecosystems. One of the goals of this program is to develop scientific hypotheses regarding biodiversity that will be discussed at three upcoming meetings: a workshop at Harvard University at the end of June, an international symposium sponsored by SCOPE to be held in Bayreuth, Germany, in October, and, later in October, an international symposium on biodiversity sponsored by UNESCO and the United Nations Environment Programme that will take place in Nalchik, USSR. An international program of research is expected to emerge from these meetings.

5.   G. Nicolis and I. Prigogine, *Exploring Complexity* (New York: Freeman Press, 1989); and O. T. Solbrig and G. Nicolis, *Perspectives on Biological Complexity* (Paris: International Union of Biological Sciences, 1991).

6.   D. L. DeAngelis and J. C. Waterhouse, "Equilibrium and Non-Equilibrium Concepts in Ecological Models," *Ecological Monographs* 57 (1987):1–21.

7.   P. Schuster, "The Interface between Chemistry and Biology: Laws Determining Regularities in Early Evolution," in G. Pifat-Mrzljak, ed., *Supramolecular Structure and Function* (Berlin: Springer, 1986); P. Schuster, "Optimization Dynamics on Valuable Landscapes: Modelling Molecular Evolution," in O. T. Solbrig and G. Nicolis, pages 115–62, note 5 above; and E. Szathmary, "The Emergence, Maintenance, and Transitions of the Earliest Evolutionary Units," *Oxford Surveys in Evolutionary Biology* 6 (1989):169–205.

8.   G. L. Stebbins, *Flowering Plants: Evolution above the Species Level* (Cambridge, Mass.: Harvard University Press, 1974); H. G. Andrewartha and L. C. Birch, *The Distribution and Abundance of Animals* (Chicago: University of Chicago Press, 1954); V. Grant, *Plant Speciation* (New York: Columbia University Press, 1981); E. Mayr, *Animal Species and Evolution* (Cambridge, Mass.: Harvard University Press, 1963); and S. Wright, *Evolution and the Genetic Populations*, 4 vols. (Chicago: University of Chicago Press, 1968–1978).

9.   G. L. Stebbins, *Variation and Evolution in Plants* (New York: Columbia University Press, 1950); and G. G. Simpson, *The Major Features of Evolution* (New York: Columbia University Press, 1953).

10.   O. T. Solbrig, "Energy, Information, and Plant Evolution," in C. R. Townsend and P. Calow, eds., *Physiological Ecology: An Evolutionary Approach to Resource Use* (Oxford, England: Blackwell Scientific Publications, 1981), 274–99; N. Eldredge, "Information, Economics, and Evolution," *Annual Review of Ecology and Systematics* 17 (1986):351–69; and J. H. Brown and B. A. Maurer, "Macroecology: The Division of Food and Space Among Species on Continents," *Science* 243 (1989):1145–50.

11.   H. Gleason, "The Individualistic Concept of the Plant Association," *Bulletin of the Torrey Botanical Club* 53 (1926):120.

12.   C. Elton, *The Ecology of Animals* (London: Methuen, 1933).

13.   J. Roughgarden, "Evolution of Niche Width," *American Naturalist* 106 (1972):638–718; and J. Roughgarden, "The Structure and Assembly of Communities," in J. Roughgarden, R. May, and S. Levin, eds., *Perspectives in Ecological Theory* (Princeton, N.J.: Princeton University Press, 1989).

14.   A. J. Cain and P. M. Sheppard, "Natural Selection in Cepaea," *Genetics* 39 (1954):89–116.

15.   I. M. Lerner, *Genetic Homeostasis* (Edinburgh: Oliver & Boyd, 1954); and I. M. Lerner, "The Concept of Natural Selection: A Centennial View," *Proceedings of the American Philosophical Society* 103 (1959): 173–82.

16.   J. F. Grassle, note 2 above.

17.   M. L. Cody, *Competition and Structure of Bird Communities* (Princeton, N.J.: Princeton University Press, 1974).

18.   R. K. Colwell and E. R. Fuentes, "Experimental Studies of the Niche," *Annual Review of Ecology and Systematics* 6 (1975):281–310.

19.   J. E. Cohen, "Food Webs and Community Structure," in Roughgarden, May, and Levin, pages 181–202, note 13 above.

20.   J. H. Brown, "On the Relationship Between Abundance and Distribution of Species," *American Naturalist* 124 (1984):255–79.

21.   F. DiCastri and T. Younès, "Ecosystem Function of Biological Diversity," *Biology International*, special issue no. 22 (1990):1–20.

Rain Forest Entrepreneurs

Cashing in on Conservation

Thomas A. Carr, Heather, L. Pedersen, and Sunder Ramaswamy

Thomas A. Carr is an assistant professor in the Economics Department at Middlebury College in Vermont. Heather L. Pedersen is a mathematics teacher at the Colorado Springs School in Colorado. Sunder Ramaswamy is an assistant professor in the Economics Department at Middlebury College.

Each year, nearly 17 million hectares of rain forest—an area roughly equal to that of Wisconsin—are lost world-wide as a result of deforestation.[1] Because more than half of all species on the planet are found in rain forests, this destruction portends serious environmental consequences, including the decimation of biological diversity.[2] Another threat lies in the fact that rain forests serve as an important sink for carbon dioxide, a greenhouse gas that contributes to global warming. The Amazon region alone stores at least 75 billion tons of carbon in its trees.[3] Furthermore, when stripped of its trees, rain forest land soon becomes inhospitable and nonarable because the soil is nutrient-poor and ill-suited to agriculture. Under current practices, therefore, the forests are being destroyed permanently.

Economic forces result in exploitation of the rain forest to extract hardwood timber and fuel and in clearcutting the land for agriculture and cattle ranching, which are primary causes of the devastation. Mounting evidence shows that these conventional commercial and industrial uses of the rain forest (see Table 1) are not only ecologically devastating but also economically unsound.[4] These findings have inspired an innovative approach to save the rain forest. Environmental groups are now targeting their efforts toward developing commercially viable and sustainable uses of the rain forest. Their strategy is to create economic incentives that encourage local inhabitants to practice efficient stewardship over the standing forests. These environmental entrepreneurs no longer view the market as their nemesis but as an instrument to bring about constructive social and environmental change. In theory, the strategy promotes win-win solutions: Environmentalists gain by preserving the rain forests, and local inhabitants gain from an improved standard of living that is generated by enlightened, sustainable development. In practice, the challenge lies in implementing such programs.

Three applications of environmental entrepreneurship in the rain forests have been particularly successful. Conservation International's "The Tagua Initiative," Shaman Pharmaceutical's search for useful drugs in the rain forest, and the management of ecotourism in Costa Rica are three projects that together provide an interesting cross section of the efforts under way to promote sustainable use of rain forest products.[5] A number of common issues and challenges confront these environmental entrepreneurs.

Responding to Deforestation

Although people everywhere may benefit from preserving the rain forest,[6] the costs of preservation are borne mainly by the local inhabitants. Usually, the inhabitants' immediate financial needs far outweigh the long-term benefit gained by forgoing the traditional extractive methods of forestry or land conversion for agriculture. In many of these countries, high levels of poverty, rapid population growth, and unequal distribution of land encourage migration into the forest regions. Local inhabitants, confronted with the tasks of daily survival, cannot be expected to respond to appeals for altruistic self-sacrifice. Consequently, forests are cut and burned for short-term economic gains. This problem is often exacerbated by misguided government policies in many countries, such as government-sponsored timber concessions that promote inefficient harvest levels, tree selection, and reforestation levels. Governments may charge a royalty far below the true economic value of the standing forest. Such low royalties and special tax breaks raise the profits of logging companies, which thereby stimulate timber booms. In addition, some governments provide special land tenure rules or tax benefits to individuals who "improve" the land by clearing the forest. These rules encourage development in the rain forest region because they impel poor settlers to seek land for agriculture and wealthy landowners to look for new investments.[7]

Environmental entrepreneurs can create commercial alternatives to the traditional damaging uses of rain for-

From *Environment*, September 1993, pp. 12-15, 33-37. Reprinted with permission of the Helen Dwight Reid Educational Foundation. Published by Heldref Publications, 1319 Eighteenth St., NW, Washington, DC 20036-1802.

est resources, but several factors must first be taken into consideration. For example, commercial development cannot be allowed to harm the ecological integrity of the ecosystem. This can be a difficult challenge as the scale of production increases for many projects. Also, if existing firms are profitable, new firms will be attracted into the industry, thus placing additional pressure on the fragile ecosystem. Of course, the product must also pass the test of the market; consumers must be willing to pay a price that covers the full cost of production. Some environmentally conscious consumers may be willing to pay a premium for sustainably harvested rain forest products. The size of this "green premium" would depend upon these consumers' willingness and ability to pay, as well as on the prices of other products competing with the rain forest products. To maintain the green premium over

time, environmental entrepreneurs need to devise a strategy that differentiates their products from others through advertising and some type of institutionalized labeling system.[8] These entrepreneurs must also anticipate the effect of expanding output on market prices. Previous studies have examined the market value of sustainable products from a single hectare. One study in the Amazonian rain forest in Peru found that sustainably harvested products such as fruit, nuts, rubber latex, and selectively logged timber yield more net value than do plantation forestry and cattle ranching.[9] If harvests are expanded, however, market prices may be pushed down, and the profitability of the program reduced. Another consideration is that entrepreneurs may be able to avoid the expense of developing extensive distribution networks and other marketing costs by forming alliances with established commercial firms. These

firms typically have retail outlets and experienced business personnel that can assist the small entrepreneur.

Finally, the environmental entrepreneur must channel income back to the effective owners of the rain forests—the local indigenous people. This return raises the issue of rain forest property rights. The property rights over rain forest resources are not well defined or enforced. Rain forest land is often held collectively, and government-owned land marked as a reserve is not always protected. Even private landowners have a difficult time preventing landless squatters from using their property. Without the enforcement of property rights, rain forests become an open-access resource that is over exploited. This result is not inevitable, however. History suggests that, when the benefits of establishing new property rights exceed the costs, societies often devise new ways to define property rights and improve the allocation of resources.[10]

In addition to the question of physical property rights, there is the problem of defining intellectual property rights. Indigenous people possess a wealth of esoteric knowledge about local plants and animals and their usages. Conservation groups argue that the wisdom of the local inhabitants must be given an economic value or else that knowledge will disappear amidst the destruction of the forest.[11] At the same time, scientists and entrepreneurs also contribute value to rain forest products by discovering useful medicinal compounds in the plants. If these interests are not protected, there will not be sufficient economic incentive to develop new products. During the Earth Summit in Rio de Janeiro last summer, the Bush administration refused to sign an international treaty on biodiversity on the grounds that it would harm the interests of biotechnology firms.[12] (The Clinton administration signed the biodiversity treaty on 4 June 1993.) A key challenge is to develop an institutional mechanism that recognizes the value of both the natives' knowledge and the scientists' and entrepreneurs' contributions,

TABLE 1
COMMERCIAL AND INDUSTRIAL PRODUCTS DERIVED FROM TROPICAL RAIN FORESTS

Product	Value of imports by region (millions of U.S. dollars)	Marketshare of rain forest products (percent)	Region receiving imports	Year of estimate
Commercial Products				
Fruit and vegetable juices	4,000	100	World	1988
Cut flowers	2,500	100	World	1985
Food additives	750	100	United States, European Community	1991
Spices	439	small	United States	1987
Nuts	216	100	World	1988
Food colorings	140	10	World	1987
Vitamins	67	small	United States	1990
Fiber	54	100	United States	1983/4
Industrial Products				
Fuel	60,000	< 1	United States	1984
Pesticides	16,000	1	World	1987
Natural rubber	666	100	United States	1978
Tannins	170	large	United States	1980
Construction material	12	1	United States	1984
Natural waxes	9.3	100	United States	1985

Note: James Duke, an economic botanist at the U.S. Department of Agriculture, has been compiling estimates of the economic value of hundreds of key commercial and industrial rain forest products. Some of the important estimates are summarized here. Although not all of the imported products are derived from tropical rain forest countries, Duke claims that they all have the potential to be sustainably harvested from these regions.

SOURCE: James Duke, "Tropical Botanical Extractives" (Unpublished manuscript, U.S. Department of Agriculture, Washington, D.C., April 1989).

and therefore rewards both types of intellectual property rights in the development of rain forest products.

The Tagua Initiative

Conservation International is an environmental organization based in Washington, D.C., that works to conserve biodiversity by supporting local rain forest communities worldwide. Through a project entitled "The Tagua Initiative," Conservation International is attempting to synthesize "the approaches of business, community development, and applied science to promote conservation through the marketing of nontimber forest products."[13] The tagua nut is an ivory-like seed that is harvested from tropical palm trees to make buttons, jewelry, chess pieces, carvings, and other arts and crafts. Conservation International links button manufacturers in the United States and other countries with rural tagua harvesters in the endangered rain forests of Esmeraldas in Ecuador. The organization works independently with participating companies to design unique marketing strategies tailored to those companies' individual images, product offerings, and marketing campaigns.

In 1990, Conservation International began expanding the market for tagua products and developing a local industry around tagua. Today, tagua buttons are being used by 24 clothing companies, including such major manufacturers as Smith & Hawken, Esprit, J. Crew, and L.L. Bean. The current distribution network links the Ecuadorian tagua producers to the clothing companies through four wholesale button manufacturers. Conservation International collects a royalty based on a percentage of sales to wholesale button manufacturers and uses the proceeds to support local conservation and community development programs in the rain forest. It has also focused its efforts on developing a viable local tagua industry that includes harvesting and manufacturing. A primary objective of The Tagua Initiative is to provide the 1,200 local harvesters with an attrac-

tive price for tagua so that they have an economic incentive to protect the standing forest. Recent figures indicate that the price paid to tagua collectors has risen 92 percent since the program began (a 32 percent real price increase after adjusting for the estimated inflation rate).[14] To increase the flow of income to the native economy, Conservation International encourages the development of new tagua products that can be manufactured locally. Currently, the tagua production line has expanded to include eight manufacturers of jewelry, arts and crafts, and other items.[15]

The Tagua Initiative provides a tremendously successful example, at least in the initial stages of development. Since February 1990, 850 tons of tagua have been delivered directly to factories, and the program has generated approximately $2 million in button sales to manufacturers in North America, Europe, and Japan.[16] According to Robin Frank, tagua product manager at Conservation International, the organization is collaborating with about 50 companies worldwide, and many others have expressed interest. Moreover, The Tagua Initiative in Ecuador has become a role model for new projects in Colombia, Guatemala, Peru, the Philippines, and a number of other countries. In all of these cases, Conservation International is working with local organizations to identify and develop sustainable commercial products in a manner that protects sensitive ecosystems. These projects are expanding the rain forest product line to Brazil nuts and pecans from Peru, fibers for textiles, and waxes and oils for the personal health and hygiene market.[17]

In addition to creating marketable rain forest products, Conservation International cooperates with conservation and community development programs, such as the Corporacion de Investigaciones para el Desarrollo Socio/Ambiental (CIDESA) in Ecuador. Ecologists, economic botanists, and conservation planners affiliated with Conservation International help

CIDESA to identify critical rain forest sites and monitor harvesting practices to ensure their sustainability, among other things. The province of Esmeraldas in Ecuador is considered a critical "hot spot" because it contains some of the highest levels of biodiversity in Latin America and harbors some of Ecuador's last remaining pristine tracts of western Andean rain forest. Coincidentally, it is one of Ecuador's poorest communities, with a meager annual average percapita income of $600, about onehalf of the national average. The community of Comuna Rio Santiago in Esmeraldas has a population of 70,000, which grows dramatically at an annual rate of 3.7 percent. Four out of every 10 children suffer from malnutrition, and the infant mortality rate is 60 per 1,000 births. There is a high level of alcoholism, and drug addiction is a growing problem. Life expectancy is just 50 years, and the illiteracy level is near 50 percent. All of these actualities indicate an urgent need to protect the natural resources found in this region, not only to maintain biodiversity but also to ensure the economic welfare of the local inhabitants. If these needs are addressed, the program will have the potential to change the current low standard of living in Ecuador by promoting both conservation and economic development.[18]

Over the next 10 years, Conservation International plans to increase the use of numerous rain forest products, such as medicines, furniture, and baskets.[19] These efforts can serve as a role model for firms in the industrialized world that seek to create rain forest products and improve the wellbeing of rain forest inhabitants.

Shaman Pharmaceuticals

Shaman Pharmaceuticals, Inc., draws its name from rain forest *shamans*, traditional medicine men who possess a vast amount of knowledge about the use of plants for medicinal purposes. The shamans' ability to cure a variety of illnesses is founded on centuries of practice and an intimate association with, and dependence upon, indigenous plants. By tapping the knowledge of the shamans, scientists hope to reduce the research costs of identifying plants with beneficial medicinal properties. Furthermore, investigating plant species already known to possess healing characteristics yields a much higher chance of success in the screening process. This ethnobotanical approach—which combines the skills of anthropology and botany to study how native peoples utilize plants[20]—is the basic premise by which Shaman Pharmaceuticals functions. By innovatively combining the disciplines of ethnobotany, isolation chemistry, and pharmacology with a keen market-driven strategy, the company hopes to create a more efficient drug-discovery program (see box on next page).

Shaman has formed strategic alliances with the pharmaceutical industry to enhance its prospects of turning a pharmaceutical discovery into a financial gain. "Shaman feels it is in a strong position to strike such alliances because the company is not only formed around a handful of products, but also around an efficient, ongoing process for generating compounds with a greater likelihood of being active in humans."[21] The company has two main objectives in building these alliances: generating research funds through cooperative arrangements and gaining access to a larger marketing network. Three major pharmaceutical manufacturers have entered into agreements with Shaman: Inverni della Beffa, an Italian manufacturer of plant-derived pharmaceuticals, has signed licensing and marketing agreements and invested $500,000 in Shaman; Eli Lilly committed $4 million to Shaman and collaborates in developing drugs for fungal infections; and Merck & Company is working with Shaman on projects targeting analgesics and medicines for diabetes.[22] (For more on this topic, see "Making Biodiversity Conservation Profitable: A Case Study of the Merck/INBio Agreement," by Elissa Blum, in the May 1993 issue of Environment.)

To address the question of intellectu-al property rights and needs of the indigenous population, Shaman Pharmaceuticals created a nonprofit conservation organization called "The Healing Forest Conservancy" to protect global plant biodiversity and promote sustainable development. The company initially donated 13,333 shares of its own stock to the conservancy and plans to channel future product profits into projects that benefit the people of the source country.[23] The first conservancy project provided health care benefits for the indigenous peoples of Amazonian Ecuador, a region that supplies valuable medicinal plants to Shaman. In return for information about these plants, physician Charles Limbach extended his medical services to three communities and treated 30 children during a whooping cough epidemic. Additionally, the conservancy seeks to create sustainable harvesting techniques for plants with commercial medicinal value. These programs have the task of reconciling the ecological constraints on plant extraction with the economic realities of producing a marketable product.[24] This strategy reflects Shaman's concern that both the physical and intellectual property rights of the indigenous population are protected and that the inhabitants benefit from the research on these products.

As a result of its research efforts over the past few years, Shaman Pharmaceuticals has a pipeline full of active plant leads. Two antiviral products are currently being tested in clinical trials and are expected to reach the market in 1996: Provir is an oral treatment for respiratory viral infections that are common in young children; Virend is a topical treatment for the herpes simplex virus. Both products use the ingredient known as SP-303, a compound that was derived from a medicinal plant that grows in South America and was isolated by the company's discovery process.[25] Patents have been filed on both the pure compounds and the methods of use for these products, which have a target market greater than $1 billion worldwide.[26] Another consequential find is an antifungal agent found in an African plant that is traditionally ingested to treat infections. Shaman is using this compound to make a product that treats thrush, a fungal infection of the mouth, esophagus, and gastrointestinal tract. Given this discovery, the company hopes to find new treatments for other types of fungal infection. Shaman has strategically targeted its product development to address problems for which few effective treatments exist, such as viral and fungal infections. Moreover, there is a growing demand to find treatments for herpes and thrush because the increasing population of immunocompromised patients (including AIDS, chemotherapy, and transplant patients) is particularly vulnerable to these ailments.[27] A third promising line of product development is in the area of analgesics. Shaman has found two plants exhibiting special binding properties that raise the prospect of creating a nonaddictive pain-relief drug. The company is conducting laboratory tests to identify the pure compounds responsible for this analgesic activity and is expanding its screening process by collaborating with Merck.[28]

The raw materials for the screening all come from plants that are either presently harvested or sustainably collected. This discovery process has been quite successful at identifying plants with potential medicinal properties. Based on thousands of field samples collected by ethnobotanical field researchers and on reviews by a scientific strategy team, the company has screened 262 plants and found 192 to be active—a "hit rate" of 73 percent in the discovery process.[29] Future products will be developed from some of these "hits."

According to company president Lisa Conte, "Shaman's well-defined strategic focus and outstanding, dedicated scientists will create a successful business by uniquely combining the newest in technology with the oldest of tribal lore."[30] As a leader in ethnobotanical investigations, Shaman hopes that its initial success will translate into the development of a market for plant-based drugs from rain forest countries. The goal here is to use the revenue generated by these medicines as an economic incentive to preserve the forests and the wisdom of the native healers.

Ecotourism in Costa Rica

Ecotourism has been defined as "purposeful travel that creates an understanding of cultural and natural history, while safeguarding the integrity of the ecosystem and producing economic benefits that encourage conservation."[31] Successful ecotourism creates economic opportunities in terms of both employment and income for the local people. These benefits furnish the local community with a strong incentive to practice good stewardship over their natural resources (see box on next page).

In Costa Rica, ecotourism has become a large and growing industry. In 1986, tourism generated $132.7 million and ranked as Costa Rica's third largest source of foreign exchange.[32] In 1989, more than 375,000 tourists visited Costa Rica, 36 percent of whom were motivated by ecotourism.[33] Tourism to Costa Rica's parks increased 80 percent between 1987 and 1990 and surged another 25 percent in 1991.[34] Costa Rica offers the ecotourist diverse rain forests, abundant biodiversity, and breathtaking scenery. To protect these valuable resources, a national park system was established in 1970, which now comprises 34 parks and covers 11 percent of the total Costa Rican land area.[35] Some of the most popular sites for ecotourism in Costa Rica, such as the Monteverde Cloud Forest Reserve and the La Selva Biological Station, are also centers for important biological research. Recently, these areas have attracted thousands of visitors each year, primarily because of the rich flora (more than 2,000 plant species) and fauna (some 300 animal species).[36]

During the mid 1980s, the Costa Rican government sought to reconcile conservation and development interests by pursuing a strategy of sustainable development. Ecotourism was viewed as a clean source of development that might facilitate the preser-

PLANT-BASED MEDICINES

There are hundreds of drugs that have been derived from tropical plants, including reserpine from the serpentine root for tranquilizers, diosgenin from the Mexican yam for antifertility drugs, quinine from the cinchona tree to combat malaria, and vincristine and vinblastine from the rosy periwinkle to fight cancer. Among the most indispensable plant-based drugs are analgesics, antibiotics, heart drugs, anticancer agents, enzymes, hormones, diuretics, antiparasite compounds, ulcer cures, dentifrices, laxatives, dysentery treatments, and anticoagulants.[1]

The value of the medicines currently derived from tropical rain forest plants is astounding. Norman R. Farnsworth of the University of Illinois calculates that 25 percent of all prescription drugs in the United States contain ingredients that are extracted from plants.[2] Given that total prescription retail sales in 1990 were $62 billion, a rough estimate of the total value of plant-derived drugs is just more than $15 billion per year. James Duke, an economic botanist for the U.S. Department of Agriculture, contends that this methodology overstates the actual value. Duke adopts more conservative assumptions and places the value of the finished pharmaceuticals containing ingredients from rain forest plants at $6.25 billion per year.[3] Peter Principe of the U.S. Environmental Protection Agency estimates the retail value of plant-based prescriptions at $8 billion per year. Principe also calculates that, given the current rates of extinction and the probability of 5 in 10,000 that any plant will be the source of a marketable drug, the forgone value of lost drug products in 1992 was $150 million.[4]

The high economic value of the medicinal properties of these plants provides a strong incentive to preserve the rain forest. A critical barrier to marketing these medicinal products, however, is the drug review and approval

process under the U.S. Food and Drug Administration (FDA). Varro E. Tyler, a professor of pharmacognosy at Purdue University, estimates that the FDA approval process imposes costs of $231 million for each new drug.[5] Under current procedures, obtaining approval for plant-derived drugs is difficult because of the complex mixture of chemicals in plant extracts. FDA requires that each constituent in these mixtures be tested for safety. As a result, pharmaceutical companies are reluctant to invest in research on tropical plants. According to Loren Israelson, an attorney involved in rain forest preservation, Europe has a different drug regulatory system that recognizes plant extract as a whole and does not require testing of its individual components. Israelson advocates that the United States adopt a "European approach, which will serve to create new incentives to both preserve and explore the medicinal plants of tropical rain forests."[6]

1. N. Myers, *A Wealth of Wild Species: Storehouse for Human Welfare* (Boulder, Colo.: Westview Press, 1983), 89-104.

2. N. R. Farnsworth, "Screening Plants for New Medicines," in E. O. Wilson, ed., *Biodiversity* (Washington, D.C.: National Academy Press, 1988), 83.

3. J. A. Duke, "Tropical Botanical Extracts" (Unpublished manuscript, Washington, D.C., April 1989), 15.

4. P. Principe, "The Economic Value of Plants as Sources of Pharmaceuticals" (Paper presented at the symposium "Tropical Forest Medical Resources and Conservation of Biodiversity" sponsored by the Rainforest Alliance, New York, 24-25 January 1992).

5. V. E. Tyler, "Success Stories in Plant-Derived Medicine" (Paper presented at the symposium "Tropical Forest Medical Resources and Conservation of Biodiversity" sponsored by the Rainforest Alliance, New York, 24-25 January 1992).

6. L. Israelson, "Creating Economic Value for Tropical Medicinal Plants and the Need for a New Drug Approval Policy" (Paper presented at the symposium "Tropical Forest Medical Resources and Conservation of Biodiversity" sponsored by the Rainforest Alliance, New York, 24-25 January 1992).

vation of the natural resource base. The actual implementation of this strategy was left to the private sector. The early environmental entrepreneurs in Costa Rica's ecotourism industry included Costa Rica Expeditions, Tikal, Horizontes, and the Organization for Tropical Studies.[37] The growth of the ecotourism industry has since put strains on the fragile resource base. For example, the large

number of visitors at popular parks is causing such problems as erosion and water pollution.[38] Given the attraction of tourist revenues and the danger of overcrowding, environmental entrepreneurs are finding it difficult to create ecotourism programs that are consistent with the principles of sustainable development. Efforts to control ecotourism in Costa Rica are still in the early stages, and more

ECOTOURISM'S BENEFITS AND PITFALLS

The World Tourism Organization estimates that tourism generates $195 billion each year in domestic and international revenues, making tourism the second largest industry in the world. The role of tourism is particularly important for developing countries, where it contributes up to one-third of the trade in goods and services.[1] Nature-oriented tourism alone accounted for between $2 billion and $12 billion in 1988.[2] A number of countries have developed particularly successful ecotourism markets: Kenya generates about $350 million in tourism revenues per year; the Galapagos Islands were responsible for most of Ecuador's $180 million in tourism receipts in 1986;[3] and the Maya Biosphere Reserve in Guatemala yielded $185 million in 1990.[4]

Elizabeth Boo of the World Wildlife Fund conducted an extensive study of ecotourism in 1990.[5] Using surveys and interviews, Boo found support for the anecdotal evidence of a surge in nature-oriented tourism in the late 1980s. At the same time, however, she found many problems in the ecotourism industry, ranging from inadequate infrastructure to a lack of trained guides. To address these problems, Boo calls for a coordinated host-government strategy to protect the environmental quality of each toured area and to implement measures that effectively capture tourist expenditures for the local economy.

1. World Tourism Organization, *Policy and Activities for Tourism and the Environment* (Madrid: World Tourism Organization, 1989), as cited in T. Whelan, "Ecotourism and Its Role in Sustainable Development," in T. Whelan, ed., *Nature Tourism* (Washington, D.C.: Island Press, 1991), 4.

2. The Ecotourism Society, "Ecotourism Statistical Fact Sheet" (Alexandria, Va.: The Ecotourism Society, 1993).

3. Whelan, note 1 above, page 5.

4. C. Reining and R. Heinzman, "Non-Timber Forest Products and the Peten, Guatemala: Why Extractive Reserves Are Critical for Both Conservation and Development" (Unpublished manuscript, Conservation International, Washington, D.C., 1992).

5. E. Boo, *Ecotourism: The Potentials and Pitfalls* (Washington, D.C.: World Wildlife Fund-U.S., 1990). See, also, E. Boo, "Making Ecotourism Sustainable: Recommendations for Planning, Development, and Management," in Whelan, note 1 above, pages 187–99.

research is needed soon if the industry is to serve its original purpose.

One firm that is striving to attain this balance is International Expeditions. This 11-year-old, Alabama-based company operates 30 travel programs on 6 continents. Company president Richard Ryel and Tom Grasse, the director of marketing and public relations, contend that the ecotourism industry needs to forgo short-run profits and adopt a four-part conservation ethic that includes increasing public awareness about the environment, maximizing economic benefits for local people, encouraging cultural sensitivity, and minimizing the negative impacts on the environment.[39] International Expeditions applies these principles to business practices: For example, to create a flow of money into the local economy, the company uses the host country's airline when possible, employs local tour operators, and uses other services within the rain forest community.[40] The company's tour of Costa Rica begins in San Jose and proceeds through the country's national parks. The tour organizers hire Costa Rican guides who are familiar with the local habitat, and both guides and tourists stay at accommodations close to the parks whenever possible. These steps are designed to prevent tourist revenue from leaking outside the local communities that live near the parks.[41]

To minimize detrimental impacts on the ecosystem and to promote respect for the rain forests, International Expeditions arranges small, manageable groups, educates participants about the ecosystem, avoids fragile habitats, and minimizes disruptions to the wildlife. In keeping with its objective of promoting natural history and conservation education, International Expeditions has designed a series of workshops in Costa Rica. The workshops are led by some of the world's leading experts on life in the rain forest, including Alwyn Gentry of the Missouri Botanical Garden, Donald Wilson of the National Museum of Natural History, and James Duke of the U.S. Department of Agriculture. Participants join in small group sessions to engage in hands-on

field experience, such as nature walks, boat trips, and bird watching, and visit such sites as the Monteverde Cloud Forest and Tortuguero National Park on the Caribbean coast. Various sites feature canopied walkways up to 125 feet off the forest floor, which allow participants to walk among the treetops and closely observe the flora and fauna. The local guides also educate tourists about the history, culture, and socioeconomic conditions of indigenous peoples.

During the 1992 season, the cost of the 10-day, general nature tour throughout Costa Rica was $1,998 per person, and the 8-day workshop cost $1,498 per person. Because roughly 50 percent of these expenditures go to Costa Rica, these trips create the dual benefits of educating the nature traveler and generating income for the local economy.

A Key to Preservation

Clearly, sustainable development of rain forest products has the potential to bring about positive change, preserve biodiversity, and improve the welfare of local communities. Because deforestation is spiraling out of control, the efforts of organizations like Conservation International, Shaman Pharmaceuticals, Inc., and International Expeditions have become imperative. E. O. Wilson of the Museum of Comparative Zoology at Harvard University calculates that deforestation of the rain forest is responsible for the loss of 4,000 to 6,000 species a year—an extinction rate 10,000 times higher than the natural extinction rate before the emergence of humans on Earth.[42] Furthermore, the unwritten knowledge of forest peoples is rapidly disappearing. Thomas Lovejoy, assistant secretary for external affairs at the Smithsonian Institution, asserts that the rain forest "is a library for life sciences, the world's greatest pharmaceutical laboratory, and a flywheel of climate. It's a matter of global destiny."[43] The need to develop methods to deal with the issue is urgent, and environmental entrepreneurs may be key to preserving the vital and fragile resources of the tropical rain forests.

NOTES

1. World Resources Institute, *World Resources 1992–93* (New York: Oxford University Press, 1992), 118, 262.

2. E. O. Wilson, "Threats to Biodiversity," *Scientific American*, September 1989, 108.

3. E. Linden, "Playing with Fire," *Time*, 18 September 1989, 78.

4. See, for example, C. M. Peters, A. H. Gentry, and R. O. Mendelsohn, "Valuation of an Amazonian Rainforest," *Nature* 339 (1989):655–56; and R. Repetto, *The Forest for the Trees? Government Policies and the Misuse of Forest Resources* (Washington, D.C.: World Resources Institute, 1988).

5. Other organizations pursuing similar types of projects are Cultural Survival; Community Products Inc., which is linked to Ben and Jerry's Ice Cream; The Body Shop; Rainforest Alliance; The Nature Conservancy; and the National Cancer Institute.

6. See, for example, N. J. H. Smith, J. T. Williams, and D. L. Plucknett, "Conserving the Tropical Cornucopia," *Environment*, July/August 1991, 7.

7. See R. Repetto, "Deforestation in the Tropics," *Scientific American*, April 1990, 39; and W. Hyde, R. Mendelsohn, and R. Sedjo, "Applied Economics of Tropical Deforestation," *Association of Environmental and Resource Economists Newsletter*, May 1991, 6–12.

8. Current examples of ecolabeling include Germany's "Blue Angel" seal, Japan's "Ecomark" label, and Canada's Environmental Choice Program. In the United States, "Green Seal" and "Green Cross" are two different ecolabel programs being developed by private, nonprofit organizations. See A. L. Salzhauer, "Obstacles and Opportunities for a Consumer Ecolabel," *Environment*, November 1991, 10.

9. Peters, Gentry, and Mendelsohn, note 4 above.

10. For example, the practice of branding cattle evolved in response to the problem of defining property rights on livestock in the open range of the U.S. West during the latter half of the 19th century. See T. Anderson and P. J. Hill, "The Evolution of Property Rights: A Study of the American West," *Journal of Law and Economics* 18 (1975):163–79. See, also, H. Demsetz, "Toward a Theory of Property Rights," *American Economic Review* 57 (1967):347–59.

11. N. Parks, "Who Owns the Rosy Periwinkle?" *Sierra*, July/August 1991, 40.

12. J. Brooke, "Britain and Japan Split with U.S. on Species Act," *New York Times*, 6 June 1992, A1.

13. Conservation International, "Seed Ventures: A Quick Overview" (Washington, D.C.: Conservation International, 1991), 1.

14. Robin Frank, product manager of Conservation International's Tagua Initiative, Washington, D.C., personal communication with the author, 12 and 19 May 1993.

15. Conservation International, note 13 above.

16. Frank, note 14 above.

17. Ibid. See, also Conservation International, note 13 above.

18. Conservation International, "Tagua Initiative: A Conservation Alternative for the Tropical Forests in Ecuador" (Washington, D.C.: Conservation International, 1992).

19. Earth Almanac, "Nuts to Ivory: Carved Seeds Help Save Forest," *National Geographic*, February 1991, 142.

20. M. J. Plotkin, "The Healing Forest," *The Futurist*, January/February 1990, 12. See, also, D. Goleman, "Shamans and Their Lore May Vanish with Forests," *New York Times*, 6 November 1991, C1.

21. Shaman Pharmaceuticals, Inc., "Company Summary" (San Carlos, Calif.: Shaman Pharmaceuticals, Inc., 1992), 3.

22. Shaman Pharmaceuticals, Inc., "Prospectus" (Prepared by S. G. Warburg Securities and The First Boston Corporation, 26 January 1993), 22–23. See, also, A. Newman, "Shaman's IPO Success Sets Example for Biotech Firms," *Wall Street Journal*, 28 January 1993, B2.

23. Shaman Pharmaceuticals, note 22 above, pages 26–27.

24. S. King, "The Source of Our Cures," *Cultural Survival Quarterly*, Summer 1991, 21.

25. Shaman Pharmaceuticals, note 22 above, pages 17–18.

26. Ibid., 23; and Shaman Pharmaceuticals, note 21 above, page 3.

27. Shaman Pharmaceuticals, note 22 above, pages 18–19.

28. Ibid., 22; and Shaman Pharmaceuticals, note 21 above, page 3.

29. Shaman Pharmaceuticals, note 22 above, page 20.

30. Shaman Pharmaceuticals, note 21 above, page 4.

31. R. Ryel and T. Grasse, "Marketing Ecotourism: Attracting the Elusive Ecotourist," in T. Whelan, ed., *Nature Tourism* (Washington, D.C.: Island Press, 1991), 164.

32. Y. Rovinski, "Private Reserves, Parks, and Ecotourism in Costa Rica," in Whelan, note 31 above, page 54.

33. Ibid., 54–55. See, also, E. Boo, *Ecotourism: The Potentials and Pitfalls* (Washington, D.C.: World Wildlife Fund, 1990).

34. The Ecotourism Society, "Ecotourism Statistical Fact Sheet" (Alexandria, Va.: The Ecotourism Society, 1993).

35. Rovinski, note 32 above, page 44.

36. Ibid., 41, 52.

37. Ibid., 46–47.

38. Ibid., 52–53.

39. Ryel and Grasse, note 31 above, pages 164–67.

40. Tom Grasse, director of marketing and public relations of International Expeditions, Inc., Helena, Ala., personal communication with the author, 14 January 1992.

41. The "leakage" refers to the amount of tourist spending that escapes the region visited. Leakages occur through such channels as national taxes, foreign tour operators, and imported luxury items for the affluent tourist.

42. Wilson, note 2 above, page 112.

43. Linden, note 3 above, page 77.

Deforestation and Public Policy

The rates of environmental changes are as important as the processes behind those changes.

*Patricia Parisi and
Michael H. Glantz*

Patricia Parisi is a research assistant with the Environmental and Societal Impacts Group at the National Center for Atmospheric Research in Boulder, Colorado. Michael H. Glantz is program director of the group.

Global environmental change has become one of the most salient scientific and political issues of the 1990s. People and their governments have begun to realize that their societies have been fouling their environments at least since the onset of the Industrial Revolution in the mid-1700s.

Global warming, stratospheric ozone depletion, deforestation, human-induced sea level rise, and desertification are some of the major environmental changes that are or could be already under way today. Making a bad situation worse are the continually expanding population numbers that are sure to burden dwindling resources. Degradation of the natural environment is increasingly being seen as an impediment to achieving sustainable economic development. Thus, interest in as well as concern about the causes and consequences of environmental change

have increased sharply in the past few years.

What makes an environmental issue a global problem? Whether environmental degradation is considered a local problem or a global problem will most likely determine how it is addressed. Some environmental changes are considered global in origin: a natural warming or cooling of the global atmosphere or a period of change in solar activity that might affect global atmospheric processes. Others are considered global in effect although they might be local in origin: volcanic eruptions that alter the chemistry of the atmosphere on a global scale for lengthy periods of time or the local emissions of heat-trapping (greenhouse) gases.

Deforestation and desertification are two environmental changes that are local in origin and impact but have been labeled as global problems because they are occurring in many countries and have attracted worldwide interest.

In the 1970s a series of UN conferences was convened to address these and other issues pertinent to the environment. The Earth Summit in Rio de

Janeiro in June can be viewed as the capstone to these many conferences, as it brought together political leaders from around the world for the purpose of addressing the global environment. From the Earth Summit came the Framework Convention on Climate Change, a biodiversity treaty, and Agenda 21, an environment and development agenda for the twenty-first century.

The catalyst for such a summit was the growing concern about a human-induced global warming of the lower atmosphere. Scenarios about the possible implications of a global warming have been generated by computer-based modeling, analyses of historical data with extrapolations to the future, and analogies of societal and environmental impacts of extreme meteorological events. The assumption is that a global warming some decades in the future would have serious implications for every region of the planet from the equator to the poles. The climate convention was designed "to protect the climate system for present and future generations" or, more pointedly, to "freeze" the global climate regime as close as possible to its present state.

Rates and processes

The processes of climate and other environmental changes are the subject of major national and global research efforts. Increased emissions of greenhouse gases, for example, are being studied in order to identify their cause-and-effect relationships to environmental changes. Chlorofluorocarbons contribute not only to ozone depletion but to global warming as well. The use of fertilizers in agriculture not only adds to groundwater contamination but also to global warming. Deforestation adds carbon to the atmosphere, reduces local rainfall, and contributes to soil erosion, which causes silt buildup in rivers. Although research on these and other environmental changes continues, and much uncertainty remains about such changes, they have been the major focus of attention by researchers and policymakers.

As important as the processes of environmental change are the rates at which those changes occur. If scientists were to suggest that global warming would occur in 1,000 years, policymakers would ask them to come back in 950 years with an update on its progress and possible impacts. If told it could be in 100 years, policymakers would consider the issue more closely. And if they were informed by a large fraction of the scientific community that such a change in global temperatures was likely in decades they would feel compelled to take action right away. High rates of change spark immediate policy responses, whereas slower rates are met with delayed political responses and lengthy studies to resolve uncertainties. Thus, policymakers and scientists must be made to realize that the rates of environmental changes are as important as the processes behind those changes. Recall that inaction is a form of action in

favor of the existing processes as well as rates of change.

In spite of their importance, rates of change are understood poorly and are more difficult to measure than the causes, as the rates vary from year to year and from region to region. Desertification (the creation of desertlike conditions where none existed) has been a major concern of governments around the world because of its devastating impact on food production potential, especially in sub-Saharan Africa. Scientific journals show wide-ranging differences (e.g., tenfold) in desertification rates. For tropical deforestation, those rates vary by a factor of six, and in the Amazon region by a factor of four. Which rates can policymakers rely on with some degree of confidence?

With respect to the tropical rain forests, if a politician favors exploitation, he or she can find the data to support their position. Similarly, policymakers wanting to preserve tropical rain forests can just as easily find data to support their alarmist position. The problem is that one can find support in the literature to justify almost any desired policy action. The wide range of estimates also confuses the media, the purveyors (as well as translators) of scientific information for the public.

The rain forest dilemma

Tropical rain forests are of special

concern to those worried about a variety of issues: loss of biodiversity, extermination of indigenous populations, and change of global climate. Latin America contains 57 percent of the world's tropical rain forests, Asia claims 25 percent, and Africa, 18 percent.

No one doubts the view that the spatial extent of the tropical rain forests, currently estimated to occupy about 13 percent of the land surface of the planet, is on the decline. The reasons for the decline, however, vary from region to region: In Latin America rain forests are being converted to pastureland; in sub-Saharan Africa they are being cleared to meet increasing demand for farmland and firewood; in Southeast Asia rain forests provide the hardwood products exported to industrialized countries. Other social factors prompting deforestation include but are not limited to: lack of equitable land reform programs, chronic poverty, large government development projects such as dam construction, poor forest management practices, demand for agricultural land, and government corruption.

The South American rain forests provide a good example of the problems that surround the rates aspect of environmental change. In fact, scientific journals and the media have focused their attention on the situation in the Amazon.

Many people in the developing world, however, see nature preservation as an obstacle to improving their economies.

6. BIOSPHERE: Plants

The Amazon forests are considered an integral part of the global climate system. They act as a major carbon sink, pulling carbon out of the atmosphere during photosynthesis and storing it. When bulldozers, chainsaws, and fires destroy the rain forest, fewer trees remain to engage in this natural process. In addition, when cut trees are burned or left in the field to decay, previously stored carbon is released back into the atmosphere as carbon dioxide, a major greenhouse gas. Scientists estimate that about 25 percent of the human contribution to atmospheric carbon dioxide comes from the effects of worldwide deforestation.

As important as rates of change are to scientists and policymakers, it is still extremely difficult to identify them. Some of the problems include the varying definitions of forests, the techniques used to interpret rates, analogies used to convey such rates to the public, and attempts to determine the original extent of forest cover.

One might be easily lulled into thinking that definitions are just words and in the world of science they might not be so important. As we have recently seen, that was not the case with respect to wetlands in the United States. With the swipe of a pen the definition of wetlands was altered by the White House Council on Economic Competitiveness. As a result, about half of the currently defined wetlands could be removed from a federally protected status in favor of land developers who lobbied for such a change. This attempt at redefining wetlands has been contested by environmental groups and the U.S. Congress.

Tropical forests encompass different kinds of forests: closed and open, wet and dry, primary and secondary. Closed forests were recently characterized by the German Parliament's Enquete Commission on Protecting the Earth as having at least 50 percent tree canopy cover, high amounts of precipitation, and minimal grass cover.

Open forests are those in which the canopy (the upper parts of a tree) covers at least 10 percent of the ground areas. Grass typically covers the land surface. Open tropical forests are also referred to as dry deciduous forests. Moist evergreen tropical forests are located close to the equator and include the Amazon and Congo basins, the land bordering Africa's Gulf of Guinea, the Indo-Malaysian Archipelago, and the eastern coast of Australia. They have also been referred to as closed dry forests, monsoon forests, seasonal forests, and semideciduous forests.

Primary forests are those that have been disturbed slightly or not at all by humans and that contain great biological diversity. They are also called climax forests, meaning that they are in a final stage of succession, which is the transformation of one ecosystem to another over long periods of time. Secondary forests are those that reestablish on sites previously cleared by logging or for shifting cultivation, cattle ranching, or infrastructure development. They reflect all stages of succession that take place on naturally bare land or on land cleared by humans. Since the composition and structure of secondary forests change over time, scientists often have difficulty distinguishing them from primary forests when using remote sensing.

Scientists as well as science writers frequently interchange such terms as deforested, degraded, and destroyed in their description of environmental changes in the rain forests. These words, however, are not synonyms. Deforested refers to the removal of trees; degraded refers to ecosystem changes that often accompany tree-cutting; destroyed generally connotes the clear cutting of trees and other vegetation in a given location.

Varying estimates

In their attempts to express the large spatial extent as well as the rapidity with which rain forests are disappearing, writers have resorted to the use of popular analogies. The following portrayals are common in the literature: an area the size of a football field is being destroyed each minute in the Amazon; an area the size of Belgium, the Netherlands, and Luxembourg is deforested globally each year; every week 113,000 soccer fields are being cleared; an area the size of Great Britain went up in smoke; and so forth. While such statements do convey a sense of urgency, they do not provide a sense of accuracy. They tend to underscore the emotional aspects of deforestation in the tropics as opposed to its scientific and objective aspects. Furthermore, they tend to confuse readers by making it difficult to compare objectively the rates and extent of deforestation provided by different writers.

Another problem related to rates is the difficulty encountered in trying to determine the original extent of forest cover in the earlier decades of this century. Different researchers have based their rate estimates on "guesstimates" about the original extent of virgin forests. This is troublesome because scientists must often rely on uncalibrated records or proxy information to calculate rates of forest loss through destruction or conversion. Some researchers have tried to reconstruct the original extent of forest cover that could be supported in theory by the existing climate of the region. In either case, determining the extent of forest cover at the beginning of the twentieth century is fraught with uncertainties.

A review of articles on tropical deforestation in what we commonly refer to as the Amazon shows that researchers have used different geographic scales on which to base their rates of change estimates. Most readers do not readily draw the necessary distinctions between the Amazon, Amazonas, Amazonia, the Brazilian Amazon, the Amazon Basin, or the Legal Amazon. Sometimes these different spatial scales are used in the same article by the same author.

The Amazon Basin encompasses parts of seven South American countries: Brazil, Ecuador, Venezuela, Peru, Colombia, Guyana, Suriname, and French Guiana; the basin covers about seven million square kilometers. It has also been referred to as Amazonia. The Brazilian Amazon, which covers about five million square kilometers, refers to the part of the basin that falls within Brazil's borders. Brazil's Legal Amazon is an administrative jurisdiction, consisting of the whole of six Brazilian states (Acre, Amapa, Amazonas, Para, Rondonia, Roraima) and parts of three others (Maranhao, Mato Grosso, and Tocantus). Comparing rates between these differently defined geographical units requires considerable expertise and care.

It is also noteworthy to point out that the high rates of deforestation in the Brazilian rain forest highlighted in the 1980s were based for the most part on spectacular changes in the Brazilian state of Rondonia. It was there that the controversial World Bank-financed highway BR 364 was constructed in the absence of environmental impact assessments. As a result, the highway served as a conduit for urban migrants seeking to improve their quality of life in the rain forest.

High rates of change spark immediate policy responses, whereas slower rates are met with delayed political responses.

The accompanying chart [on the next page] depicts yearly changes in deforestation in the Brazilian state of Rondônia during the 1980s and clearly shows how rates of change are not constant but can vary over time. Charts like these can be very instructive and can be used to encourage scientists to seek to identify and understand the natural and human processes behind those annual changes in rates.

Finally, any specific deforestation figure is questionable because of the inadequacies of current technology to monitor environmental changes in forested regions; such technology is either too costly or cannot provide the details required by local, regional, and national policymakers. NOAA's weather satellites, for example, pass over the earth every day and are able to distinguish forested from nonforested areas, as well as detecting fires and plumes. However, with a resolution of one kilometer, the satellites gather relatively crude data over large areas, without the ability to differentiate gradual changes in forest types. Even with the more expensive, higher resolution (30-meter) LANDSAT satellites, scientists often have trouble distinguishing between primary forest areas and those areas that are in a state of regeneration. Fires further complicate evaluations by distorting satellite images, as the smoke appears on the images beyond where the fires are actually occurring.

Conclusions

Decisionmakers have difficulty gaining an accurate picture of where and to what extent tropical deforestation is occurring given this state of confusion. Such discrepancies within the scientific literature enable policymakers to pursue their hidden agendas, be they conservation or all-out exploitation. The rates of environmental changes in general, and of deforestation in particular, are often as important to decisionmakers as are the processes of changes themselves.

Governments in the tropics, like others around the globe, are pursuing development strategies designed to improve the quality of life of their citizens. Often, these strategies come into conflict with the desires by some members of the international community eager to preserve the global environment. A major reason for this conflict is the fact that many in the industrialized world are beginning to realize that they can no longer enjoy long-term sustained economic growth unless they maintain the health of their local, regional, and global environments.

Many people in the developing world, however, see nature

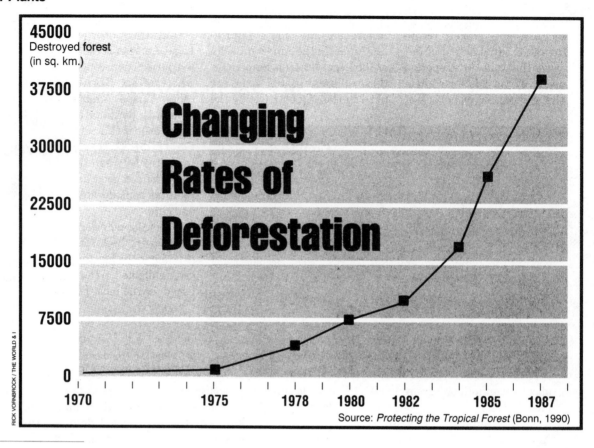

45000
Destroyed forest
(in sq. km.)

37500

30000

22500

15000

7500

0

Changing Rates of Deforestation

1970   1975   1978   1980   1982   1985   1987

RICK VORNBROCK / THE WORLD & I

Source: *Protecting the Tropical Forest* (Bonn, 1990)

■ The rate of deforestation is not easy to calculate, as the rate often changes from year to year. This chart shows how the deforestation rate has changed from year to year for the Brazilian state of Rondônia.

preservation as an obstacle to improving their economies. In addition, policymakers in the Third World view the high level of industrial development within rich countries as having been achieved at the expense of the environment. Now that the developed countries have achieved their high standards of living, they can turn their attention and resources to the restoration of their trashed environments.

The ongoing disappearance of tropical forests provides an example of how urgent and nec-essary it is to reduce the confu-sion surrounding the rates of deforestation so that rational pol-icy decisions can be made before any ecosystem is completely sacrificed. The importance of sus-taining the earth's tropical forests—their biodiversity, their indigenous inhabitants, their renewable products—cannot be overstated. This same urgency applies to the need to reassess the rates of wetland loss, desertifi-cation, ozone depletion, mangrove and coral reef destruction, and sea level rise.

The effects of misguided poli-cies toward the environment based on insufficient, if not incor-rect, scientific information will be felt immediately at the local, regional, and national levels, while the effects of climate change resulting from people degrading the earth's atmosphere and ecosytems will most likely be felt after many decades. The con-sequences of today's environmen-tal policies must be acknowledged and dealt with now for the benefit of present and future genera-tions.

OUT OF THE WOODS

Women are the real caretakers of the world's forests, yet they are being driven from the forests by development plans that ignore this role. Much more than just the trees are suffering.

JODI L. JACOBSON

Jodi L. Jacobson is a senior researcher at the Worldwatch Institute and author of the just-released Worldwatch Paper Gender Bias: Roadblock to Sustainable Development.

I n areas of India where the forests have been ravaged, women and children have to spend three to five hours each day just to gather enough fuel to cook the evening meal. Women there have taken to saying, "It's not what's in the pot, but what's under it, that worries you."

Their concerns are well-founded. The scarcity of leaves, twigs, branches, grasses, and other materials used for cooking fuel now rivals the scarcity of food itself as a cause of malnutrition in parts of sub-Saharan Africa, Haiti, India, Mexico, Nepal, and Thailand. The rampant loss of forests and other lands has a devastating effect on women and children throughout the developing world.

For cash-poor women, healthy forests are a savings bank from which they draw the interest—in the form of fuel, food, fodder, and countless other goods—that their families depend on to survive. As these lands become debased, the principal in the account declines, and women are forced to borrow against the future. For example, they compensate for the scarcity of fuel by, among other things, cutting live trees instead of dead ones, and cooking fewer meals. And

because women are largely responsible for feeding their families, they are forced to work longer hours to make ends meet, a situation that further compromises their health and well-being and that of their children.

Given women's vast potential for helping to reverse the loss of forests, it's hard to imagine why most development programs are actually *reducing* women's access to and ownership of land. The prominent role women play in using and maintaining forests makes it not only logical, but critical, that they be included in managing forest ecosystems. But throughout Africa, Asia, and Latin America, women are being muscled out of forests—and off croplands and grasslands—by governments and private interests looking to make a quick buck through the development of "cash crops."

Women are rarely included in these plans. And as more and more of the land on which they depend is made off-limits to them, their circumstances are becoming more tenuous than ever before. Only when development strategies recognize, and are geared toward supporting the role of women in conserving forests, can we begin to solve many of the economic and environmental problems that otherwise promise to spin out of control.

Survival Stores

Forests and other land resources are to women in the Third World what grocery

From *World Watch,* Vol. 5, No. 6, November/December 1992, pp. 26-31. Copyright © 1992 by Worldwatch Institute. Reprinted by permission.

stores and utilities are to most women in industrialized nations. Forests, for example, are a major source of fuel for Third World homes. Without wood, women can't cook the food they've grown and harvested, or do simple things like boil water or heat their homes. Much of the Third World still depends on wood, leaves, and other forms of "biomass" for its domestic energy needs, and it usually falls to women and girls to collect this fuel—as in rural Africa, where mothers and their daughters gather 60 to 80 percent of all the fuel their families need.

Women look to the forest for other needs, too. In the woods they collect plant fibers to make cloth and thread, plants and herbs used for medicines, seeds used in condiments and oils, and many other materials. In India, women use the bark of the tendu tree to treat diarrhea.

Foods gathered from trees and bushes also add critical nutrients to the grain-based diets of people throughout the Third World. Fruits, vegetables, and nuts—important sources of vitamins, minerals, proteins, and fats—are widely used to supplement staple crops. In many areas it is the job of women to protect and manage these trees. Indonesian women harvest bananas, mangos, guavas, and avocados from 37 species of trees growing around their homes. Women often use these and other forest goods to make money for their families. In Senegal, for example, shea-nut butter made from the fruit of local trees is one of many marketable goods women produce to earn cash. Muslim women in the Middle East living in purdah (seclusion) often cultivate trees within the confines of their gardens, sending their children out to sell the fruit.

But because they are not considered major cash crops, these foods women gather and the lands that produce them often are overlooked and undervalued by development experts. A study by the United Nations Food and Agriculture Organization in northeast Zambia found that what was officially cataloged as "useless" forest land was actually a major source of leafy wild vegetables and mushrooms, as well as caterpillars, eaten by local families. Women there rely on these items as sources of protein and of cash income. In times of scarcity, tree foods often make the difference between life and death. In Tanzania, the fruits and nuts from just three types of trees provide some

food for rural women every month of the year. The United Nations found that these trees were used more intensively during famines and droughts.

Forests also provide fodder for the livestock rural women keep to add milk and meat to their families' diets. In most regions, women are solely responsible for finding food for their animals.

Finally, the forests give women wood and thatch to repair homes, and the branches and logs they use to build fencing or to carve spoons, bowls, and other utensils.

The bottom line is that much of what women need to help their families survive comes from forests. When they can't turn to the land, it's anyone's guess how they will make a living.

The Real Forest Workers

Forests offer women plenty of resources for everyday life, but they also constitute a source of jobs and income. Forget Paul Bunyan and other conventional images of macho, flannel-shirted lumberjacks; in many rural areas in the Third World, women make up a big share of the labor force in logging, wood processing, nurseries, plantations, and small-scale forest products industries. They log trees to make charcoal, grow seedlings for sale, and harvest wood for carving, among other things.

And while logging is usually considered the big-money forest industry, products that contain no wood at all often play an even bigger part in local and national economies throughout the Third World. World Bank researchers Augusta Molnar and Gotz Schreiber estimate that in India, 40 percent of revenue from forests—and 75 percent of the net export earnings from forestry products—come from so-called non-wood resources. The same is true in Thailand, Indonesia, and Burma.

Poor, rural women can make a large chunk of their income, for instance, by collecting raw materials and using them to make saleable items out of bamboo, rattan, and rope. In one province in Egypt, 48 percent of the women make their living through such "minor" forest products industries. In hard times, women who don't own land and can't make enough as field workers often fall back on collecting these resources. In India, for example, about 600,000 women harvest tendu leaves in the wild for use as wrappings for domestic cigarettes.

Robbed of Resources

Imagine an American woman trying to scrape together food for herself or her children if she suddenly had no access to grocery stores or markets, and the modern dilemma of women in subsistence communities may be easier to understand. Traditionally, even though African and Asian women could rarely "own" land, they usually had equal rights—as members of a community—to use community, or "commons," land according to their families' needs. However, they have always had to face restrictions on the use of land that have never applied to men.

But now even women's limited access to land is rapidly eroding, despite the fact that they depend on forests and other land resources for survival.

The "commons" lands formerly open to women increasingly are being converted to private property or turned into "open access" systems that *no one* controls. Investments by governments, donor agencies, and multinational corporations are directly encouraging communities to shift land use from community-controlled "subsistence" activities to raising cash crops. In many countries, large areas that were once communal forest have been privatized and set aside for agriculture or plantation forestry, resulting in widespread deforestation and a blow to women's ability to provide for their families.

In Zanzibar, Tanzania, commercial clove tree plantations began to replace once-communal natural forests in the late 19th century. One hundred years later, the spread of commercial agriculture continues and is creating a crisis in poor households, which now have to spend up to 40 percent of their income on the fuel they once were able to find in nearby forests. In Western Kenya, more and more land is now privately owned and thus off-limits to the poor.

And in India, much of the commons land now disappearing into government and private hands was once used by village women to secure fuel under a community system. "Contrary to the popular belief that...the gathering of wood for fuel...is [primarily] responsible for deforestation and fuelwood shortages," says Bina Agarwal, professor of agricultural economics at the Institute of Economic Growth in Delhi, India. "Evidence [in India] points to past and ongoing state policies and schemes as significant causes."

Agarwal contends that what remains of India's forests will disappear within 45 years if the widespread conversion of sacred groves and other communal land for cash crops, dams, and commercial timber continues at today's pace.

Women won't benefit from these changes in land use, because privatization favors male landholders. Legal and cultural obstacles prevent women from obtaining titles to land, and without these titles, they can't be included in the cash crop schemes affecting forests and other lands. (Land titles invariably are given to men because governments and international agencies routinely identify them as heads of their households, regardless of whether or not they actually support their families.)

Women's rights to land are subject to the wishes of their husbands or the whims of male-dominated courts and community councils. In one region of Kenya, where women cannot own land, they also are restricted from planting trees. According to custom, control over land is determined by the ownership or planting of trees on it. Not surprisingly, men in the area have opposed women's attempts to plant more trees and other sources of fuel.

In northern Cameroon, some men only allow their wives to plant papaya trees, which are short-lived and do not count toward establishing land rights. In Nigeria, women have rights to the kernel, but not the oil, of the palm tree, which is sold by men as a cash crop.

Unfortunately, when men make more money, it rarely comes back to their families. It goes instead to unessential purchases such as radios, clothes, and alcohol. Family welfare continues to depend on what women can come up with, according to studies in every region of the Third World—making women and their children the big losers in the transition to a cash crop economy.

Forest Keepers

Women in Third World subsistence economies are the unacknowledged experts on the use and management of trees and other forest products. They know how important forests are in preserving ecosystems (forests play a critical role in replenishing fresh water supplies, for instance) and appear to be as careful in conserving forests as they are reliant on using them.

Women are more effective at protecting

and regenerating the environment than either the state or private landowners, according to research on common lands conducted by Babjik Kudar, director of the Indian State Common Lands College. The reason is obvious enough: When you depend on something, you learn to take care of it.

Beyond their experience at managing forests, women's knowledge of forest resources constitutes a vast mental store of information on species that scientists regularly lament being unable to catalog. Tribal women in India, for example, know of medicinal uses for some 300 forest species. A survey in Sierra Leone found that women could name 31 products they gathered or made from nearby trees and bushes, while men could name only eight.

Unfortunately, "most forest policies and most foresters continue to overlook or ignore this diversity of knowledge," says Paula Williams, a forest and society specialist at the Institute for Current World Affairs in London. Tree-planting campaigns and international investments to stem deforestation have all but ignored women. Of 22 social forestry projects appraised by the World Bank from 1984 through 1987, only one mentioned women as a project beneficiary. And only four of 33 rural development programs that involved forestry funded by the World Bank over the same period included women in some way.

Undervaluing women's social and economic contributions is certain to slow progress toward broad social and environmental goals, such as preserving biodiversity and protecting the role played by forests in water cycles. The only real hope of saving forests and other resources is to begin providing greater economic opportunity to the women who depend on them for survival.

The Population Trap

It's an axiom of the environmental movement that population growth is a big obstacle to preserving the environment: the more people there are, the greater the demands on the Earth's resources. But few development experts consider that when a woman's basic resources are taken away, she is actually likely to want *more* children to help her shoulder the increased workload—even when the region where she lives is already home to more people than the land can realistically support.

From the perspective of a poor, rural woman, children represent an investment in the future. With few opportunities to save cash and no chance to own land, women too old to fend for themselves turn to their children for support. Children are a ready labor force. The time constraint imposed on women by the longer hours they must work to make ends meet means they lean more heavily on the contributions of their children—especially girls.

Often, mothers are forced by circumstance to keep their daughters out of school so that they might help with the housework—ensuring that another generation of females will grow up with fewer prospects than their brothers. In Africa, for example, "more and more girls are dropping out of both primary and secondary school or just missing school altogether due to increasing poverty," says Madame Phoebe Asiyo, forest specialist at the United Nations Fund for Women.

This is the population trap. Many of the policies and programs carried out in the name of development actually increase women's dependence on children as a source of status and security. And the continued rapid population growth in Third World communities just compounds the destruction of the local environment.

To fail to recognize women's roles in forest management and as income earners is to fail in the fundamental purpose of development itself. If women in subsistence economies are the major suppliers of food, fuel, and water for their families, and yet their access to productive resources is declining, then more people will suffer from hunger, illness, and loss of productivity.

If women have learned ecologically sustainable methods of forestry, and have extensive knowledge about genetic diversity, yet are denied partnership in development, then this wisdom will be lost.

Without addressing issues of equity and justice, then, development goals that are ostensibly universal—such as the alleviation of poverty, the protection of ecosystems, and the creation of a balance between human activities and environmental resources—simply cannot be achieved.

"Many people believe," says Paula Williams, "that first we should save the world's tropical forests, then we can worry about women and children. Unless we work with women and children, however, it will be impossible to 'save' the tropical forests."

KILLED BY KINDNESS

We nearly wiped out the beasts of the field. Then we tried to save them. Now greed, corruption and stupidity are finishing the job that bullets began.

Sharon Begley

Han-Han was beautiful, the "most perfect" panda that wildlife biologist George Schaller had ever seen. With luminous fur and paws so clumsy-looking they had no business peeling bamboo as delicately as a finicky dieter peels celery, she sat for hours in the forest, her rounded back snuggled against a log, her seat cradled in the soft humus. With big black eye patches and a "rolling sailor's gait," Han-Han evoked in the humans who admired her "a desire to hug and protect," says Schaller.

The humans failed her, badly. The 785-square-mile Wolong wildlife reserve in China's Sichuan province, where Han-Han lived, is hardly guarded; Wolong police say they don't have the time and Beijing doesn't force the issue. So on the fogbound, snowy morning of Jan. 24, 1983, Han-Han strangled to death in a poacher's snare. The hunter-peasant carried the carcass and skin back to his hut. His wife cooked some of the meat with turnips, the peasant later said. "It did not taste good. So we fed it to the pigs." The unspeakable waste, the assault on the creatures he studied and loved, made Schaller rage—at "the officials who talked and talked while ever more pandas slipped silently into darkness. Damn. Damn. Damn them all."

Have we bungled the jungle? In "The Last Panda" *(291 pages. University of Chicago Press. $24.95),* to be published next week, Schaller lets loose with a shattering attack on the sham called

panda conservation. In doing so, he has opened a new front in the wildlife wars. Taking on zoos, governments and international conservation groups that sanctimoniously swear fealty to the ideal of saving species, Schaller charges that stupidity, greed and indifference are causing mankind to hasten the loss of the world's wildlife. The panda may be the most egregious example, but there are, tragically, others. Many others. From crocodile reserves in Colombia that have been turned into poachers' paradises to international whaling bans that save one species only to keep others on the cusp of extinction, the message is sobering. "If most wildlife groups were a business," fumes Earth Island Institute biologist Sam LaBudde, whose clandestine filming of dolphins caught in nets led to the 1988 tuna boycott, "they would be sued for fraud."

Species: **GIANT PANDA**
Habitat: **SICHUAN PROVINCE, CHINA**
Population in the Wild: **1,000**
Estimated Population in the Year 2025: **5,000 to 0**

A constellation of forces conspires against efforts to save animals, starting with poaching and the destruction of species' habitat. But corrupt governments, inefficient bureaucracies and the muddling of even well-intentioned green

groups add to the mix. Most disturbing is conservationists' willingness to spend money to appease local government honchos. "You have to buy your way into a country—with vehicles, equipment, something," says Alan Rabinowitz, senior zoologist at The Wildlife Conservation Society. China, he discovered, actually has a formal list of the prices it charges foreigners to study particular species. Wildlife managers are often overmatched by corruption. Just last month a four-zoo consortium, formed in 1984 to capture Sumatran rhinos and breed them in captivity, gave up in frustration over spiraling payments that seemed not to be helping the hairy rhinos. Under its agreement with Indonesia, the $3.5 million Sumatran Rhino Trust had to pay the government $60,000 for each animal captured, ostensibly to protect the remaining creatures' forest habitat against loggers. "We know we gave the Indonesians that money for rhino conservation," says Jim Doherty of the Bronx zoo. He pauses. "That's all we know." Dead-pans president Kathryn Fuller of the World Wildlife Fund, "Working in conservation can be a fascinating multicultural experience."

Schaller, scientific director of WCS (the field biology arm of the Bronx zoo) and National Book Award winner for "The Serengeti Lion" in 1972, details countless such frauds, cover-ups, lies and fatal mistakes that may yet doom the giant panda to extinction. "I am placing the blame on every institution, whether American zoos or the Chinese government or World Wildlife Fund—anyone who did not think of the panda first," Schaller told NEWSWEEK. Among his charges:

■ WWF had cosponsored Schaller's research, but he pulls no punches in his criticism of the well-respected group whose very logo is the giant panda. WWF failed to "push an agenda on behalf of the panda in the wild," charges Schaller. Instead it acquiesced in Beijing's demand for the fanciest panda-research lab anywhere on earth. The new Wolong research and breeding facility got crammed with expensive equipment donated by WWF. But it was barely used. In WWF's defense, a fat check may have been the only way to get Beijing's permission to work in China. All but the most naive groups budget for "consulting fees" to foreign officials. The difficulty of working abroad is underscored by the small number that do it. "WWF has more failures because it's made more effort," says Michael Bean of the Environmental Defense Fund. But Schaller isn't the only on-the-ground biologist disenchanted with WWF. A new joke asks, Why did the dinosaurs go extinct? Because WWF had a program to save them.

Species: **BENGAL TIGER**
Habitat: **INDIAN JUNGLE**
Population in the Wild: **2,300**
Estimated Population in the Year 2025: **0**

■ Zoos "vie for status, publicity and profit" through "rent a panda" deals, says Schaller. Beijing reaps millions of dollars in fees; zoos rake in more through increased attendance and T shirt sales. But panda rentals disrupt the breeding of a species that reproduces poorly in captivity even without transoceanic sabbaticals. When they take up long-term residence, any cubs they have rarely survive: since 1972 panda lovers had followed the tragic courtship of Washington National Zoo's Ling-Ling, who had five cubs with her mate Hsing-Hsing, but for various reasons, not one lived. (Ling-Ling, whose picture is on the cover of this issue, died last December.) Despite zoos assurances, they "are not cooperating in seeing that conservation money goes toward saving the panda," says Schaller. Like WWF, zoos have failed to press China to use the

millions to truly help wild pandas. And announcements that the loans are for "breeding programs" are often no more than a cover. In May, Adventure World Zoo in Japan will get one male and one female panda from China, paying $10 million for the 10-year loan. China says it will use the money for panda conservation. Japan says it will try to breed the pair. But the zoo has no breeding program or experts, says WWF. "The loan appears to be mainly a commercial undertaking," says WWF's Stuart Parkins.

■ Beijing exerts little pressure on the provinces to cut down snares in the bamboo forests and comb the panda reserves for poachers. (Dealers and collectors in Japan, Hong Kong and Taiwan pay $40,000 for a panda skin good for nothing but hanging on a wall.) When Schaller emphasized the need for patrols, he was told the Wolong police were "too busy with other duties . . . though it was my impression that their daily task consisted chiefly of reading newspapers and drinking tea." Others echo Schaller's charge of misguided priorities. "The main thing being preserved are not animals but the lifestyles of officials," says an animal specialist in Beijing who asked to remain anonymous: many local conservation officials consider a new Jeep Cherokee their top conservation priority. Some of the $4 million it gave China for the Wolong reserve went to build a hotel and a school for the children of resident Chinese scientists.

Schaller particularly faults Beijing for seeking any excuse to capture wild pandas, even though they die in captivity faster than they reproduce. Unlike other male mammals that are perpetually randy, male pandas come into heat only once a year. In captivity, it tends not to be when females are receptive. Even worse, babies taken to the Wolong center sometimes fall ill and die because of inexperienced and incompetent vets. (Once a Communist Party official performed a lethal feeding experiment on a cub.) During the bamboo die-off of 1983, China made a great show of rescuing "starving pandas"—even though Schaller discovered that the pandas had plenty to eat. Of 108 pandas captured, 33 died, 35 were released into a different forest and 40 were sent to zoos or holding stations, like the one where they lived in "dark, cold [cages] encrusted with frozen urine and feces [where they] suffered silently," relates Schaller. China had its most successful captive-breeding year ever in

1992—a record 11 of 13 newborns survived—but few biologists view that as the norm. "We don't believe captive breeding works," admits one Beijing scientist. "If we depend on zoos, the panda has no hope."

Species: **AFRICAN ELEPHANT**
Habitat: **BUSH**
Population in the Wild: **6,000, most in southern Africa**
Estimated Population in the Year 2025: **Unknown**

Schaller's broadside has forced environmentalists to rethink the accommodations they make with host governments, and a similar revisionism has struck even cases deemed unqualified conservation successes. Perhaps the animal lovers' greatest recent triumph was the 1989 ban on trade in ivory, a move intended to stop the elephant slaughter in east Africa. Eliminating the ivory market has curtailed poaching even in countries, like Tanzania, that can't afford to police their herds. In the early 1980s, poaching claimed 3,000 to 4,000 elephants a year in Kenya, and ivory brought $150 a pound. Now, with the black-market price down to $5 a pound, fewer than 50 elephants are poached annually (see box).

But in a contrariant book being published this month, journalist Raymond Bonner, who lived in Nairobi for several years, attacks the "ecocolonial" conservation groups that rammed through the ban. He charges that their strategy portrayed anyone who favored a less absolute stance as heartless advocates of elephant killing. Such billing would have crippled fund raising for groups like WWF and zoos. But the herds of Botswana, Zimbabwe and South Africa can sustain selective culling and ivory harvesting, he argues in "At the Hand of Man" *(302 pages. Knopf $24).* The ban deprives African nations of an estimated $50 million they could spend on wildlife reserves. (They could make it all up, though, by raising admission prices to their game parks $2: currently, Kenya charges $15; Zimbabwe, nothing.) The growing elephant population, charges Bonner, "has meant the destruction of the habitat for other species: . . . impala, giraffes, bush-babies and monkeys; kudus, bushbuck and genet." Even ivory-trade foes admit that, in perhaps 10 years, east Africa's herds may be large enough to allow legal ivory trade. The catch: making sure legal sales don't provide a laundering system for illegal poaching.

But there are perils in legalizing what had once been contraband. In Colombia, poachers had nearly wiped out the Orinoco crocodile, American crocodile and black caiman (a croc cousin) in order to illegally export thousands of hides a year in the 1980s. To save the prehistoric reptiles, the government in 1987 began overseeing licensed breeding areas, where ranchers may raise caiman for profit as long as they release 5 percent of them to the wild. The idea was to control poaching by providing a legal alternative source. But the program has proven much better at making ranchers rich than making caiman safe: the farms serve as a cover for poaching. Tens of thousands of caiman hides labeled LEGAL EXPORTS were in fact captured illegally in the wild and laundered through the ranches. Not one of the 32 farms has released 5 percent of its production to the wild. And many farmers lie about the size of their stocks and hunt thousands more illegally: one farm reported having 15,000 caiman eggs, but an inspector counted only 3,000.

What happened to the other 12,000? The shortfall would be made up with caiman poached from the wild. Officials of Colombia's Natural Resources Institute, which runs the ranch program, don't deny the accusations. They just blame it on the lack of money and personnel to enforce regulations.

There is no clearer sign of rampant revisionism than the challenge to that very rallying cry of green activism, "Save the whales." The International Whaling Commission banned whaling seven years ago, but last year Norway resumed its hunt and this month its whalers are going back. Their rationale: catching a few hundred of the 86,000 minkes in the eastern North Atlantic is the essence of sustainable development, that newest eco-buzzphrase meaning "using a resource without using it up." "If you say that one animal, however abundant, has an absolute right to life, you have harpooned the principle of sustainability," says Georg Blichfeldt, secretary of a coastal alliance. "What you get are the principles of Disney World: whales become 'fun animals'." Even

Norway's environmentalists support the limited hunt in the interest of research. Is the whaling ban really a bungle? It is if the failure to harvest the minke harms other species. The minke may compete for the same turf and food as the blue whale; if the minke retains its huge population lead, the blue may never recover.

Sometimes the veneer of success can hide a disaster. Take Project Tiger, in which India set aside tiger reserves, trained wardens and equipped patrols. By most accounts, this saved the Bengal tiger from extinction. But, say critics like Wildlife Conservation's Rabinowitz, the tiger's numbers have been regularly inflated by reserve managers whose pay is tied to their performance and whose deception is winked at by conservation groups. "They needed a success story because these organizations can't keep saying things are getting worse and worse," charges Rabinowitz. "If they do, the public asks, 'Why should I give money to a lost cause?'" The result, Peter Jackson of the International Union for the Conservation of Nature told reporters last year, is less pressure to stop poaching and the illegal sale of tiger skins and bones, which are used in Chinese medicine. After years of seeing reports from Project Tiger that the number of great cats in India was rising, to 4,600, biologists now believe there are fewer than half that. The deception may have helped allow the population to shrink to a size that will doom it to extinction before 1999.

Species: **BLACK RHINO**
Habitat: **AFRICAN BUSH**
Population in the Wild: **2,400**
Estimated Population in the Year
  2025: **Possibly 0**

It's easy to lay blame, whether on corrupt politicians or blinded-by-idealism wildlifers. But at least when fingers get pointed, solutions sometimes follow: in December China's State Council approved a $52.6 million panda-conservation plan that doubles the number of reserves and creates "corridors" between them, moves out lumber firms and resettles local peasants. Beijing hopes to have one anti-poaching patrol per 2,500 acres of reserve (it's 1 per 7,400 to 12,350 now). In the truly disheartening cases, the forces of extinction are so

overwhelming that there seems to be no solution at all. NEWSWEEK's Joseph Contreras traveled to Rhodes Matopos National Park in Zimbabwe last week to see that impoverished nation's desperate—and seemingly futile—attempt to save the black rhino. His report:

Since daybreak, six trackers in military uniforms and two veterinarians had been stalking a pair of black rhinos through the dense bush. They located their quarry: the crack of a dart gun rang out, and the 1,600-pound cow slumped against the bank of a muddy creek. After making sure she was in stable condition, Dr. Mark Atkinson fired up a chain saw and amputated the anesthetized animal's horns—horns so highly prized by poachers from Zambia that now, only 305 black rhinos remain in Zimbabwe. Last May, to deter poachers who sell the horns to China, Taiwan and South Korea for their alleged pharmaceutical properties, Zimbabwe began amputating horns from black rhinos. (In 1991 it had begun dehorning white rhinos, of which only 198 remain in Zimbabwe and 6,000 in all of Africa.) So far, 123 whites and 140 blacks have been dehorned. And now Zimbabwe has grasped the full horror of the poaching problem. In the last 20 months, 10 dehorned whites and six dehorned blacks have been slaughtered by Zambian poachers intent on demonstrating that they can slip into the country and kill rhinos with impunity. Says Tony Ferrar of the Wildlife Society of Southern Africa, "It's taught the Zimbabweans how futile their whole campaign against poaching has been." The parks and wildlife department is broke; 250 scouts have been laid off and anti-poaching patrols scaled way back.

Is it all hopeless? Sam LaBudde of Earth Island Institute charges that the hundreds of millions of dollars spent on wildlife conservation have "maintained animals for the benefit of poachers and smugglers." Only massive international pressure can choke both the supply of and the demand for endangered beasts. LaBudde calls for withholding foreign aid until a country implements tough anti-poaching and anti-smuggling programs. In some places, at least, the strategy has worked for other contraband: "It's very difficult to get guns or heroin in Taiwan," says LaBudde. "But it's simple to get rhino horn, bear gallbladder or tiger bone." Does the world have the resolve? "It's going to take all the tigers disappearing, all the rhinos, all the elephants, all the mountain gorillas," laments Ronald Tilson of the Minnesota Zoo. "Then maybe there will be a true war declared on behalf of endangered species."

Until then, more pandas follow Han-Han's snowy footprints into the snares of Wolong, and poachers steal across the Zambian border to kill rhinos for the bragging rights. Zoos try frantically to breed endangered beasts, banking their eggs and sperm cryogenically in frozen Noah's arks. Someday, maybe there will again be a place on earth for the beasts of the field. But that day may be long in coming. As Leonardo da Vinci observed 500 years ago, "All the animals languish, filling the air with lamentations."

With DANIEL GLICK *in Washington,* BROOK LARMER *in Buenos Aires,* KARI HUUS *in Beijing,* WILLIAM UNDERHILL *in London and* JOHN STEVENSON *in Nairobi*

The War to Save the Tsavo Reserve

Few places stand as more powerful symbols of wildlife's fragility—and humans' greed—than Tsavo National Park in Kenya. A reserve of sweeping savannas and dramatic mesas roughly the size of El Salvador, Tsavo was once home to 35,000 elephants. But during the booming ivory trade of the 1980s, poachers swept down from Somalia in an unstoppable wave, reducing the herds to 4,300 by 1988. Says Steven Gichange, Tsavo's chief warden, "We believed the elephants were doomed."

Today they have a new lease on life. In 1989, under pressure from conservationists and Western donors, Kenyan President Daniel arap Moi appointed Richard Leakey director of the Kenya Wildlife Service. The son of anthropologists Louis and Mary Leakey, he has created a model conservation program. He cleaned up corruption and helped triple the park's budget to $20 million. He was also a key supporter of global legislation to ban commerce in ivory. "The results are encouraging," says Mark Stanley Price, director of African operations for the African Wildlife Foundation. The herds are growing by 15 percent a year; tourism is flourishing.

Leakey's first priority was improving Tsavo's lax security. Using funds from foreign donors, including the United States and Japan, he beefed up the ranger force from 30 to 130 and tripled salaries. He augmented Tsavo's one broken-down Land Rover and one truck with 40 new vehicles. Wardens built observation posts, patrolled round the clock and scattered bases through the park. The wildlife service hired 30 rangers from the Orma tribe, Somali-speaking Kenyans who live between Tsavo and the Somali border. Their main mission: to recruit spies among the Orma nomads. Somali poachers have traditionally employed the Orma to guide them through the unfamiliar countryside. "The Orma know every movement the Somalis make," says Gichange. "That's where we're winning the war."

The wildlife service is trying to gain the sympathy of villagers. The subsistence farmers and herders at the edge of the park have bitterly complained for years about rogue elephants that destroy their crops and lions that kill their cattle and a few of their neighbors. Leakey initiated a revenue-sharing program that designates 25 percent of entrance fees collected in Kenya's game parks for local populations; around Tsavo, money has gone for dispensaries, health clinics, wells and scholarships.

Tsavo's staff remains wary; five elephants killed in February sent a signal that poaching won't go away. And rangers continue to uncover stockpiles of ivory buried outside Tsavo, stashed in anticipation of a relaxation of the ban. "That tells us," says Gichange, "that we can never let down our guard."

JOSHUA HAMMER
in Tsavo National Park

New-Species Fever

To preserve biodiversity, scientists must discover new forms of life fast—before they're gone.

Patrick Huyghe

Patrick Huyghe is a science journalist whose work has appeared in The New York Times Magazine, Discover, *and* The Sciences. *His latest book is* Columbus Was Last.

Seated in her elegant office at the Smithsonian's National Zoological Park, in Washington, D.C., Devra Kleiman is making animal sounds. "*Doo, doo, doo, doo, doo, doo, doo, doot,*" she says rapidly, then slows to a whine for another "*doo, doo, doo, doo,*" and ends with a cluck, short and sharp: "*Chuck, chuck, chuck, chuck, chuck, chuck.*" That is the closest she can come to the sound of a caissara, the lion tamarin monkey discovered in 1990 on a small forested island off the coast of Brazil. Called black face by the local people, it is the fourth species of lion tamarin known to science, joining the golden lion tamarin, the golden-headed lion tamarin, and the black lion tamarin.

From a file, Kleiman pulls out a chart with a choppy rising and falling line—a diagram that represents the monkey's call. Since lion tamarins show little genetic variability, Kleiman and other researchers are trying to differentiate the species by noting variations in their behavior. "I'm interested in how different the vocalizations are in the four species," says Kleiman, who has studied golden lion tamarins for two decades, "and whether you can use the structure of the vocalizations to differentiate the species taxonomically."

The lucky discovery of this new primate is credited to Maria Lucia Lorini and Vanessa Guerra Persson, who had just taken their first jobs as professors with Brazil's Capão da Imbuia Natural History Museum. While conducting a bird survey on the island of Superagüi, they had gotten wind of the monkeys' existence from a local fisherman; their description of the new species is based on a well-preserved skin provided by the fisherman and on a month spent observing the animals in the wild.

The find stunned the scientific world. Imagine a species of primate unknown to science roaming the densely populated Brazilian coast less than 200 miles away from São Paulo. Russell A. Mittermeier, chairman of a primate-specialist group for the International Union for the Conservation of Nature and president of Conservation International, likened the discovery to finding a major new species in the suburbs of Los Angeles.

But what was unknown to science was not unknown to everyone. Local people are keenly aware of the animals in their midst; the *caissara*, the traditional coastal fishermen of southern Brazil, were quite familiar with this particular lion tamarin. In their honor, the new species was named *Leontopithecus caissara*.

In June 1990 a report by Lorini and Persson in the journal of the National Museum of Rio de Janeiro announced that the caissara was a new and distinct species, citing its fur coloring—gold, except for its face, forearms, feet, and tail, which are black. But disputes over whether an animal is a new species or a subspecies can simmer among scientists for years. In fact, since the caissara was found on a small island and its population was estimated in the dozens, some suspected it might be a subspecies. But during the two and a half years since its discovery, the animals have been spotted on the mainland, and recent population estimates suggest there may be several hundred in existence—strong support for the black-faced lion tamarin being a new species.

Not a year goes by without the discovery of thousands of new species, whether they are found in the potted plant of a Washington, D.C., office building, where a new species of *Pheidole* ant was discovered just a few years ago, or in remote jungles. Most discoveries are of small creatures—ants and beetles and invertebrates. But since the start of the 1990s several new species of birds—the best-known group of animals—have been described, including a Peruvian parrotlet, a flightless rail from the Solomon Islands, and a striped babbler from the mountains of the central Philippines. In addition, three new species of monkey have been discovered, the most recent being a pocket-size marmoset with a koala-like face that was described in October 1992. And just the year before, a new species of whale entered the scientific inventory.

"The interest in new species," says Michael Smith, a senior research scientist at the Center for Marine Conservation, in Washington, D.C., "has in large part come from the biodiversity crisis, the realization that we are now losing species at

6. BIOSPHERE: Animals

enormously high rates and that to do anything about it, we first have to know what species are out there. . . . We have to know what biodiversity is and where it is in order to conserve it rationally. So getting new species described has become critically important."

According to Edward O. Wilson, the Pulitzer Prize–winning Harvard biologist and author of *The Diversity of Life*, scientists don't have time to officially describe more than a small fraction of the species discovered each year. "It's a myth," he says, "that scientists break out the champagne when a new species is discovered." But as the search for new forms of life heats up, some critics regard the enterprise—which in haste may be carried out carelessly—as something of a disease. They call it new-species fever.

THE UNKNOWN WORLD
Since 1758, when the biologist Linnaeus first established a classification system for living things, about 1.4 million living species have been discovered and described. Insects constitute more than half of all named species, of which about 40 percent are beetles—a fact that led British biologist J. B. S. Haldane to comment that the Creator must have had "an inordinate fondness for beetles." Other arthropods, such as spiders and crustaceans, account for another 125,000 species. Flowering plants number about 250,000 and fungi about 70,000. Vertebrates account for about 45,000 species, of which almost half are fish, a quarter are reptiles and amphibians, a quarter are birds, and a tenth are mammals. Algae and microorganisms make up the remaining hundred thousand or so known species.

The concept of species is more than an attempt by humans to impose an artificial order on the world. Species distinctions are real—or so argued zoologist Ernst Mayr in the 1920s, when he found that the natives of New Guinea recognized the same species of birds as Western scientists did.

But just what constitutes a species has been the subject of considerable debate since Linnaeus. For nearly 200 years, the notion of species was based almost entirely on physical characteristics. Then, in 1942, Mayr introduced the Biological Spe-

cies Concept, which holds that the only way to tell one species from another is to determine whether or not they have interbred. Though the concept does not hold true for all species—many plants, invertebrates, certain mammals—it was widely accepted.

Taxonomic methodology was again called into question in the 1980s. This time, geneticists argued that the most reliable indicator of the uniqueness of a species is mitochondrial DNA, which is inherited only from the mother and is therefore the best record of the animal's evolution. Although there is still considerable disagreement as to how DNA tests should be interpreted, the introduction of genetic characters for species determination is generally viewed as a healthy sign. "It means," says Troy Best, a mammalogist at Auburn University, "that we're still pushing back the frontiers of one of the oldest types of science—systematics and taxonomy. We're still learning new ways of looking at relationships among animals."

However they are defined, though, most of the world's species remain unknown. Until about a decade ago, the best estimates of the number of species remaining to be discovered stood at about 1.5 million, for a grand total of about 3 million species. But according to Frederick Grassle, director of Rutgers University's Institute of Marine and Coastal Sciences, new studies of the "lost world" of the deep sea, which until recently was thought to be almost devoid of life, suggest that it alone may harbor another 10 million undiscovered species. Although that figure is a controversial one—other experts put the estimate at no more than half a million—what seems certain is that the deep sea is a habitat teeming with life.

And in the last decade, the canopy of the tropical rainforest has been opened to science, revealing an unexpected number of new species. So dense and so varied is life in this habitat—which provides a home for the majority of the world's species—that one entomologist, Terry Erwin, has proposed that there must be some 30 million species in the world. That figure is based on recent discoveries in the rainforest; other experts, who base their estimates on the total number of

species named so far, claim it is too high, although even they have revised their numbers upward.

But whether the total of life on earth is 3 million or 100 million is less important than is the extent of our ignorance. "What's important," according to Smith, "is that the number still to be discovered is greater than the number we've been able to process in the two hundred and fifty years since Linneaus."

SPECIES DIPLOMACY
Discovery is not always a "happy accident," as it was for the caissara tamarins. Sometimes it's deliberate. But, says Smith, "finding the species is never the object." For him, the primary motive is conservation. "There are as many species of fish in the world as all birds, amphibians, reptiles, and mammals put together," he says. "But virtually no protected areas have yet been set up specifically for fish. I'm saying that the larger part of biodiversity is being overlooked."

One sure way to make a discovery is to find a gap in knowledge, explains Smith. He found one such gap in the West Indies with the help of a biological rule of thumb that states that for each 10-fold increase in an island's area you can expect roughly double the number of species. When Smith compared the number of endemic species of vascular plants, ants, and fish in Cuba, Hispaniola, Jamaica, and Puerto Rico, he found that the number of fish species in Cuba, the largest island, fell far short of expectations—by 25 to 50 species.

Smith set his sights on Cuba in the late 1980s, though given the diplomatic climate, his task was not without difficulty. Scientists trying to get there had to be invited, but to be invited they first had to know someone there. The same was true

> Since 1990 several new species of birds and monkeys—as well as a new whale—have been discovered.

of Cuban scientists trying to get into the United States. Smith got around this catch-22 by taking a visiting professorship at the University of Santo Domingo, in the Dominican Republic, which had an exchange program with Cuba. In 1989 he learned that Havana's National Museum of Natural History had just begun a biological inventory of the island; eventually, he teamed up a number of Cuban and American scientists to help uncover the island's hidden cornucopia of life. Says Smith, "For me, ichthyology in Cuba was like ping-pong with China. You do what you do and use it to change the world."

So far, the collaboration has paid off handsomely. Smith has already found four new species of small freshwater fish, all in the order *Cyprinodontiformes*. They aren't a very charismatic group of organisms, he admits, but in terms of biodiversity, "a species is a species." By the time the Cuban survey is complete, the team expects to have doubled the known number of fish species on the island, where the head scientist of the national museum estimates that about 40 percent of the fauna are yet to be discovered. According to Orlando Garrido, curator of birds and reptiles at Havana's National Museum, the last two years have produced 19 new reptiles—3 geckos, 4 snakes, 5 frogs, and 7 lizards. "We get a kick out of every discovery," says Garrido. "To be in the field is a thrill already. To discover a new thing is a bigger thrill. It doesn't matter what it is."

NAMING NAMES

For biologists, a species doesn't exist until it's been scientifically described. But although the challenge of discovery has fueled the quest for new species, the formal process of description is so long and drawn out that some biologists end up keeping a new species in their back pocket for years before announcing a discovery.

James Mead, curator of marine mammals at the Smithsonian's National Museum of Natural History, collected the remains of his find—a new species of beaked whale—in February of 1976. But since the official date of discovery for a new species is tied to the publication date of its description in a refereed scientific journal, the world did not learn of Mead's discovery for 15 years.

"I was lucky," says Mead, recalling his original find. He had been beachcombing near the fish market of San Andrés on the coast of Peru when he spotted a skull in the sand. Though the back of the head and the snout were missing, he recognized it as a species of *Mesoplodon*, or beaked whale—his specialty. Then just 10 feet away, a colleague picked a whale vertebra out of the sand. It belonged to a mature animal. In talking with the local fishermen, Mead learned that the parts belonged to a whale that had been brought in about a year and a half earlier.

The whale, later described as *Mesoplodon peruvianus*, is the smallest of 13 known species of *Mesoplodon*. Deep gray on its back and light gray from midside to belly, it is about 5 feet long at birth, 12 feet long as an adult. What distinguishes it from other species are the size, shape, and position of its teeth, as well as its strikingly small cranium.

Although Mead "strongly suspected" he had found a new species back in 1976, nine years passed before another specimen showed up—a female calf also caught off the Peruvian coast. The specimen was acquired by Julio Reyes, a university student Mead had been in touch with. "But that still wasn't enough for a complete description," says Mead. "Most *Mesoplodon* are diagnosed on characters that occur in adult males, and we didn't have one. So we sat and waited."

Three years and six specimens later, Mead still lacked an adult male. "At that point," he recalls, "we just threw up our hands and decided to describe the species anyway." But in November of 1988, while Reyes was in Washington working up the description with Mead, an American named Tony Luscombe—a contact for biological expeditions from the United States—found an adult male stranded at Playa Paraíso, near where the first find was made. As the only adult male found of the species, the specimen became the official example of the species. The following year, Reyes examined a 10th specimen, a young female taken by fishermen in the town of Pucusana. So Mead and Reyes, along with Koen Van Waerebeek of the Peruvian Center for Cetalogic Studies, revised their species description and sent it off to the journal *Marine Mammal Science* in October of 1989.

They nearly got scooped. While the paper was being prepared for publication, Mead received a manuscript for review from the *Journal of Mammalogy* that described a new species of *Mesoplodon* from Baja California, Mexico. "It was our species," said Mead, astonished. When the other researchers were shown a preprint of Mead's description they withdrew their manuscript, leaving Mead and his colleagues the honor of discovery. Says Mead, "That we can still find an animal as large as a whale—even a relatively small one—serves to point out that we don't know a lot about the oceans."

THE FERTILE SEA

Rutgers University scientist Frederick Grassle and his co-workers were not really looking for new life when he took 233 box-core samples of the sea bottom off New Jersey and Delaware in 1984 and 1985. They were mainly trying to describe the diversity of deep-sea communities. But what they found in those samples eventually led them to propose a dramatic revision of the total number of species likely to inhabit the oceans.

Widely regarded as being short of nutrients and thinly populated, the deep sea has been largely ignored in estimates of the total number of species on earth. It is generally believed that the 160,000 or so named and recorded marine species are representative of a true total that some put at no more than half a million.

Grassle and his colleagues collected samples that would cover an area no larger than two tennis courts and found 90,677 small invertebrates living on or in the sediments. The creatures were tiny but larger than half a millimeter—big enough to see. Most of this macrofauna were polychaete worms, crustaceans, and mollusks, but they represented more than 14 different phyla, 171 different families, and 798 species, of which 460, or 58 percent, had never before been seen by science.

Such diversity is unheard of on terra firma. "If you took the same area of rainforest and came up with a hundred and seventy-one families," says Grassle, "people would be startled." The oceans are home to 28 phyla in all, 13 of which are found neither on land nor in fresh water. But there are no more than 11 phyla in all terrestrial habitats, of which only one is unique to the land. Even then, most terrestrial species belong to just two phyla. "So at the phylum or family level," says Grassle, "the sea is a lot more diverse than the rainforest. But in terms of species it's still an open question."

If the rest of the ocean bottom is as bountiful as the area Grassle sampled—given that some areas are more so and others certainly less so—"there have to be at least one million species out there," he concludes. "And ten million is not an unreasonable number."

PHANTOM OF THE HIGHLANDS

Somewhere in the wilds of Madagascar lurks a gigantic moth with a six-inch wingspan and a proboscis of astonishing size—more than a foot long. No one has reported seeing it, nor has it ever been described in scientific literature, but Gene Kritsky, an entomologist at the College of Mount St. Joseph-on-the-Ohio, near Cincinnati, is certain it exists.

Kritsky's quest began in the spring of 1991, when he came upon the description of the *Angraecum longicalcar* 'Bosser,' a rare Madagascan orchid with a 16-inch-deep nectar tube. Since only moths can pollinate this group of orchids, Kritsky realized that for the moth to reach the inch-deep pool of nectar at the bottom of such an orchid, it would have to have a proboscis 15 inches long.

"We've got the orchid," says Kritsky, "now we've got to find the moth." He knows the effort will not be easy. "These moths are nocturnal pollinators, and it's a two-day hike from the main power source to set up infrared lighting to watch this thing come swooping in." Nor is it possible to pinpoint a 48-hour period when the orchids will be open and the moth has to come in. Unlike other orchids, which stay open only a day or two, these angraecoids stay open and fresh for weeks.

So whoever seeks out the giant moth may have a long wait—assuming, of course, that the orchid itself can be found. Horticultural importers have not seen a Madigascan longicalcar for several

Population Explosion

On an ordinary expedition to the rainforest, Terry Erwin, curator of entomology at the National Museum of Natural History, brings back 500,000 insect specimens at a time. In fact, since 1974 his fogging for arthropods—everything from mites to beetles—in the rainforest canopies of Brazil, Peru, Bolivia, Panama, and Venezuela has nabbed some 5 or 6 million specimens. He estimates that about 80 percent of the creatures falling into his aluminum funnels below the canopy are new species.

On each trip to the rainforest Erwin does anywhere from 300 to 400 foggings. After each one he sorts through the four pints or so of mixed insects and pulls out all the beetles, which are his specialty. Sometimes a colleague will pull out the spiders. All the rest go to the museum sorting center, where they are broken down into orders, sorted, and stored. Erwin then examines all the beetles with his microscope, pulls out one specimen of every species, labels it, codes it, and enters the data into his computer. This general cut tells him how many species there are in each sample of each microhabitat. For some families, such as carabid beetles, Erwin may then do the species description himself.

"Everybody thinks of the tropics as these big insects crawling around," says Erwin, "and sure, that's what you see when you walk through. But as many as two-thirds of the arthropod species in the tropics are in the canopy. What a lot of people don't appreciate is that everything up on the top of the trees is only about—this is beetles again—three millimeters long. That's the magic of the tropics, and it's obvious to any taxonomist that that's why so much is new. People just haven't dealt with all the little things."

The diversity of insects is such, says Erwin, that it doesn't matter that new species be named. What matters is that they exist. So for the past five years he has tried to develop a short cut—an interim taxonomy—to catalogue his enormous number of finds.

The ground rules of nomenclature

require going back to the Latinized names and type specimens, but the sheer numbers Erwin brings back make this well-nigh impossible. And even if it were possible, it wouldn't be a good idea, says Erwin, who considers most of the original literature on entomology in the tropics "worthless." Three-quarters of the tropical species of insects currently catalogued were discovered in the 1800s, and most were described on the basis of just one or two specimens, when in fact 30 or so are usually needed to properly establish the variations in the species. Many of the descriptions also lack such crucial information as the precise habitat in which the specimen was found.

It must all be checked eventually, says Erwin. But in the meantime he uses an alphanumeric code in lieu of an official Latinized taxonomic name. Every family of beetles, for instance, gets a four-letter code, followed by a number that he uses only for that species. Later the specimens may be named by a specialist, but until then Erwin can do his bioinventory without having to travel all over Europe looking at museum type specimens and assigning the proper Latinized name to each species. "I hope to show that we can do bioinventories by assessing the number of microhabitats and the complexity of their biological architecture," he says, "rather than by counting every living species of insect."

In 1982 this kind of work led Erwin to propose a dramatic increase in the estimated number of species in the world. On the basis of the beetles he has collected in the canopies of 19 individuals of one species of tree in Panama, Erwin estimated that there may be 30 million species worldwide. And he's talking insects only. When he is asked for a grand total that includes other forms of life, Erwin shrugs. "Well," he says, "that would make it about thirty million, four hundred thousand."

Erwin's 30-million-species estimate is subject to heated debate. But even those who dispute it as too high agree that on this planet at least, insects are clearly in the majority—and are likely to suffer the greatest losses as the world's tropical rainforests steadily disappear. —*P.H.*

years. They fear that it may no longer be around. If so, it's likely that the giant moth Kritsky has posited may have vanished before ever being seen.

In that sense, the Madagascar moth may not be so unusual. We are losing species faster than they are being described. Conservative estimates place the yearly loss at about 27,000 species a year in the rainforests alone. Harvard's Wilson has argued that what is needed is a massive inventory effort on the scale of the human genome project. It could be completed in 50 years, says Wilson, and would be worth every cent.

But unfortunately, with the diversity of life largely locked up in small creatures like insects, calls for their preservation have not engendered the same public sympathy as the battle to save endangered whales or eagles. Few seem to realize that small, obscure creatures are the foundation of intricate webs that support all life. Or that, as Wilson puts it, "each species is a masterpiece, regardless of size. Species are not interchangeable. Each is exquisitely adapted to the environment, and each is a potential reservoir of vast new scientific knowledge."

Born Free

Our knowledge of the world's birds is more complete than that of any other class of animals. A recent tally recognized 9,881 species, 43 of which were first described in the 1980s. But the most notorious avian discovery of late is the rare shrike found in Somalia and described in 1991.

The extraordinary saga began in Bulo Burti, Somalia, on August 27, 1988, when Edmund Smith, a British entomologist and avid birdwatcher, spotted a "weird" black and white bird he could not identify.

Normally the bird would have been killed, to become the official species representative. But Smith was reluctant to kill a bird of a species that might be near extinction. So with the help of a Somalian, he captured the lone bird and placed it in the care of a West German ornithologist, Jan-Uwe Heckel, then living in Mogadishu.

Over six months Heckel carefully noted the bird's behavior. Blood samples, along with a few feathers the bird had lost in captivity, eventually reached the University of Copenhagen, where a comparison of the bird's DNA with that of other shrikes proved that it was a new species.

Meanwhile, following an outbreak of civil unrest in Somalia, Heckel flew to Germany with the bird, then returned to Somalia some months later. But as the bird's original habitat was still in the grip of a civil war, Heckel took the shrike to a nearby nature reserve. And there he released it.

That controversial action brought to the forefront fears of conservationists, who think that when it comes to rare new species, the collecting and killing of even a single animal could push the species to extinction. Some ornithologists agree, but many others find the release of this bird sentimental and shortsighted.

Writing in the centenary volume of the *Bulletin of the British Ornithologists' Club*, Mary LeCroy and François Vuilleumier deplored the waste of the type specimen, doubting that the bird could survive after a year in captivity, especially in a strange area. And without a specimen for biologists to study, they noted, much about the shrike would remain unknown.

—*P. H.*

Barnyard Biodiversity

Agricultural breeds around the world—like wild species—are vanishing and taking with them a rich genetic heritage.

Verlyn Klinkenborg

Verlyn Klinkenborg is working on a book about the Indian Health Service. He is also the author of The Last Fine Time *and* Making Hay. *His essay on biopolitics appeared in the January–February 1992 issue of* Audubon.

The next time you drive past a barnyard full of Holsteins—the familiar black and white dairy cows—pull over and watch them for a moment. They will most likely be standing up to their dewclaws in mud or dust and will either be eating for the sake of eating or eating while waiting to be milked while eating. Holsteins are large, hungry beasts with great veined orbs for udders, highly successful by today's standards of automated, pneumatic dairying. In the rural landscape, the ubiquity of Holsteins somehow makes them seem permanent, as durable and timeless as the stone walls that once divided the farms of New England, before they became suburbs.

But Holsteins aren't timeless, either as individuals or as a breed. Average modern Holsteins are allowed to live only 4 1/2 years before they're sold at cull-cow sales and turned into lean hamburger. If they lived any longer their legs would begin to give way, as would the sinews supporting their udders, which may weigh as much as a full-grown man. Except in the sentimental minds of city dwellers touring the countryside, there's nothing pastoral about Holsteins. To produce a pound of milk, a Holstein needs a pound of high-protein food. They're industrial animals, no more fuel efficient than a mammoth tractor, suited to the richness of an industrial economy. And though they may seem insouciant in the barnyard or the pasture, their angular rumps turned to the wind, they're quite delicately positioned at the acme of agricultural fashion. Holsteins have had a good run of it, nearly 60 years of popularity, with more to come. But agricultural fashions don't last forever.

What, for instance, if you were a farmer who wanted a dairy cow that thrived on rough forage—mediocre grassland, say—rather than on alfalfa hay and expensive high-grade feeds? What if you didn't need 20,000 pounds of milk a year from each of your animals? What if you wanted a cow that didn't require a steady supply of vaccines, drugs, and hormones to reach its expected level of milk production? What if you just wanted a cow that lived longer without breaking down?

Losing a breed is like losing a language or any other unique, complex relic of human culture.

You might look for other dairy breeds, but soon enough you'd find that many other dairy breeds have become uncommon, endangered, or extinct. In the United States and Great Britain, you don't see many Jerseys or Guernseys these days. You see almost no Ayrshires or Milking Shorthorns or Brown Swiss or Milking Devons or Red Polls or Gloucesters or Kerrys or Shetlands or Dexters or Canadiennes or Dutch Belted or Irish Moiled cows, much less Irish Duns, Alderneys, Sheeted Somersets, Glamorgans, and Castlemartins, which are extinct. What you see are Holsteins.

But I *have* seen a mature Kerry cow, one of the rarest dairy breeds. There are only a few dozen Kerrys in the United States and only about 300 in their native Ireland. The one I saw was standing with her younger sister and a Kerry bull calf in a stone-walled paddock at the Museum of American Frontier Culture in Staunton, Virginia, a living history museum that, in the interest of historical and geographical accuracy, raises other rare breeds as well, like Percheron horses, Milking Shorthorn cows, and Hog Island sheep.

The Kerry's name was Patty. She was a small, graceful, light-limbed cow, with delicate upswept horns, a solid-black coat, and the ability to thrive on inferior pasture where a Holstein would soon go hungry. Patty may also calve until she's 14 or 15 years old. Like all purebred Kerrys she is, according to the British conservationist Lawrence Alderson, "one of the purest surviving descendants of cattle brought to Britain in the early part of the second millennium B.C." and is thus of especially high genetic value. At Patty's feet a chicken – a White Wyandotte – picked at her hoardings. Kit Nicholson, the museum's livestock expert, and I leaned against the stone wall and watched as the Kerrys grazed and the Wyandotte scratched. "That," Nicholson observed, "is a nineteenth-century chicken." That, she might as well have added, is a pre-Christian cow.

Not far from the Kerrys' paddock stood a giant tent that temporarily housed the American Minor Breeds Conservancy, an organization founded in 1977 and devoted to preserving endangered agricultural breeds. Don Bixby, the AMBC director, and his staff had driven up from their offices in Pittsboro, North Carolina, to the Museum of American Frontier Culture's annual Fall Festival, held in September. There, under the big top, they had assembled a menagerie of rare breeds of cattle, sheep, and poultry.

Bixby, an affable veterinarian, led me around the pens. There were Lincoln Longwool, Cotswold, and Tunis sheep. There were chickens of every feather: White Silkies, Black Rosecombs, Buff Leghorns, Black Cochins, Old English Gamecocks, and Dominiques. There was a Scotch Highland cow, a Belted Galloway, and a deep-red bull whose sign read "Milking Devon Bull." "There's one for Ripley," said an old man walking past the sign. Bixby explained that in the 1960s the Devon breed split into two types, the Beef Devon and the Milking Devon. This was a bull of the Milking Devon type. The Devon is the oldest recognized breed in North America, Bixby noted, dating to 1623 in the records of the Plymouth Plantation.

It takes some getting used to, the idea that time can be yoked so firmly to an animal. After all, wild creatures are timeless until they become extinct, and a date is fixed to their disappearance. But farm animals belong to the culture of mankind, and they're touched, like humans, by time and by the succession of whims that so helpfully mark time's passing.

Since animals were first domesticated some 10,000 years ago, several thousand breeds of buffalo, horses, cattle, swine, sheep, goats, and fowl have arisen through human and environmental selection. Breeds, which are roughly equivalent to subspecies in the wild, were at first, and in many cases still are, kept distinct mainly by geographical isolation, though there's plenty of evidence to suggest that early breeders had a sound practical grasp of selective breeding. The peculiarly Western concept of breed, which requires a stud, flock, or herd registry book with detailed pedigrees, actually emerged in early 18th-century England, where some of

the most dramatic breakthroughs in breed improvement occurred.

Since then, breeders in Europe and America have tended to select for production-oriented traits, like high-protein-and-butterfat milk or marbled beef. Multipurpose cattle breeds, like the Devon, which once provided milk, draft power, and meat, were split into separate strains and then nearly disappeared as other, more specialized breeds came into fashion. The number and quality of breeds has always ebbed and flowed. But what has changed in modern agriculture, with new tools like artificial insemination and embryo transplants, is the speed and efficiency with which unfashionable breeds can be replaced.

"In popular breeds," Bixby said, "certain bulls can now not only be prepotent, they can be ubiquitous" – like the British Holstein bull Alsopdale Sunbeam II, who "provided two hundred and fourteen thousand, two hundred and ninety-three first inseminations" and left another 200,000 doses of semen in storage before he died, in 1979.

In fact, the rate at which agricultural breeds are vanishing is nearly as alarming as the rate at which wild subspecies are going up in smoke. "In 1947," wrote conservationist Alderson in *The Chance to Survive*, "there were 42 Italian breeds of cattle; by 1974 only 22 breeds remained, and of these 13 were gravely endangered." In 1984, according to the Food and Agriculture Organization of the United Nations (FAO), a "survey of five livestock species in Europe showed that, of 700 breeds reviewed, a third were in danger of extinction. The same pattern has been repeated in countries where intensive livestock breeding rules the day.

It seems oddly poignant that animals that coevolved so closely with humans and that contributed so largely to the success of the human species should fall away under the pressures of modern agricultural selection, only to be replaced by animals whose breeders have, in many cases, sacrificed efficiency and the health of the creature itself to obtain industrial levels of production.

But since all breeds emerged under the hand of man to meet human needs, and since those needs have changed dramatically over time, why should losing a breed matter now? There are two answers. The diversity of breeds – the profuse adaptations that have occurred among domesticated animals throughout history – is itself one of the most sophisticated accomplishments of the human race. Losing a breed is like losing a language or any other unique, complex relic of human culture. But like losing a wild subspecies, losing a breed also means losing genetic diversity. The 20 or so major domesticated animal species represent only a tiny part of earth's genetic resources (there are probably more than 30 million species), but it's an extremely important part.

In agricultural circles, there's general agreement about the value of genetic conservation. But there's also real dispute about how best to conserve the diversity of genes in both major and minor breeds. In a sense, the idea of genetic conservation conflicts with the whole thrust of agricultural research, which is geared toward constant improvement of agricultural animals—however "improvement" happens to be defined at the moment.

One steamy day last September, Don Bixby took me to a farm outside Pittsboro, North Carolina, that is owned by Clarence Durham, a retired village postal worker. Clarence is president of the Tamworth Swine Association, which promotes that breed and elects a (human) National Tamworth Queen each year. He guided us through cobwebbed hog houses filled with Tamworth pigs, a rare hog that, according to the AMBC, is "one of the oldest pig breeds known." Clarence raises purebred Tamworths for sale to other breeders—and for barbecue—and he also crosses his boars and sows with Swedish Landrace and Large White hogs, two modern breeds.

Domesticated species are a tiny part of the earth's genetic resources—but an extremely important part.

In the farrowing house, the air was filled with the sound of blowing fans and the high-pitched clamor of Tamworth and Tamworth-cross piglets, some only a day or two old. Clarence, who moves among his hogs with remarkable stateliness, reached down into the mass of russet-colored piglets swarming at their mother's teats, lifted one out by the trotters, and handed him to Don Bixby, who has a good sense of humor but couldn't quite muster the look of pride that was expected of him. Then, while we waited at the fence line, Clarence walked down into the brushy bottom at the back of his property and led his Red Poll cattle into view. As he sauntered among his herd, some dozen or so in number, Clarence said, or seemed to say, in a slow, even, North Carolina drawl, "Cooowww-baby . . . cooowwwbaby . . . cooowwwbaby."

Like the Tamworths, there was nothing exotic in the appearance of the Red Polls—nothing but the deep saturation of oxblood red in their coat, and their quiet demeanor and modest size. The rarity of the cows standing in the open around Clarence wasn't apparent, nor was their historicity. But only about 2,000 Red Polls, first bred in England in the early 19th century, were registered in 1991 (there are nearly 1.5 million registered Holsteins in the United States). Red Polls are now on the AMBC's watch list, which means they were in steady decline and still have not fully recovered—despite the fact that Red Polls live longer than

Holsteins, are more efficient foragers, and yield higher quality milk.

Keeping the breeds themselves intact, as Clarence Durham and the other 3,800 members of the American Minor Breeds Conservancy are trying to do, is one way to ensure genetic diversity. Like the Rare Breeds Survival Trust in Great Britain and similar organizations in Switzerland and Hungary, the AMBC tries to preserve minor breeds at population levels high enough to avoid excessive inbreeding or the loss of bloodlines or genetic drift, which happens when populations get too small and a gene simply disappears through random events. The AMBC maintains a breeders' directory and a census of livestock, and it has been instrumental in recovering a number of breeds, including the Wilbur-Cruces strain of Spanish mustang, the milking type of Devon cow, the Florida Cracker cattle, and the Piney-woods cattle of Georgia. It also establishes registries and herd books for breeds that have fallen on hard times, and it stores semen at depositories in New York, Ohio, Illinois, California, and Louisiana.

Mike Strauss, an expert on the conservation of genetic resources who is affiliated with the American Association for the Advancement of Science, in Washington, D.C., calls the AMBC strategy preservationist—conserving genes by preserving breeds. But, says Strauss, who is also a member of the AMBC board, "for some people, collecting rare breeds is like collecting stamps—they want every individual variety and imperfection. You have to be careful about building a case for the potential genetic utility of a single breed, much less a single animal. The genetic value lies in the whole group of breeds."

Even when it comes to storing semen, the strategies for preservation aren't simple, according to James Womack, who chaired the United States Department of Agriculture committee on genome mapping in poultry, swine, and cattle. "What do you do?" he asked me. "Do you preserve the germ plasm of every individual, every line, every breed, every group of breeds?"

Many scientists and breeders, especially those at the USDA and the nation's agricultural colleges, don't think it's necessary to preserve the minor breeds themselves. Strauss calls this group utilizationist. In their view, as he describes it, "breed identity is not considered critical . . . so long as the desirable genes from available breeds are retained in the commercial populations. Emphasis is placed on retaining useful genes and genetic variation, not on retaining breeds per se." The hidden dynamic behind the utilizationist view is the power of market forces. Because economic incentives drive breed improvement, economic forces—the market itself—will always ensure that the commercial animal population contains the genes essential for future improvement. So the argument goes.

But there are several problems with the utilizationist point of view. For one thing, it's never clear where the livestock market is headed or what traits will be valuable in the future. And in the meantime the number of people who make decisions about livestock breeding has been rapidly shrinking over the past 50 years. Once local stockmen raised local breeds to suit local demand. "But decisions about what the market wants are made by many, many, many fewer people than used to be the case," Don Bixby said as we talked at the AMBC office, overlooking central Pittsboro. "Livestock is scarcely held by stockmen any longer; it's held by syndicates, which distribute animals to contract growers. The poultry industry is controlled by very few people, and the genetic plan for industrial poultry stretches only about three years into the future.

"I'm really impatient with agricultural experts who say we've already got everything we'll ever need. In the first place, we don't know what we've got, and in the second place, we don't know what we'll need."

Even as scientists gain a foothold in new fields of biotechnology, old breeds continue to slip away.

In the past, too, breeders in Europe and America often imported exotic breeds like Zebu cattle or the Chinese pig to improve domestic breeds. "The old assumption," said Bixby, "was that we could always get genetic resources from somewhere else." But as one FAO expert explains, in the Third World "many local breeds of undistinguished appearance are disappearing without it even being known whether or not they have unique adaptive genetic traits." These days agricultural fashion is a worldwide phenomenon. Industrial animals like Holsteins have been exported to regions in the Third World where it makes little economic sense to raise high-input cattle and where for many reasons, almost none of which have to do with efficiency, Western animals tend to quickly replace indigenous species—usually just long enough to doom them.

This is a complex tragedy. Because Western breeds have been so carefully and intensely bred, they tend to have a high level of genetic redundancy: They form sets of genetically consistent information. But in most of Africa and Asia, where breed associations and registry books are nonexistent, agricultural breeds are defined much more loosely—"by regional isolation, color pattern, or other morphological or behavioral characteristics" rather than by strict pedigree, notes Mike Strauss. As a result, such breeds usually contain far more genetic diversity within themselves than Western breeds do. "If you lose an African breed," says Strauss, you may be losing a large quantity of genetic diversity."

Even as scientists gain a foothold in new fields of biotechnology, old breeds continue to slip away. The USDA has shown little interest in breed preservation, while investing heavily in gene mapping, which is still in its infancy. Gene mapping, transgenic experiments, and single-gene storage all offer great promise for genetic conservation in a still distant future. But as Strauss remarks, "a breed cannot be reconstituted from a DNA library."

Bixby takes a more philosophical attitude. "We think we're at the apogee of our knowledge," he says. "And we never are. We're always stupid, and we don't have to look very far back to see how stupid we were. What's happening with contemporary agriculture so far is selection for uniformity, because we've been led to believe that the ultimate goal in agriculture is to have uniform animals—completely interchangeable. But the imposition of an industrial model on a biological system isn't going to be stable in the long run.

In fact, there has been a strong surge of interest worldwide—even within the USDA—in sustainable agriculture, which the FAO calls "environmentally nondegrading, technically appropriate, economically viable, and socially acceptable." It may well happen that there will soon come a time when it makes no sense to have long lines of Holsteins standing between concrete stanchions with their heads dunked in a manger full of high-protein concentrate while their milk is sucked out of them pneumatically. There may come a time when it makes sense again to raise dairy animals that thrive on available feeds, feeds they harvest themselves on sound, long-lived legs.

Don Bixby says, "What's happening with contemporary agriculture so far is selection for uniformity."

Lawrence Alderson calls this "the ecological approach," by which he means taking "the natural environment as the starting point, selecting breeds that are suited to it . . . with emphasis on efficiency rather than maximum production." But that, of course, is how it always used to be done, in the days when wild species were first being domesticated.

As for the preservationists, they've had the goal of sustainable agriculture in mind all along. "We're future utilizationists," says Bixby. "Sometime in the future any of these minor breeds might be useful. But we don't know and we don't presume to know which ones will be needed, so we're interested in diversity, not only in genetics, but also in the ways these animals are conserved."

6. BIOSPHERE: Animals

And that's where the members of the American Minor Breeds Conservancy weigh in. For them, in the end, it comes down not to economics or genetics but to the animals themselves. One day in mid-September I drove into the rolling countryside near Staunton to the home of Margot and Phil Case. The Cases breed a rare horse called the Akhal-Teke, which originated on the Akhal oasis in southern Turkmenistan and is directly descended from an ancient type of Turkoman war-horse used by Alexander the Great. There are only 100 or so Akhal-Tekes in this country, and their numbers in Turkmenistan are unknown but presumed to be less than 1,000 and declining.

There was a horse show that day at the Case farm, and alongside the show-jumping ring, amid the familiar collection of chestnut, bay, and black animals, there stood an Akhal-Teke stallion with a young woman astride him. He was more lightly framed than the horses around him, with a delicate face and a sense of intense concentration in his manner. But what distinguished him most of all—and it's a characteristic of the breed—was his coat. It glowed in the sun with a metallic sheen, as if he were a sculpture beaten from copper or gold. In the pastures on the surrounding hills, more Akhal-Tekes, nearly 40 of them, gleamed in the autumn sun.

A week or two later, I sought out one last rare breed. I drove to Scotland, Connecticut, to visit Penelope de Peyer at the farm she calls Chakola's Place. De Peyer is a brisk, direct woman who was asked, at age 13, to hold a horse for someone at a show in Windsor, England. That horse was part Cleveland Bay, the oldest established breed in England, originally from Yorkshire, and the only remaining purebred native horse of the United Kingdom. "It changed my life," she said.

There are just 25 Cleveland Bays in the United States, and in 1962 the breed had declined to such an extent that there were only three purebred stallions in England. (Their numbers remain low largely because they make such good sport horses when crossed with Thoroughbreds that they're rarely raised as purebreds.) In the words of historian Elwyn Hartley Edwards, the Cleveland Bay "hauled heavy loads in testing conditions; it carried large men to hounds and was in no way deterred by being asked to jump out of deep clay, and it was also a coach horse of great strength."

"I always say," said De Peyer as she led out her 12-year-old stallion, Fryup Marvel, "that Cleveland Bays throw sanity, size, and substance." Marvel displays all the traits of this distinguished breed. He's a large, quiet horse with a beautifully dappled bay coat and an exceptionally kind eye, a defining characteristic of the breed. But the big news at Chakola's Place was in the paddock behind us. Lying in the sunshine beneath Ariadne, a Cleveland Bay mare, was a three-week-old purebred colt, still in his milk hairs and as soft as angora. His name was Advocate, the 26th Cleveland Bay in America.

262

ENVIRONMENTAL INFORMATION RETRIEVAL—ON FINDING OUT MORE

There is probably more printed information on environmental issues, regulations, and concerns than on any other major topic. So much is available from such a wide and diverse group of sources, that the first effort at finding information seems an intimidating and even impossible task. Attempting to ferret out what agencies are responsible for what concerns, what organizations to contact for specific environmental information, and who is in charge of what becomes increasingly more difficult.

To list all of the governmental agencies private and public organizations, and journals devoted primarily to environmental issues is, of course, beyond the scope of this current volume. However, we feel that a short primer on environmental information retrieval should be included in order to serve as a springboard for further involvement; for it is through informed involvement that issues, such as those presented, will eventually be corrected.

I. Selected Offices Within Federal Agencies and Federal-State Agencies for Environmental Information Retrieval

Appalachian Regional Commission
Public Information: 1666 Connecticut Avenue, NW, Washington, DC 20235 (202) 673-7968

Council on Environmental Quality
722 Jackson Place, NW, Washington, DC 20503 (202) 395-5750

Delaware River Basin Commission
Main Interior Building, Washington, DC 20240 (202) 343-5761

Department of Agriculture
14th Street and Independence Avenue, SW, Washington, DC 20250 (202)447-2798

Department of the Army (Corps of Engineers)
Office of the Chief of Engineers, U.S. Army Corps of Engineers, Washington, DC 20314 (202) 272-0001

Department of Commerce
Main Commerce Building, 14th Street and Constitution Avenue, NW, Washington, DC 20230 (202) 377-2000

Department of Defense
The Pentagon, Washington, DC 20301 (703) 545-6700

Department of Health and Human Services
Department of Health and Human Services, 200 Independence Avenue, SW, Washington, DC 20201 (202) 245-6343

Department of the Interior
Interior Building, 1849 C Street, NW, Washington, DC 20410 (202) 708-0980
- Bureau of Indian Affairs (202) 208-3711
- Bureau of Land Management (202) 208-5717
- Bureau of Outdoor Recreation (202) 343-1005
- National Park Service (202) 208-7394
- United States Fish and Wildlife Service (202) 208-5634

Department of State, Bureau of Oceans and International Environmental and Scientific Affairs
2201 C Street, NW, Washington, DC 20520 (202) 647-3686

Department of the Treasury, U.S. Customs Service
1301 Constitution Avenue, NW, Washington, DC 20229 (202) 566-2000

Energy Department
1000 Independence Avenue, SW, Washington, DC 20585 (202) 586-5806

Energy Information Administration
1000 Independence Avenue, SW, Washington, DC 20585 (202) 586-8800

Environmental Protection Agency
Director, Office of Federal Activities, Environmental Protection Agency, 401 M Street, SW, Washington, DC 20460 (202) 382-4355

Regional Administrator I, U.S. Environmental Protection Agency
Room 2303, John F. Kennedy Federal Building, Boston, MA 02203 (617) 565-3417
(Connecticut, Maine, Massachusetts, New Hampshire, Rhode Island, Vermont)

Regional Administrator II, U.S. Environmental Protection Agency
26 Federal Plaza, New York, NY 10007 (212) 264-2515
(New Jersey, New York, Puerto Rico, Virgin Islands)

Regional Administrator III, U.S. Environmental Protection Agency
841 Chestnut Street, Philadelphia, PA 19107 (215) 597-9370
(Delaware, Maryland, Pennsylvania, Virginia, West Virginia, District of Columbia)

Regional Administrator IV, U.S. Environmental Protection Agency
345 Courtland Street, NE, Atlanta, GA 30365 (404) 347-3004
(Alabama, Florida, Georgia, Kentucky, Mississippi, North Carolina, South Carolina, Tennessee)

Regional Administrator V, U.S. Environmental Protection Agency
230 South Dearborn Street, Chicago, IL 60604 (312) 353-2072
(Illinois, Indiana, Michigan, Minnesota, Ohio, Wisconsin)

Regional Administrator VI, U.S. Environmental Protection Agency
1445 Ross Avenue, Dallas, TX 75202 (214) 655-2200
(Arkansas, Louisiana, New Mexico, Texas, Oklahoma)

Regional Administrator VII, U.S. Environmental Protection Agency
726 Minnesota Avenue, Kansas City, KS 66101 (913) 551-7003
(Iowa, Kansas, Missouri, Nebraska)

Regional Administrator VIII, U.S. Environmental Protection Agency
999 18th Street, Denver, CO 80202 (303) 293-1119
(Colorado, Montana, North Dakota, South Dakota, Utah, Wyoming)

Regional Administrator IX, U.S. Environmental Protection Agency
75 Hawthorne Street, San Francisco, CA 94105 (415) 744-1020
(Arizona, California, Hawaii, Nevada, American Samoa, Guam, Trust Territories of Pacific Islands, Wake Island)

Regional Administrator X, U.S. Environmental Protection Agency
1200 Sixth Avenue, Seattle, WA 98101 (206) 442-1465
(Alaska, Idaho, Oregon, Washington)

Federal Energy Regulatory Commission
825 North Capitol Street, NE, Washington, DC 20426 (202) 208-1088

Interstate Commission on Potomac River Basin
6110 Executive Boulevard, Rockville, MD 20852 (301) 984-1908

Nuclear Regulatory Commission
Director, Office of Public Affairs, Washington, DC 20555 (202) 492-0240

Susquehanna River Basin Commission
1100 L Street, NW, Washington, DC 20240 (202)343-4091

Tennessee Valley Authority
412 1st Street, SE, Washington, DC 20444 (202) 479-4412

II. Selected State, Territorial, and Citizens' Organizations for Environmental Information Retrieval

A. GOVERNMENT AGENCIES

Alabama:
Department of Conservation and Natural Resources, 64 North Union Street, Room 702, Montgomery, AL 36104 (205) 261-3486

Alaska:
Department of Environmental Conservation, PO Box O, Juneau, AL 99811 (907) 465-2600

Arizona:
Department of Water Resources, 15 South 15th Avenue, Phoenix, AZ 85007-3226 (602) 542-1554

Natural Resources Division, 1616 West Adams Street, Phoenix, AZ 85007-2600 (602) 542-4625

Arkansas:
Department of Pollution Control and Ecology, 8001 National Drive, Little Rock, AR 72209-4800 (501) 562-7444

Energy Office, 1 State Capitol Mall, Room 4B-215, Little Rock, AR 72201-1012 (501) 682-1370

California:
Environmental Affairs Agency, 1102 Q Street, Sacramento, CA 95814-6511 (916) 322-4203

Colorado:
Department of Natural Resources, 131 Sherman Street, Room 718, Denver, CO 80203-2239 (303) 866-3311

Connecticut:
Department of Environmental Protection, 165 Capitol Avenue, Hartford, CT 06106-1600 (203) 566-2110

Delaware:
Natural Resources and Environmental Control Department, 89 Kings Highway, Dover, DE 19901-3816 (302) 736-4403

District of Columbia:
Department of Energy, Environmental Safety and Health, 1000 Independence Avenue, SW, Room 7B-058, Washington, DC 20585 (202) 586-6151

Florida:
Department of Natural Resources, 3900 Commonwealth Boulevard, Tallahassee, FL 32399-3000 (904) 488-1554

Georgia:
Natural Resources Department, 205 Butler Street, SW, Atlanta, GA 30334-4100 (404) 656-3530

Guam:
Department of Agriculture, Agana, Guam 96910 (617) 734-3941

Environmental Protection Agency, PO Box 2999, Agana, Guam 96910 (617) 646-8863

Hawaii:
Land and Natural Resources Department, 1151 Punchbowl Street, Room 130, Honolulu, HI 96813-2407 (808) 548-2544

Idaho:
Department of Health and Welfare, 450 West State Street, Boise, ID 83720 (208) 334-5500

Department of Lands, State Capitol Building, Room 201, Boise, ID 83720 (208) 334-3284

Department of Water Resources, 1301 North Orchard, Boise, ID 83720 (208) 327-7900

Illinois:
Department of Energy and Natural Resources, 325 West Adams Street, 3rd Floor, Springfield, IL 62704 (217) 785-2800

Indiana:
Department of Natural Resources, State Office Building, Room 608, Indianapolis, IN 46204 (317) 232-4020

Iowa:
Department of Natural Resources, 900 East Grand Avenue, Des Moines, IA 50319 (515) 281-5145
- Air Quality and Solid Waste Protection Bureau, (515) 281-8852
- Energy Bureau, (515) 281-8681

Kansas:
Division of Environment, Forbes Field Building 740, Topeka, KS 66620 (913) 296-1535

Kentucky:
Department for Environmental Protection, 18 Reilly Road, Frankfort, KY 40601-1139 (502) 564-2150

Louisiana:
Department of Natural Resources, 625 North 4th Street, Baton Rouge, LA 70802-5364 (504) 342-4503

Environmental Quality, 11720 Airline Highway, Baton Rouge, LA 70817-4401 (504) 295-8900

Maine:
Department of Environmental Protection, State House, Suite 17, Augusta, ME 04333 (207) 289-7688

Maryland:
Department of Natural Resources, 580 Taylor Avenue, Annapolis, MD 21401 (301) 974-3041

Massachusetts:
Department of Environmental Management, 100 Cambridge Street, Boston, MA 02202 (617) 727-3159

Michigan:
Department of Natural Resources, PO Box 30028, Lansing, MI 48909-7528 (517) 373-1220

Minnesota:
Department of Natural Resources, 500 Lafayette Road, St. Paul, MN 55155 (612) 296-2549

Mississippi:
Department of Environmental Quality, PO Box 20305, Jackson, MS 39289-1305 (601) 961-5000

Missouri:
Department of Natural Resources, 205 Jefferson Street, Jefferson City, MO 65101-2981 (314) 751-4422

Montana:
Department of Natural Resources and Conservation, 1520 E 6th Avenue, Helena, MT 59601-4541 (406) 444-6699

Nebraska:
Department of Environmental Control, PO Box 98922, Lincoln, NE 68509-8922 (402) 471-2186

Nevada:
Department of Conservation and Natural Resources, 123 West Nye Lane, Room 201, Carson City, NV 89710 (702) 885-4360

New Hampshire:
Department of Environmental Services, 6 Hazen Drive, Concord, NH 03301 (603) 271-3503

Department of Resources and Economic Development, 105 Loudon Road, Concord, NH 03301-5601 (603) 271-3727

New Jersey:
Department of Environmental Protection, 401 West State Street, CN-402, Trenton, NJ 08625 (609) 292-2885

New Mexico:
Environmental Improvement Agency, 1190 St. Francis Drive, Santa Fe, NM 87503 (505) 827-2850

New York:
Department of Environmental Conservation, 50 Wolf Road, Room 604, Albany, NY 12233 (518) 457-3446

North Carolina:
Department of Environmental Health and Natural Resources, 512 North Salisbury Street, Raleigh, NC 27611 (919) 733-4984

North Dakota:
Game & Fish Department, 100 North Bismarck Expressway, Bismarck, ND 58501-5086 (701) 221-6300

Ohio:
Department of Natural Resources, 1939 Fountain Square, Building E-3, Columbus, OH 43224 (614) 265-6722

Oklahoma:
Conservation Commission, 2800 North Lincoln Boulevard, Suite 160, Oklahoma City, OK 73105-4210 (405) 521-2384

Oregon:
Department of Environmental Quality, 811 SW 6th Avenue, Portland, OR 97204-1334 (503) 229-5696

Pennsylvania:
Department of Environmental Resources, 3rd & Locust Streets, Harrisburg, PA 17120 (717) 787-2814

Puerto Rico:
Department of Natural Resources, PO Box 5887, Puerta de Tierra Station, San Juan, Puerto Rico 00906 (809) 722-8774

Rhode Island:
Department of Environmental Management, 83 Park Street, Providence, RI 02908 (401) 277-3437

South Carolina:
Department of Environmental Quality Control, 2600 Bull Street, Columbia, SC 29201-1708 (803) 734-5360

Land Resources Conservation Commission, 2221 Devine Street, Suite 222, Columbia 29205-2474 (803) 734-9100

South Dakota:
Department of Water and Natural Resources, 523 East Capitol Avenue, Pierre, SD 57501-3182 (605) 773-3151

Tennessee:
Department of Conservation, 701 Broadway, Nashville, TN 37243 (615) 742-6749

Texas:
Environmental Protection Division, PO Box 12548, Austin, TX 78711 (512) 463-2012

Utah:
Department of Natural Resources, 1636 West North Temple, Suite 316, Salt Lake City, UT 846116-3193 (801) 538-7200

Vermont:
Agency of Natural Resources, 103 South Main Street, Waterbury, VT 05676-1534 (802) 244-7347

Virgin Islands:
Department of Conservation and Cultural Affairs, PO Box 4399, St. Thomas, Virgin Islands 00801 (809) 774-3320

Virginia:
Council on the Environment, 202 North 9th Street, Suite 900, Richmond, VA 23219-3402 (804) 786-4500

Washington:
Department of Ecology, Abbott Raphael Building, St. Martins College, Campus PV-11, Olympia, WA 98504 (206) 459-6168

Department of Natural Resources, 201 John A. Cherberg Building, Olympia, WA 98504 (206) 753-5327

West Virginia:
Department of Natural Resources, 1900 Kanawha Boulevard East, Room 669, Charleston, WV 25305-0660 (304) 348-2754

Wisconsin:
Department of Natural Resources, 101 South Webster Street, Madison, WI 53702 (608) 266-2121

Wyoming:
Environmental Quality Department, 122 West 25th Street, Herschler Building, 4th Floor West, Cheyenne, WY 82002 (307) 777-7938

B. CITIZENS' ORGANIZATIONS

Advancement of Earth & Environmental Sciences
International Association for Northeastern Illinois University, Geography and Environmental Studies Department, 5500 North St. Louis Avenue, Chicago, Illinois 60625 (312) 794-2628

Air Pollution Control Association
4400 Fifth Avenue, PO Box 2861, Pittsburgh, PA 15213 (412) 578-8111

American Association for the Advancement of Science
1333 H Street, NW, Washington, DC 20005-4707 (202) 326-6400

American Chemical Society
1155 16th Street, NW, Washington, DC 20036 (202) 872-4600

American Committee for International Conservation
Center for Marine Conservation, 1725 DeSales Street, NW, Washington, DC 20036 (202) 429-5609

American Farm Bureau Federation
225 Touhy Avenue, Park Ridge, IL 60068 (312) 399-5700

American Fisheries Society
5410 Grosvenor Lane, Bethesda, MD 20014 (301) 897-8616

American Forest Council
1250 Connecticut Avenue, NW, Suite 320, Washington, DC 20036 (202) 463-2455

The American Forestry Association
1516 P Street, NW, Washington, DC 20005-1932 (202) 667-3300

American Institute of Biological Sciences, Inc.
730 11th Street, NW, Washington, DC 20001-4521 (202) 628-1500

American Museum of Natural History
Central Park West at 79th Street, New York, NY 10024-5192 (212) 769-5100

American Petroleum Institute
1220 L Street, NW, Washington, DC 20005 (202) 682-8000

American Rivers
801 Pennsylvania Avenue, SE, Suite 400, Washington, DC 20034 (202) 547-6900

Audubon Society, National
950 Third Avenue, New York, NY 10020 (212) 832-3200

Boone and Crockett Club
241 South Fraley Boulevard, Dumfries, VA 22026 (703) 221-1888

Canada–United States Environmental Council
Canada: c/o Canadian Nature Federation, 75 Albert Street, Ottawa, Ontario K1P 6G1 (613) 238-6154
United States: c/o Defenders of Wildlife, 1244 19th Street, NW, Washington, DC 20036 (202) 659-9510

Center for Environmental Education, Inc.
1725 DeSales Street, NW, Suite 500, Washington, DC 20036 (202) 429-5609

Children of the Green Earth
PO Box 31550, Seattle, WA (206) 781-0852

Citizens for a Better Environment
407 South Dearborn, Suite 1775, Chicago, IL 60605 (312) 939-1530

Coastal Conservation Association
4801 Woodway, Suite 220W, Houston, TX 77056 (713) 626-4222

Conservation Engineers, Association of:
•Alabama Department of Conservation, 64 North Union Street, Montgomery, AL 36130 (205) 261-3476
• Missouri Department of Conservation, PO Box 180, Jefferson City, MO 65076 (314) 751-4115

Conservation Foundation
1250 24th Street, NW, Suite 400, Washington, DC 20037 (202) 293-4800

Conservation Fund
1800 North Kent Street, Suite 1120, Arlington, VA 22209 (703) 525-6300

Conservation Information, Association for
PO Box 10678, Reno, NV 89520 (702) 688-1500

Conservation International, 1015 18th Street, NW, Suite 1000, Washington, DC 20036 (202) 429-5660

Defenders of Wildlife
1244 19th Street, NW, Washington, DC, 20036 (202) 659-9510

Ducks Unlimited, Inc.
One Waterfowl Way, Long Grove, IL 60047 (708)438-4300

Earth First
PO Box 5871, Tucson, AZ 85703

Earthwatch
680 Mt. Auburn Street, Box 403N, Watertown, MA 02272
(617) 926-8200

Ecology and Environment, Inc.
368 Pleasant View Drive, Lancaster, NY 14086-1397
(716) 684-8060

Environmental Action Foundation, Inc.
1525 New Hampshire Avenue, NW, Washington, DC 20036
(202) 745-4870

Environmental Health Association, National
720 South Colorado Boulevard, Suite 970, Glendale, CO 80222
(303) 756-9090

Fish and Wildlife Service
1849 C Street, NW, Room 3240, Washington, DC 20240-0001
(202) 208-4131

Fisheries Institute, National
200 M Street, Washington, DC 20036

Food and Agriculture Organization of the United Nations (FAO)
Via delle Terme di Caracalla, 00100, Rome, Italy

Foresters, Society of American
5400 Grosvenor Lane, Bethesda, MD 20814-2198 (301) 897-8720

Friends of the Earth
218 D Street, SE, Washington, DC 20003 (202) 544-2600

Greenpeace U.S.A.
1436 U Street, NW, Washington, DC 20009 (202) 462-1177

Human Environment Center
1001 Connecticut Avenue, NW, Suite 827, Washington, DC
20036 (202) 331-8387

International Association of Fish and Wildlife Agencies
444 North Capitol Street, NW, Suite 534, Washington, DC 20001
(202) 624-7890

International Fund for Agricultural Development (IFAD)
107 Via del Serafico, 00142, Rome, Italy

Keep America Beautiful
Mill River Plaza, 9 West Broad Street, Stamford, CT 06902
(203) 323-8987

Marine Conservation, Center for
1725 De Sales Street, NW, Suite 500, Washington, DC 20036
(202) 429-5609

National Association of Conservation District
509 Capitol Court, NE, Washington, DC 20002-4937
(202) 547-6223

National Audubon Society
950 Third Avenue, New York, NY 10022 (212) 832-3200

National Geographic Society
1145 17th Street, NW, Washington, DC 20036 (202) 857-7000

National Institute for Urban Wildlife
10921 Trotting Ridge Way, Columbia, MD 21044 (301) 596-3311

National Rifle Association of America
1600 Rhode Island Avenue, NW, Washington, DC 20036
(202) 828-6000

National Wildlife Federation
1400 16th Street, NW, Washington, DC 20036 (202) 797-6800

Natural Resources Council of America
801 Pennsylvania Avenue, NW, No. 410, Washington, DC 20003
(202) 547-7553

Nature Conservancy
1815 North Lynn Street, Arlington, VA 22209 (703) 841-5300

Parks and Conservation Association, National
1015 31st Street, NW, Washington, DC 20007 (202) 944-8530

Population Association of America
1429 Duke Street, Alexandria, VA 22314 (703) 684-1221

Population Conservation, Center for
1725 De Sales Street, NW, Suite 500, Washington, DC 20036
(202) 429-5609

Rainforest Alliance
270 Lafayette Street, Suite 512, New York, NY 10012
(212) 941-1900

Sierra Club
730 Polk Street, San Francisco, CA 94109 (415) 981-8634

Smithsonian Institution
900 Jefferson Drive, Washington, DC 20560 (202) 357-1300

Society of American Foresters
5400 Grosvenor Lane, Bethesda, MD 20814 (301) 897-8720

Sport Fishing Institute
1010 Massachusetts Avenue, NW, Suite 100, Washington, DC
20001 (202) 737-0668

United Nations Educational, Scientific, and Cultural Organization
(UNESCO)
UNESCO House, Place de Fontenoy, 7e, Paris, France

Wilderness Society
900 17th Street, NW, Washington, DC 20006 (202) 293-2732

Wildlife Federation, National
1400 16th Street, NW, Washington, DC 20036-2266 (202)
797-6800

World Wildlife Fund
1255 23rd Street, NW, Washington, DC 20037 (202) 293-4800

Zero Population Growth
1400 16th Street, NW, Suite 320, Washington, DC 20036
(202) 332-2200

III. Canadian Agencies and Citizens' Organizations

A: GOVERNMENT AGENCIES

Alberta:
Energy and Natural Resources, Main Floor, North Tower,
Petroleum Plaza, 9945 108th Street, Edmonton, Alberta T5K
2G6
(403) 427-3674

British Columbia:
Ministry of Environment and Parks, Parliament Building,
Victoria, British Columbia V8V 1X5 (604) 387-5429

Manitoba:
Department of Natural Resources, Legislative Building,
Winnipeg, Manitoba R3C 0V8 (204) 945-3730

New Brunswick:
Department of Natural Resources and Energy, PO Box 6000,
Fredericton, New Brunswick E3B 5H1 (506) 453-2510

Newfoundland:
Department of Culture, Recreation, and Youth, Wildlife Division,
Building 810, Pleasantville, PO Box 4650, St. John's,
Newfoundland A1C 5T7

Northwest Territories:
Department of Economic Development and Tourism,
Government of the NWT, Yellowknife, Northwest Territories X1A
2L9 (403) 873-7115

Nova Scotia:
Department of Lands and Forests, PO Box 698, Halifax, Nova
Scotia B3J 2T9 (902) 424-5935

Ontario:
Ministry of Natural Resources, Toronto, Ontario M7A 1WS
(416) 965-1301

Prince Edward Island:
Department of Community and Cultural Affairs, PO Box 2000, Charlottetown, Prince Edward Island C1A 7N8 (902) 892-0311

Quebec:
Department of Recreation, Fish, and Game, Place de la Capitale, 150 East, St. Cyrille Boulevard, Quebec City, Quebec G1R 2B2 (418) 643-6527

Saskatchewan:
Department of Parts and Renewable Resources, 3211 Albert Street, Regina, Saskatchewan S4S 5W6 (306) 787-9130

Yukon Territory:
Department of Renewable Resources, Box 2703, Whitehorse, Yukon Territory Y1A 2C6 (403) 667-5460

B. CITIZENS' GROUPS

Alberta Wilderness Association
Box 6398, Station D, Calgary, Alberta T2P 2E1 (403) 283-2025

Algonquin Wildlands League
69 Sherbourne Street, Suite 313, Toronto, Ontario M5A 3X7 (416) 366-3494

British Columbia Wildlife Federation
5659 176th Street, Surrey, British Columbia V3S 4C5 (604) 576-8288

Ducks Unlimited Canada
954A Lavel Crescent, Kamloops, British Columbia V2C 5P5 (604) 374-8307

Federation of Ontario Naturalists
FON Conservation Centre, Moatfield Park, 355 Lesmill Road, Don Mills, Ontario M3B 2W8 (416) 444-8419

Manitoba Wildlife Federation
1770 Notre Dame Avenue, Winnipeg, Manitoba R3E 3K2 (201) 633-5967

Newfoundland Labrador Wildlife Federation
(709) 364-8415

Nova Scotia Forestry Association
64 Inglis Place, Suite 202, Truro, Nova Scotia B2N 4B4 (902) 893-4653

Quebec Forestry Association, Inc.
110, 915 St. Cyrille Boulevard West, Quebec City, Quebec G1S 1T8 (418) 681-3588

Quebec Wildlife Federation
319 St. Zotique East, Montreal, Quebec H2S 1L5 (514) 271-2487

Saskatchewan Natural History Society
Box 414, Raymore, Saskatchewan S0A 3J0 (306) 746-4544

IV. Selected Journals and Periodicals of Environmental Interest

American Forests
The American Forestry Association, 1319 Eighteenth Street, NW, Washington, DC 20036

American Scientist
PO Box 13975, Research Triangle, PA 27709-3975

Annual Report on the Council of Environmental Quality
Superintendent of Documents, U.S. Government Printing Office, Washington, DC 20401

Audubon
National Audubon Society, 950 Third Avenue, New York, NY 10022

BioScience
American Institute of Biological Science, 730 11th Street, NW, Washington, DC 2001-4584

BUZZWORM, The Environmental Journal
2305 Canyon Boulevard, Suite 206, Boulder, CO 80302

California Environmental Directory
California Institute of Public Affairs, Box 10, Claremont, CA 91711

The Canadian Field-Naturalist
Box 3264, Postal Station C, Ottawa, Ontario K1Y 4J5, Canada

Conservation Directory
1400 16th Street, NW, Washington, DC 20036

E Magazine (The Environmental Magazine)
28 Knight Street, Norwalk, CT 06851

Earth
Kalmbach Publishing Company, 21027 Crossroads Circle, Waukesha, WI 53187

Ecology USA
Business Publishers Inc., 951 Pershing Drive, Silver Spring, MD 20910-4464

Environment
1319 Eighteenth Street, NW, Washington, DC 20036

Environment Action Bulletin
Rodale Press Inc., 33 East Minor Street, Emmaus, PA 18098

Environment Reporter
Bureau of National Affairs Inc., 1231 25th Street, NW, Washington, DC 20037

Environmental Information Handbook
Simon and Schuster Inc., 1230 Avenue of the Americas, New York, NY 10020

Environmental Science and Technology
American Chemical Society Publications, 1155 16th Street, NW, Washington, DC 20036

The Futurist
7910 Woodmont Avenue, Suite 450, Bethesda, MD 20814

Greenpeace Magazine
1436 U Street, NW, Washington, DC 20009

Journal of Soil and Water Conservation
7515 Northeast Ankeny Road, Ankeny, IA 50021

Journal of Wildlife Management
The Wildlife Society, 5410 Grosvenor Lane, Bethesda, MD 20814

Mother Earth News
80 Fifth Avenue, New York, NY 10011

National Wildlife
8925 Leesburg Pike, Vienna, VA 22184

Natural Resources Journal
University of New Mexico, School of Law, 1117 Stanford, NE, Albuquerque, NM 87131

Nature
MacMillan Journals Ltd., 4 Little Essex Street, London WC2R 3LF, England

Nature Canada
Canadian Nature Federation, 453 Sussex Drive, Ottawa, Ontario K1P 5K6, Canada

Nature Conservancy Magazine
1815 North Lynn Street, Arlington, VA 22209

Oceans
2001 West Marin Street, Stamford, CT 06902

Omni
Omni Publications International Ltd., 1965 Broadway, New York, NY 10023-5965

Pollution Abstracts
7200 Wisconsin Avenue, 6th Floor, Bethesda, MD 20814

Science
1333 H Street, NW, Washington, DC 20005

Sierra Club Bulletin
730 Polk Street, San Francisco, CA 94109

Smithsonian
900 Jefferson Drive, Washington, DC 20560

Technology Review
M.I.T. Alumni Association, W59-200, Cambridge, MA 02139

U.S. News and World Report
2400 N Street, NW, Washington, DC 20037-1196

Wilderness
Wilderness Society, 900 17th Street, NW, Washington, DC 20036

The World & I
2800 New York Avenue, NE, Washington, DC 20002

World Wildlife Fund and The Conservation Foundation
1250 24 Street, SW, Washington, DC 20037

SOURCES *used to compile this list: Conservation Directory, National Wildlife Federation, 3rd Edition; Encylopedia of Associations, 26th Edition, 1992; Information Please World Almanac, 1992; National Directory of Addresses and Telephone Numbers, 1991; The World Almanac, 1992.*

Glossary

This glossary of 168 environment terms is included to provide you with a convenient and ready reference as you encounter general terms in your study of environment which are unfamiliar or require a review. It is not intended to be comprehensive but taken together with the many definitions included in the articles themselves it should prove to be quite useful.

Abiotic Without life; any system characterized by a lack of living organisms.

Abortion Expulsion of a fetus from the uterine cavity prior to birth.

Acid Any compound capable of reacting with a base to form a salt; a substance containing a high hydrogen ion concentration (low pH).

Acid Rain Precipitation containing a high concentration of acid.

Acre Foot Unit used to measure the volume of water equal to the quantity of water required to cover one acre to a depth of one foot; equal to 325,851 gallons.

Adaptation Any characteristic that aids an organism to survive and reproduce in its environment.

Additive A substance added to another in order to impart or improve desirable properties or suppress undesirable properties.

Aerobic Environmental conditions where oxygen is present; aerobic organisms require oxygen in order to survive.

Aesthetic Pertaining to a sense or feeling of the beautiful.

Age Distribution The proportion of a population in each age class.

Agriculture Production of crops, livestock, or poultry.

Air Quality Standard A prescribed level of a pollutant in the air that should not be exceeded.

Alkali Soil A soil of such a high degree of alkalinity (high pH) that growth of most crop plants is reduced.

Alpha Particle A positively charged particle given off from the nucleus of some radioactive substances: it is identical to a helium atom that has lost its electrons.

Anaerobic Without oxygen; environmental conditions where oxygen is absent.

Ammonia A colorless gas composed of one atom of nitrogen and three atoms of hydrogen; liquified ammonia is used as a fertilizer.

Anthropocentric Considering man to be the central or most important part of the universe.

Atom The smallest particle of an element, composed of electrons moving around an inner core (nucleus) of protons and neutrons. Atoms of elements combine to form molecules and chemical compounds.

Atomic Energy Energy released by changes in the nucleus of an atom, either by the splitting of the nucleus or by the joining of nuclei.

Atomic Reactor A structure fueled by radioactive materials which generates energy usually in the form of electricity; reactors are also utilized for medical and biological research. Plutonium, a radioactive substance, is also produced by reactors and has been used for the production of atomic devices.

Autotrophic Organisms capable of using chemical elements in the synthesis of larger compounds; green plants are autotrophic.

Background Radiation The normal radioactivity present; coming principally from outer-space and naturally occurring radioactive substances in the earth.

Bacteria One-celled microscopic organisms found in the air, water, and soil. Bacteria cause many diseases of plants and animals; they also are beneficial in agriculture, decay of dead matter, in food industries and in chemical industries.

Biochemical Oxygen Demand (BOD) The oxygen utilized in meeting the metabolic needs of aquatic organisms.

Biodegradable Capable of being reduced to simple compounds through the action of biological processes.

Biogeochemical Cycle The cyclical series of transformations of an element through the organisms in a community and their physical environment.

Biological Control The suppression of reproduction of a pest organism utilizing other organisms rather than chemical means.

Biomass The weight of all living tissue in a sample.

Biome A major climax community type covering a specific area on earth.

Biota The flora and fauna of any region.

Biotic Biological; relating to living systems.

Biotic Potential Maximum possible growth rate of living systems under ideal conditions.

Birth Rate Number of live births in one year per 1,000 midyear population.

Breeder Reactor A nuclear reactor in which the production of fissionable material occurs.

Carbon One of the most common elements; compounds of carbon are the chief constituents of living systems.

Carbon Cycle Process by which carbon is incorporated into living systems, released to the atmosphere and returned to living organisms.

Carbon Dioxide A gas, CO_2, making up about 0.03 of the earth's atmosphere. It is an end product of burning (oxidation) of organic matter or carbon-containing substances.

Carbon Monoxide A gas, poisonous to most living systems; formed when burning occurs in the absence of much oxygen.

Carcinogen Any substance capable of producing cancer.

Chlorinated Hydrocarbon Insecticide Synthetic organic poisons containing hydrogen, carbon, and chlorine. Because they are fat soluble they tend to be recycled through food chains eventually affecting non-target systems. Damage is normally done to the organism's nervous system. Examples include DDT, Aldrin, Deildrin and Chlordane.

Clear-Cutting The practice of removing all trees in a specific area.

Climax Community Terminal state of ecological succession in an area; the redwoods are a climax community.

Coal Gasification Process of converting coal to gas; the resultant gas, if used for fuel, sharply reduces sulfur oxide emissions and particulates that result from coal burning.

Commensalism The relationship between two different species in which one benefits while the other is neither harmed nor benefited.

Community All organisms existing in a specific region.

Competition The struggle between individuals of the same or different species for food, space, mates or other limited resource.

Competitive Exclusion Resulting from competition; one species forced out of part of an available habitat by a more efficient species.

Conservation The planned management of a natural resource to prevent over-exploitation, destruction, or neglect.

Contraception Process of preventing conception.

Crankcase Smog Devices (PCV System) A system, used principally in automobiles, designed to prevent discharge of combustion emissions to the external environment.

Death Rate Number of deaths in one year per 1,000 midyear population.

Decomposer Any organism which causes the decay of organic matter; bacteria and fungi are two examples.

Demography The statistical study of populations; related principally to human populations.

Desert An arid biome characterized by little rainfall, high daily temperatures, and low diversity of animal and plant life.

Detergent A synthetic soap-like material that emulsifies fats and oils and holds dirt in suspension; some detergents have caused pollution problems because of certain chemicals used in their formulation.

DNA (Deoxyribonucleic Acid) One of two principal nucleic acids, the other being RNA (Ribonucleic Acid). DNA contains information used for the control of a living cell. Specific segments of DNA are now recognized as genes, those agents controlling evolutionary and hereditary processes.

Dominant Species Any species of plant or animal that is particularly abundant or controls a major portion of the energy flow in a community.

Dust Bowl The name usually associated with the south-central region of the United States; applied during the 1930s during periods of droughts and dust storms that occurred in Colorado, Kansas, New Mexico, Texas, and Oklahoma.

Ecological Density The number of a singular species in a geographical area; including the highest concentration points within the defined boundaries.

Ecology Study of the interrelationships between organisms and their environments.

Ecosystem The organisms of a specific area, together with their functionally related environments; considered as a definitive unit.

Effluent A liquid discharged as waste.

Electron Small, negatively charged particle; normally found in orbit around the nucleus of an atom.

Eminent Domain Superior dominion exerted by a governmental state over all property within its boundaries that authorizes it to appropriate all or any part thereof to a necessary public use, reasonable compensation being made.

Environment The physical and biological aspects of a specific area.

Environmental Protection Agency (EPA) Federal agency responsible for control of air and water pollution, radiation and pesticide problems, ecological research and solid waste disposal.

Erosion Progressive destruction or impairment of a geographical area; wind and water are the principal agents involved.

Estuary Area formed where streams, or rivers, enter oceanic zones.

Eutrophic Well nourished; refers to aquatic areas rich in dissolved nutrients.

Evolution A change in the gene frequency within a population; sometimes involving a visible change in the population's characteristics.

Fallow Cropland that is plowed but not replanted; left idle in order to restore productivity mainly through water accumulation, weed control, and buildup of soil nutrients.

Fauna The animal life of a specified area.

Feral Animals or plants that have reverted to a noncultivated or wild state.

Fertilizer Any natural or artificial substance added to soil to promote growth.

Fission The splitting of an atom into smaller systems.

Flora The plant life of an area.

Food Additive Substance added to food, usually added to improve color, flavor, or shelf life.

Food Chain The sequence of organisms in a community, each of which uses the lower source as its energy supply. Green plants are the ultimate basis for the entire sequence.

Fossil Fuel Coal, oil, natural gas, and/or lignite; those fuels derived from former living systems; usually called nonrenewable fuels.

Fuel Cell Manufactured chemical systems capable of producing electrical energy; usually derive their capabilities via complex reactions involving the sun as the driving energy source.

Fusion The formation of a heavier atomic complex brought about by the addition of atomic nuclei; during the process there is an attendant release of energy.

Gamma Ray A ray given off by the nucleus of some radioactive elements. A form of energy similar to X-rays.

Greenhouse Effect The effect noticed in greenhouses when shortwave solar radiation penetrates glass, is converted to longer wavelengths, and is blocked from escaping by the windows. It results in a temperature increase. The earth's atmosphere acts in a similar manner.

Ground Water All water located below the earth's surface.

Habitat The natural environment of a plant or animal.

Heterotrophic Obtaining nourishment from organic matter.

Herbicide Any substance used to kill plants.

Hydrocarbon Organic compounds containing hydrogen, oxygen and carbon. Commonly found in petroleum, natural gas, and coal.

Hydrogen Lightest known gas; major element found in all living systems.

Hydrogen Sulfide Compound of hydrogen and sulfur; toxic air contaminant, smells like rotten eggs.

Immigration Movement into a new area of residence.

Ion An atom or group of atoms, possessing a charge; brought about by the loss or gain of electrons.

Ionizing Radiation Energy in the form of rays or particles which have the capacity to dislodge electrons and/or other atomic particles from matter which is irradiated.

Irradiation Exposure to any form of radiation.

Isotope Two or more forms of an element having the same number of protons in the nucleus of each atom but different numbers of neutrons.

Kilowatt Unit of power equal to 1,000 watts.

Leaching Dissolving out of soluble materials by water percolating through soil.

Limnologist Individual who studies the physical, chemical, and biological conditions of aquatic systems.

Malthusian Theory The theory that populations tend to increase by geometric progression (1, 2, 4, 8, 16, etc.) while food supplies increase by arithmetic means (1, 2, 3, 4, 5, etc.).

Metabolism The chemical processes in living tissue through which energy is provided for continuation of the system.

Methane Often called marsh gas (CH_4); an odorless, flammable gas that is the major constituent of natural gas. In nature it develops from decomposing organic matter.

Migration Periodic departure and return of organisms to and from a population area.

Monoculture Cultivation of a single crop, such as wheat or corn, to the exclusion of other land uses.

Mutagen Any agent capable of causing a mutation.

Mutation Change in genetic material (gene) that determines species characteristics; can be caused by a number of agents including radiation and chemicals.

Natural Selection The agent of evolutionary change by which organisms possessing advantageous adaptations leave more offspring than those lacking such adaptations.

Niche The unique occupation or way of life of a plant or animal species; where it lives and what it does in the community.

Nitrate A salt of nitric acid. Nitrates are the major source of nitrogen for higher plants. Sodium nitrate and potassium nitrate are used as fertilizers.

Nitrite Highly toxic compound; salt of nitrous acid.

Nitrogen Oxides Common air pollutants. Formed by combination of nitrogen and oxygen; often the products of petroleum combustion in automobiles.

Oil Shale Rock impregnated with petroleum. Regarded as a potential source of future petroleum products.

Oligotrophic Most often refers to those lakes with a low concentration of organic matter. Usually containing considerable oxygen; Lakes Tahoe and Baikal are examples.

Organic Derived from living systems.

Organophosphates A large group of nonpersistent synthetic poisons used in the pesticide industry; include parathion and malathion.

Ozone Molecule of oxygen containing three oxygen atoms; shields much of the earth from ultraviolet radiation.

Particulate Existing in the form of small separate particles; various atmospheric pollutants are industrial produced particulates.

Peroxyacyl Nitrate (PAN) Compound making up part of photochemical smog and the major plant toxicant of smog type injury; levels as low as 0.01 ppm can injure sensitive plants. Also causes eye irritation in man.

Pesticide Any material used to kill rats, mice, bacteria, fungi, or other pests of man.

Petrochemical Chemicals derived from petroleum bases.

pH Scale used to designate the degree of acidity or alkalinity; ranges from 1-14; a neutral solution has a pH of 7; low pHs are acid in nature while pHs above 7 are alkaline.

Phosphate A phosphorous compound; used in medicine and as fertilizers.

Photochemical Pertaining to the chemical effects of light.

Photochemical Smog Type of air pollution; results from sunlight acting with hydrocarbons and oxides of nitrogen in the atmosphere.

Photosynthesis Formation of carbohydrates from carbon dioxide and hydrogen in plants exposed to sunlight; involves a release of oxygen through the decomposition of water.

Physical Half-Life Time required for half of the atoms of a radioactive substance present at some beginning to become disintegrated and transformed.

Plutonium A heavy, radioactive, manmade, metallic element. Used in weapons and as a reactor fuel. Highly toxic to life forms and possesses an extremely long physical half-life.

Pollution The process of contaminating air, water, or soil with materials that reduce the quality of the medium.

Polychlorinated Biphenyls (PCBs) Poisonous compounds similar in chemical structure to DDT. PCBs are found in a wide variety of products ranging from lubricants, waxes, asphalt, and transformers, to inks and insecticides. Known to cause liver, spleen, kidney, and heart damage.

Population All members of a particular species occupying a specific area.

Predator Any organism that consumes all, or part, of another system; usually responsible for death of the prey.

Primary Production The energy accumulated and stored by plants through photosynthesis.

Rad (Radiation Absorbed Dose) Measurement unit relative to the amount of radiation absorbed by a particular target, biotic or abiotic.

Radioactive Waste Any radioactive by-product of nuclear reactors or nuclear processes.

Radioactivity The emission of electrons, protons (atomic nuclei), and/or rays from elements capable of emitting radiation.

Recycle To reuse; usually involves manufactured items, such as aluminum cans, being restructured after use and utilized again.

Redwood *Sequoia sempervirens;* world's tallest tree. Used extensively for lumber; grows in coastal areas of California and Oregon.

Riparian Water Right Legal right of an owner of land bordering a natural lake or stream to remove water from that aquatic system.

RNA Ribonucleic acid; nucleic acid most often located in the cytoplasm of cells; used principally in the manufacture of proteins.

Scrubber Anti-pollution system which uses liquid sprays in removing particulate pollutants from an air stream.

Sediment Soil particles moved from land into aquatic systems as a result of man-caused activities.

Seepage Movement of water through soil.

Selection The process, either natural or artificial, of removing or selecting the best or less desirable members of a population.

Selective Breeding Process of selecting and breeding organisms containing traits considered most desirable.

Selective Harvesting Process of taking specific individuals from a population; removal of trees in a specific age class would be an example.

Sewage Any waste material coming from domestic and industrial origins.

Smog A mixture of smoke and air; now applies to any type of air pollution.

Solid Waste Unwanted solid materials usually resulting from industrial processes.

Species Populations, or a population, capable of interbreeding and producing viable offspring.

Species Diversity A ratio between the number of species in a community and the number of individuals in each species. Generally, the greater the species diversity comprising a community, the more stable is the community.

Strip Mining Mining in which the earth's surface is removed in order to obtain sub-surface materials.

Strontium-90 Radioactive isotope of strontium; results from nuclear explosions and is dangerous, especially for vertebrates, because it is taken up in the construction of bone.

Succession Change in the structure and function of an ecosystem; replacement of one system with another through time.

Sulfur Dioxide (SO₂) Gas produced by burning coal and as a byproduct of smelting and other industrial processes. Very toxic to plants.

Sulfuric Acid (H₂SO₄) Very corrosive acid; produced from sulfur dioxide; found as a component of acid rain.

Sulfur Oxides (SOₓ) Oxides of sulfur produced by the burning of oils and coal which contain small amounts of sulfur. Common air pollutants.

Technology Applied science; application of knowledge for practical application.

Tetraethyl Lead Major source of lead found in living tissue; produced to reduce engine knock in automobiles.

Thermal Pollution Unwanted heat; the result of ejection of heat from various sources into the environment.

Thermocline The layer of water in a body of water that separates an upper warm layer from a deeper colder zone.

Threshold Effect The situation in which no effect is noticed, physiologically or psychologically, until a certain level or concentration is reached.

Tolerance Limit The point at which resistance to a poison or drug breaks down.

Toxic Poisonous; capable of producing harm to a living system.

Toxicant Any agent capable of producing toxic reactions.

Trophic Relating to nutrition; often expressed in trophic pyramids in which organisms feeding on other systems are said to be at a higher trophic level; an example would be carnivores feeding on herbivores which, in turn, feed on vegetation.

Turbidity Usually refers to the amount of sediment suspended in an aquatic system.

Uranium 235 An isotope of uranium that when bombarded with neutrons undergoes fission resulting in radiation and energy. Used in atomic reactors for electrical generation.

Zero Population Growth The condition of a population in which birthrates equal death rates; results in no growth of the population.

Index

Credits/ Acknowledgments

Cover design by Charles Vitelli

1. The Global Environment
Facing overview—Courtesy of NASA.

2. The World's Population
Facing overview—United Nations photo by Paul Heath Noeffel.
67, 69, 71—Art direction by Suzanne Morin.

3. Energy
Facing overview—Solar Products Manufacturing Corp.,
Cromwell, CT.

4. Pollution
Facing overview—EPA Documerica.

5. Resources
Facing overview—USDA-SCS photo.

6. Biosphere
Facing overview—U.S. Fish & Wildlife Service photo by Galen
Rathbun.

ANNUAL EDITIONS ARTICLE REVIEW FORM

■ NAME: _____ DATE: _____

■ TITLE AND NUMBER OF ARTICLE: _____

■ BRIEFLY STATE THE MAIN IDEA OF THIS ARTICLE: _____

_____

_____

_____

■ LIST THREE IMPORTANT FACTS THAT THE AUTHOR USES TO SUPPORT THE MAIN IDEA:

_____

_____

_____

_____

_____

_____

■ WHAT INFORMATION OR IDEAS DISCUSSED IN THIS ARTICLE ARE ALSO DISCUSSED IN YOUR
TEXTBOOK OR OTHER READING YOU HAVE DONE? LIST THE TEXTBOOK CHAPTERS AND PAGE
NUMBERS:

_____

_____

_____

_____

_____

_____

■ LIST ANY EXAMPLES OF BIAS OR FAULTY REASONING THAT YOU FOUND IN THE ARTICLE:

_____

_____

_____

_____

■ LIST ANY NEW TERMS/CONCEPTS THAT WERE DISCUSSED IN THE ARTICLE AND WRITE A
SHORT DEFINITION:

_____

_____

_____

_____

*Your instructor may require you to use this Annual Editions Article Review Form in any number of ways:
for articles that are assigned, for extra credit, as a tool to assist in developing assigned papers, or simply
for your own reference. Even if it is not required, we encourage you to photocopy and use this page;
you'll find that reflecting on the articles will greatly enhance the information from your text.

ANNUAL EDITIONS: ENVIRONMENT 94/95
Article Rating Form

Here is an opportunity for you to have direct input into the next revision of this volume. We would like you to rate each of the 35 articles listed below, using the following scale:

1. **Excellent: should definitely be retained**
2. **Above average: should probably be retained**
3. **Below average: should probably be deleted**
4. **Poor: should definitely be deleted**

Your ratings will play a vital part in the next revision. So please mail this prepaid form to us just as soon as you complete it.
Thanks for your help!

Rating	Article	Rating	Article
	1. The World Transformed		21. Common Threads: Research Lessons From Acid Rain, Ozone Depletion, and Global Warming
	2. The Mirage of Sustainable Development		
	3. The GATT: Menace or Ally?		
	4. The Environment of Tomorrow		22. 25th Environmental Quality Index: A Year of Crucial Decision
	5. How Many Is Too Many?		
	6. A New Strategy for Feeding a Crowded Planet		23. Beyond the Ark: A New Approach to U.S. Floodplain Management
	7. Population: The Critical Decade		24. Desktop Farms, Backyard Farms, or No Farms?
	8. The Landscape of Hunger		
	9. The Great Energy Harvest		25. 20 Years of the Clean Water Act
	10. What Would It Take to Revitalize Nuclear Power in the United States?		26. Redeeming the Everglades
			27. Global Warming on Trial
	11. Here Comes the Sun		28. Exploring the Links Between Desertification and Climate Change
	12. Energy Crops for Biofuels		
	13. Tilting Toward Windmills		29. The Origin and Function of Biodiversity
	14. All the Coal in China		30. Rain Forest Entrepreneurs: Cashing in on Conservation
	15. Facing Up to Nuclear Waste		
	16. A Place for Pesticides?		31. Deforestation and Public Policy
	17. Stewing the Town Dump in Its Own Juice		32. Out of the Woods
	18. Where the Air Was Clear		33. Killed by Kindness
	19. Ravaged Republics		34. New Species Fever
	20. Chernobyl's Lengthening Shadow		35. Barnyard Biodiversity

(Continued on next page)

ABOUT YOU

Name_____ Date_____

Are you a teacher? ☐ Or student? ☐

Your School Name _____

Department _____

Address _____

City_____ State _____ Zip _____

School Telephone #_____

YOUR COMMENTS ARE IMPORTANT TO US!

Please fill in the following information:

For which course did you use this book? _____

Did you use a text with this Annual Edition? ☐ yes ☐ no

The title of the text? _____

What are your general reactions to the Annual Editions concept?

Have you read any particular articles recently that you think should be included in the next edition?

Are there any articles you feel should be replaced in the next edition? Why?

Are there other areas that you feel would utilize an Annual Edition?

May we contact you for editorial input?

May we quote you from above?

ANNUAL EDITIONS: ENVIRONMENT 94/95

BUSINESS REPLY MAIL

First Class   Permit No. 84   Guilford, CT

Postage will be paid by addressee

**The Dushkin Publishing Group, Inc.
Sluice Dock**
DPG **Guilford, Connecticut 06437**